Earthquakes

Book Series: Statistical Physics of Fracture and Breakdown
Editors: Bikas K. Chakrabarti and Purusattam Ray

Why does a bridge collapse, an aircraft or a ship break apart? When does a dielectric insulation fail or a circuit fuse, even in microelectronic systems? How does an earthquake occur? Are there precursors to these failures? These remain important questions, even more so as our civilization depends increasingly on structures and services where such failure can be catastrophic. How can we predict and prevent such failures? Can we analyze the precursory signals sufficiently in advance to take appropriate measures, such as the timely evacuation of structures or localities, or the shutdown of facilities such as nuclear power plants?

Whilst these questions have long been the subject of research, the study of fracture and breakdown processes has now gone beyond simply designing safe and reliable machines, vehicles and structures. From the fracture of a wood block or the tearing of a sheet of paper in the laboratory, the breakdown of an electrical network on an engineering scale, to an earthquake on a geological scale, one finds common threads and universal features in failure processes. The ideas and observations of material scientists, engineers, technologists, geologists, chemists and physicists have all played a pivotal role in the development of modern fracture science.

Over the last three decades, considerable progress has been made in modeling and analyzing failure and fracture processes. The physics of nonlinear, dynamic, many-bodied and non-equilibrium statistical, mechanical systems, the exact solutions of fibre bundle models, solutions of earthquake models, numerical studies of random resistor and random spring networks, and laboratory-scale innovative experimental verifications have all opened up broad vistas of the processes underlying fracture. These have provided a unifying picture of failure over a wide range of length, energy and time scales.

This series of books introduces readers – in particular, graduate students and researchers in mechanical and electrical engineering, earth sciences, material science, and statistical physics – to these exciting recent developments in our understanding of the dynamics of fracture, breakdown and earthquakes.

Earthquakes

Models, Statistics, Testable Forecasts

Yan Y. Kagan
Department of Earth and Space Sciences, University of California, Los Angeles, USA

This work is a co-publication between the American Geophysical Union and Wiley

AGU American Geophysical Union WILEY

This edition first published 2014 © 2014 by John Wiley & Sons, Ltd
This work is a co-publication between the American Geophysical Union and Wiley

Registered office: John Wiley & Sons, Ltd, The Atrium, Southern Gate, Chichester, West Sussex, PO19 8SQ, UK

Editorial offices: 9600 Garsington Road, Oxford, OX4 2DQ, UK
The Atrium, Southern Gate, Chichester, West Sussex, PO19 8SQ, UK
111 River Street, Hoboken, NJ 07030-5774, USA

For details of our global editorial offices, for customer services and for information about how to apply for permission to reuse the copyright material in this book please see our website at www.wiley.com/wiley-blackwell.

Library of Congress Cataloging-in-Publication Data
Kagan, Yan Y., author.
Earthquakes : models, statistics, testable forecasts / Yan Y. Kagan.
pages cm – (Statistical physics of fracture and breakdown)
Includes bibliographical references and index.
ISBN 978-1-118-63792-0 (hardback)
1. Earthquake prediction. 2. Earthquake hazard analysis. I. Title.
QE538.8.K32 2014
551.2201′12 – dc23

2013033255

A catalogue record for this book is available from the British Library.

Wiley also publishes its books in a variety of electronic formats. Some content that appears in print may not be available in electronic books.

Cover image: Earthquake. Men looking at cracks in the ground caused by a magnitude 7.1 earthquake that struck near the city of Van, Turkey, on 23rd October 2011. Photographed in near Alakoy, Turkey, on 30th November 2011. Ria Novosti/Science Photo Library.
Structural damage on an apartment building during the earthquake of February 27, 2010 in Chile (Santiago). © iStockphoto.com/erlucho (Front cover).
Location of shallow earthquakes in the Global Centroid Moment Tensor (GCMT) catalog, 1976/1/1–2012/12/31. Courtesy of Göran Ekström and the GCMT project (Back cover).
Cover design by Steve Thompson

Set in 10.5/12.5pt DanteMT by Laserwords Private Limited, Chennai, India
Printed and bound in Malaysia by Vivar Printing Sdn Bhd

1 2014

To my wife Olga, always my first reader, editor and critic

Contents

Preface

Quantitative prediction is the aim of every science. As Ben-Menahem (1995, p. 1217) puts it:

> *[T]he ultimate test of every scientific theory worthy of its name, is its ability to predict the behavior of a system governed by the laws of said discipline.*

Accordingly, the most important issue in earthquake seismology is *earthquake prediction*. This term, however, has been been the topic of scientific debate for decades. For example, Wood and Gutenberg (1935) write:

> *To have any useful meaning the prediction of an earthquake must indicate accurately,* ***within narrow limits****, the region or district where and the time when it will occur – and, unless otherwise specified, it must refer to a shock of important size and strength, since small shocks are very frequent in all seismic regions.*

Because earthquake prediction is complicated by a number of factors, Wood and Gutenberg propose the term *earthquake forecast*, as an alternative, where in effect the earthquake occurrence rate is predicted.

Long-term studies, however, indicate that the prediction of individual earthquakes, as suggested in the first definition by Wood and Gutenberg, is impossible (Geller 1997; Geller *et al.* 1997; Kagan 1997b). Furthermore, as we show in Chapters 2 and 3, even the notion of individual earthquakes or individual faults cannot be properly defined because of earthquake process fractality. Therefore, below we treat the terms *earthquake prediction* and *earthquake forecast* as synonyms.

Available books on seismology primarily discuss the problems of elastic wave propagation and study the Earth structure. This book will take a different approach, focusing instead on earthquake seismology, defined as rigorous quantitative study of the earthquake occurrence Even though several books on earthquake seismology and some books on earthquake prediction are available, there are no in-depth monographs considering the stochastic modeling of the fractal multidimensional processes and the rigorous statistical analysis of the earthquake occurrence. In this book the results of modeling and statistical analysis are applied to evaluate the short- and long-term occurrence rates of future earthquakes, both regionally and globally, and, most importantly, to test these forecasts according to stringent criteria.

The subject of this book could therefore be roughly defined as "Statistical Seismology" (Vere-Jones 2009, 2010). There has been significant interest in the problems of statistical seismology recently: since 1998, the International Workshops on Statistical Seismology (Statsei2–Statsei7) have provided researchers with an opportunity to evaluate recent developments in statistical seismology, as well as define future directions of research (see http://www.gein.noa.gr/statsei7/). Problems explored in these meetings include the statistical behavior of earthquake occurrence and patterns, time-dependent earthquake forecasting, and forecast evaluations. In addition, in this book we investigate geometrical properties of the earthquake fault system and the interrelations of earthquake focal mechanisms.

Thus, this book is a comprehensive and methodologically rigorous analysis of earthquake occurrence. Earthquake processes are inherently multidimensional: in addition to the origin time, 3-D locations, and the measures of size for each earthquake, the orientation of the rupture surface and its displacement requires for its representation either second-rank symmetric tensors or quaternions. Models based on the theory of stochastic multidimensional point processes were employed here to approximate the earthquake occurrence pattern and evaluate its parameters. The terms "moment" or "moment tensor" used in seismology to signify "the seismic moment" or "the seismic moment tensor" (see Section 2.2), will throughout this book be distinguished from moments used in statistics.

Adequate mathematical and statistical techniques have only recently become available for analyzing fractal temporal, spatial, and tensor patterns of point process data generally and earthquake data in particular. Furthermore, only in the past 20–30 years have the processing power of modern computers and the quality, precision, and completeness of earthquake datasets been sufficient to allow a detailed, full-scale investigation of earthquake occurrence.

Since the early nineteenth century, the Gaussian (normal) distribution was used almost exclusively for the statistical analysis of data. However, the Gaussian distribution is a special, limiting case of a broad class of *stable* probability distributions. These distributions, which, with the exception of the Gaussian law, have a power-law (heavy) tail, have recently become an object of intense mathematical investigation. These distributions are now applied in physics, finance, and other disciplines. One can argue that they are more useful in explaining natural phenomena than the Gaussian law. For the stable distributions with the power-law tail exponent $1.0 < \beta < 2.0$, the variance is infinite; if $1.0 \geq \beta$, the mean is infinite (see Section 5.4). The application of these distributions to the analysis of seismicity and other geophysical phenomena would significantly increase our quantitative understanding of their fractal patterns.

After careful analysis of systematic and random effects in earthquake registration and interpretation of seismograms, we show that most of these statistical distribution parameters have universal values. These results help explain such classical distributions as Omori's law and the Gutenberg-Richter relation, used in earthquake seismology for many decades. We show that the parameters of these distributions are universal constants defined by simple mathematical models. We derived a negative-binomial distribution for earthquake numbers, as a substitute for the Poisson distribution. The fractal correlation dimension for spatial distributions of earthquake hypocenters was determined. We also investigated the disorientation of earthquake focal mechanisms and showed that it follows the rotational Cauchy distribution. We evaluated the parameters of these distributions in various earthquake zones, and estimated their systematic and random errors.

These statistical and mathematical advances made it possible to produce quantitative forecasts of earthquake occurrence. The theoretical foundations for such forecasts based on multidimensional stochastic point processes were first proposed by Kagan (1973). Later we showed how the long- and short-term forecasts can be practically computed and how their efficiency can be estimated. Since 1999, daily forecasts have been produced, initially for several seismically active regions and more recently expanded to cover the whole Earth. The recent mega-earthquake in Tohoku, Japan, which caused many deaths and very significant economic losses, demonstrates the importance of forecasts in terms of a possible earthquake size, its recurrence time, and temporal clustering properties.

An important issue in the study of earthquake occurrence and seismic hazard is the verification of seismicity models. Until recently seismic event models and predictions were based exclusively on case histories. It was widely believed that long-term earthquake occurrence, at least for large earthquakes, was quasi-periodic or cyclic (seismic gap and characteristic earthquake hypotheses). The Parkfield earthquake prediction experiment and many other forecasts were therefore based on these models. However, when we tested the seismic gap models against the earthquake record, it turned out that the performance of the gap hypothesis was worse than a similar earthquake forecast (null hypothesis) based on a random choice (temporal Poisson model). Instead of being quasi-periodic, large earthquakes are clustered in time and space (Section 2.4). The Tohoku event consequences underscore that all statistical properties of earthquake occurrence need to be known for correct prediction: the extent of the losses was to a large degree due to the use of faulty models of characteristic earthquakes to evaluate the maximum possible earthquake size (Section 5.6).

Earthquake occurrence models that are too vague to be testable, or are rejected by rigorous objective statistical tests (see above) are not discussed in detail here. In our opinion, the only models worthy of analyses, are those which produce testable earthquake forecasts.

Since this book is an initial attempt to thoroughly and rigorously analyze earthquake occurrence, many unresolved issues still remain. In the final Section (12.3), we list some challenging questions that can now be addressed by thorough theoretical studies and observational statistical analysis. There is, of course, the possibility that some of these problems have been solved in other scientific disciplines; in this case, we will need to find out how to implement these solutions in earthquake science.

Acknowledgments

This book is the result of my work over many years. I am grateful to the various scientists throughout the world with whom I have collaborated. There have been far too many people for me to list them individually here. There are nevertheless a few individuals I do want to especially thank.

First of all, I would like to mention the seismologists and mathematicians I worked with in the former Soviet Union from 1957–1974: Anatoliy A. Zhigal'tsev, Stanislav S. Andreev, Yuriy N. Godin, Michail S. Antsyferov, Nina G. Antsyferova (Goykhman), Igor M. Lavrov, Zinoviy Gluzberg (Zinik), Yuriy V. Riznichenko, Alexandr A. Gusev, and George M. Molchan. The work of some of these individuals sparked my interest in earthquake seismology and applying mathematical and statistical methods to the solution of seismological problems.

I wish to express deep gratitude to Leon Knopoff, who brought me to UCLA in 1974 and who was my coauthor for many years. I would also like to gratefully acknowledge my long-term collaborator Dave Jackson. About half of my papers in the United States were developed in cooperation with these colleagues, and I learned a lot from them.

I was also greatly influenced by my collaboration with statisticians David Vere-Jones and George Molchan, from whom I learned about many issues in mathematical statistics. Their recommendations have been used throughout my work over the years.

I have also benefited from the advice of and very useful discussions with many coauthors of my papers, including Peter Bird, Frederick Schoenberg, Robert Geller, Heidi Houston, Max Werner, Agnés Helmstetter, Didier Sornette, Zhengkang Shen, Paul Davis, Ilya Zaliapin, Francesco Mulargia, Qi Wang, Silvia Castellaro, and Yufang Rong among others.

Several individuals have, through my reading of their work and through conversations with them, significantly influenced my approach to solving the problems described in this book. Of these I would like to note with special gratitude Benoit Mandelbrot, Per Bak, Akiva Yaglom, George Backus, Vladimir Zolotarev, Adam Morawiec, Yosi Ogata, Cliff Frohlich, Andrei Gabrielov, Fred Schwab, Vlad Pisarenko, Philip Stark, Tokuji Utsu, Göran Ekström, Jiancang Zhuang, Ritsuko Matsu'ura, Jeremy Zechar, Yehuda Ben-Zion, William Newman, David Rhoades, Danijel Schorlemmer, David Harte, and Peiliang Xu.

I am grateful to Kathleen Jackson who helped me to become a better writer by editing many of my published papers.

I also want to offer profound thanks to several computer specialists who helped me in my calculations and in other computer-related tasks: John Gardner, Robert Mehlman, Per Jögi, Oscar Lovera, and Igor Stubailo.

Reviews by an anonymous reviewer and by Peter Bird have been very helpful in revising and improving the book manuscript.

Finally, I would like to thank several people from John Wiley & Sons publishing company for guiding me through the process of the book production. I am grateful to Ian Francis, Delia Sandford, and Kelvin Matthews, all of Oxford, UK. I thank Alison Woodhouse (Holly Cottage, UK) for tracking down permissions for my previously published figures and tables. The copy-editing work by Susan Dunsmore (Glasgow, UK) is appreciated. I am also grateful to Production Editor Audrie Tan (Singapore) and Project Manager Sangeetha Parthasarathy (Chennai, India) for their work in producing and typesetting the book.

List of Abbreviations

AIC	Akaike Information Criterion
ANSS	Advanced National Seismic System (catalog)
CBM	Critical Branching Model
CDF	Cumulative distribution function
CLVD	Compensated Linear Vector Dipole
CSEP	Collaboratory for Study of Earthquake Predictability
DC	Double-couple
ED	Error diagram
ETAS	Epidemic Type Aftershock Sequence
GCMT	Global Centroid Moment Tensor (catalog)
GPS	Global Positioning System
G-R	Gutenberg-Richter (relation)
GSRM	Global Strain Rate Map
IGD	Inverse Gaussian Distribution
i.i.d.	independent identically distributed
INSAR	Interferometric Synthetic Aperture Radar
MLE	Maximum likelihood estimate
NBD	Negative binomial distribution
PDE	Preliminary Determinations of Epicenters (catalog)
PDF	Probability density function
PF	Probability function
RELM	Regional Earthquake Likelihood Models
ROC	Relative Operating Characteristic
SCEC	Southern California Earthquake Center
SOC	Self-Organized Criticality
STF	Source Time Function
TGR	Tapered G-R (distribution)
USGS	U.S. Geological Survey

List of Mathematical Symbols

3-D	three-dimensional
CLVD	Γ-index, Eq. 8.15
DC1	double-couple earthquake source with no symmetry, p. 160
DC2	double-couple source with C_2, order 2 cyclic symmetry, p. 160
DC4	double couple source with D_2, order 2 dihedral symmetry, p. 160
I	Information score, Eqs. 11.11, 11.13
I_0	Forecast Information score (specificity), Eq. 11.14
I_1	Information score (success) for cell centers of forecasted events, Eq. 11.17
I_1'	Information score for earthquakes which occurred in the training period
I_2	Information score for forecasted events, Eq. 11.18
I_3	Information score for simulated events, Eq. 11.19
I_4	Information score based on forecasted events curve, Eq. 11.23
m_b	body-wave magnitude
m_L	local magnitude
M_S	surface-wave magnitude
M_t	seismic moment detection threshold of a seismographic network
m_t	magnitude threshold
$\mathbf{R}^3$ or $\mathbf{R}^2$	Euclidian space
S^2	two-dimensional (regular) sphere
S^3	three-dimensional sphere
$SO(3)$	group of 3-D rotations
Tr(.)	square matrix trace

Part I

Models

1 Motivation: Earthquake science challenges

Our purpose is to analyze the causes of recent failures in earthquake forecasting, as well as the difficulties of earthquake investigation. It is widely accepted that failure has dogged the extensive efforts of the last 30 years to find "reliable" earthquake prediction methods, the efforts which culminated in the Parkfield prediction experiment (Roeloffs and Langbein 1994; Bakun *et al.* 2005 and their references) in the USA and the Tokai experiment in Japan (Mogi 1995). Lomnitz (1994), Evans (1997), Geller *et al.* (1997), Jordan (1997), Scholz (1997), Snieder and van Eck (1997), and Hough (2009) discuss various aspects of earthquake prediction and its lack of success. Jordan (1997) comments that

> *The collapse of earthquake prediction as a unifying theme and driving force behind earthquake science has caused a deep crisis.*

Why does theoretical physics fail to explain and predict earthquake occurrence? The difficulties of seismic analysis are obvious. Earthquake processes are inherently multidimensional (Kagan and Vere-Jones 1996; Kagan 2006): in addition to the origin time, 3-D locations, and measures of size for each earthquake, the orientation of the rupture surface and its displacement requires for its representation either second-rank tensors or quaternions (see more below). Earthquake occurrence is characterized by extreme randomness; the stochastic nature of seismicity is not reducible by more numerous or more accurate measurements. Even a cursory inspection of seismological datasets suggests that earthquake occurrence as well as earthquake fault geometry are scale-invariant or fractal (Mandelbrot 1983; Kagan and Vere-Jones 1996; Turcotte 1997; Sornette 2003; Kagan 2006; Sornette and Werner 2008). This means that the statistical distributions that control earthquake occurrence are power-law or stable (Lévy-stable) distributions. See also http://www.esi-topics.com/earthquakes/interviews/YanYKagan.html.

After looking at recent publications on earthquake physics (for example, Kostrov and Das 1988; Lee *et al.* 2002; Scholz 2002; Kanamori and Brodsky 2004; Ben-Zion 2008), one gets the impression that knowledge of earthquake

Earthquakes: Models, Statistics, Testable Forecasts, First Edition. Yan Y. Kagan.

process is still at a rudimentary level. Why has progress in understanding earthquakes been so slow? Kagan (1992a) compared the seismicity description to another major problem in theoretical physics: turbulence of fluids. Both phenomena are characterized by multidimensionality and stochasticity. Their major statistical ingredients are scale-invariant, and both have hierarchically organized structures. Moreover, the scale of self-similar structures in seismicity and turbulence extends over many orders of magnitude. The size of major structures which control deformation patterns in turbulence and brittle fracture is comparable to the maximum size of the region (see more in Kagan 2006).

Yaglom (2001, p. 4) commented that turbulence status differs from many other complex problems which twentieth-century physics has solved or has considered.

> *[These problems] deal with some very special and complicated objects and processes relating to some extreme conditions which are very far from realities of the ordinary life ... However, turbulence theory deals with the most ordinary and simple realities of the everyday life such as, e.g., the jet of water spurting from the kitchen tap. Therefore, the turbulence is well-deservedly often called "the last great unsolved problem of the classical physics."*

Although solving the Navier-Stokes equations, describing turbulent motion in fluids is one of the seven millennium mathematical problems for the twenty-first century (see http://www.claymath.org/millennium/), the turbulence problem is not among the ten millennium problems in physics presented by the University of Michigan, Ann Arbor (see http://feynman.physics.lsa.umich.edu/-strings2000/millennium.html), or among the 11 problems by the National Research Council's board on physics and astronomy (Haseltine 2002). In his extensive and wide-ranging review of current theoretical physics, Penrose (2005) does not include the turbulence or Navier-Stokes equations in the book index.

Like fluid turbulence, the brittle fracture of solids is commonly encountered in everyday life, but so far there is no real theory explaining its properties or predicting outcomes of the simplest occurrences, such as a glass breaking. Although computer simulations of brittle fracture (for example, see O'Brien and Hodgins 1999) are becoming more realistic, they cannot yet provide a scientifically faithful representation. Brittle fracture is a more difficult scientific problem than turbulence, and while the latter has attracted first-class mathematicians and physicists, no such interest has been shown in the mathematical theory of fracture and large-scale deformation of solids.

In this book we first consider multidimensional stochastic models approximating earthquake occurrence. Then we apply modern statistical methods to investigate distributions of earthquake numbers, size, time, space, and focal mechanisms. Statistical analysis of earthquake catalogs based on stochastic point process theory provides the groundwork for long- and short-term forecasts. These forecasts are rigorously tested against future seismicity records. Therefore, here statistical study of earthquake occurrence results in verifiable earthquake prediction.

The book has 12 chapters. In this chapter, we discuss the fundamental challenges which face earthquake science. In Chapter 2 we review the seismological background information necessary for further discussion as well as basic models of earthquake occurrence. Chapter 3 describes several multidimensional stochastic models used to approximate earthquake occurrence. They are all based on the theory of branching processes; the multidimensional structure of earthquake occurrence is modeled. Chapter 4 discusses the distribution of earthquake numbers in various temporal-spatial windows. In Chapters 5–8 some evidence for the scale-invariance of earthquake process is presented, in particular, one-dimensional marginal distributions for the multidimensional earthquake process are considered. Fractal distributions of earthquake size, time intervals, spatial patterns, focal mechanism, and stress are discussed. Chapter 9 describes the application of stochastic point processes for statistical analysis of earthquake catalogs and summarizes the results of such analysis. Chapter 10 describes the application of the results of Chapter 9 for long- and short-term prediction of an earthquake occurrence. Methods of quantitative testing of earthquake forecasts, and measuring their effectiveness or skill are discussed in Chapter 11. The final discussion (Chapter 12) summarizes the results obtained thus far and presents problems and challenges still facing seismologists and statisticians.

2 Seismological background

2.1 Earthquakes

This chapter discussion mainly follows Kagan (2006). Since this book is intended for seismologists, as well as statisticians, physicists, and mathematicians, we briefly describe earthquakes and earthquake catalogs as primary objects of the statistical study. A more complete discussion can be found in Bullen (1979), Lee *et al.* (2002), Scholz (2002), Bolt (2003), Kanamori and Brodsky (2004). As a first approximation, an earthquake may be represented by a sudden shear failure – appearing as large quasi-planar dislocation loop in rock material (Aki and Richards 2002).

Figure 2.1a shows a fault-plane diagram. Earthquake rupture starts on the fault-plane at a point called the "hypocenter" (the "epicenter" is a projection of the hypocenter on the Earth's surface), and propagates with a velocity close to that of shear waves (2.5 – 3.5 km / s). The "centroid" is in the center of the ruptured area. Its position is determined by a seismic moment tensor inversion (Ekström *et al.* 2012, and references therein). As a result of the rupture, two sides of the fault surface are displaced by a slip vector along the fault-plane. For large earthquakes, such displacement is on the order of a few meters.

The earthquake rupture excites seismic waves which are registered by seismographic stations. The seismograms are processed by computer programs to obtain a summary of the earthquake's properties. Routinely, these seismogram inversions characterize earthquakes by their origin times, hypocenter (centroid) positions, and second-rank symmetric seismic moment tensors.

Equivalent to the earthquake focus is a quadrupole source of a particular type (Fig. 2.1b) known in seismology as a "double-couple" or *DC* (Burridge and Knopoff 1964; Aki and Richards 2002; Kagan 2005b; Okal 2013). Figure 2.1c represents a "beachball" – the quadrupolar radiation pattern of earthquakes. The focal plots involve painting on a sphere the sense of the first motion of the far-field primary, P-waves: solid for compressional motion and open for dilatational. Two orthogonal planes separating these areas are the fault and the auxiliary planes.

Earthquakes: Models, Statistics, Testable Forecasts, First Edition. Yan Y. Kagan.

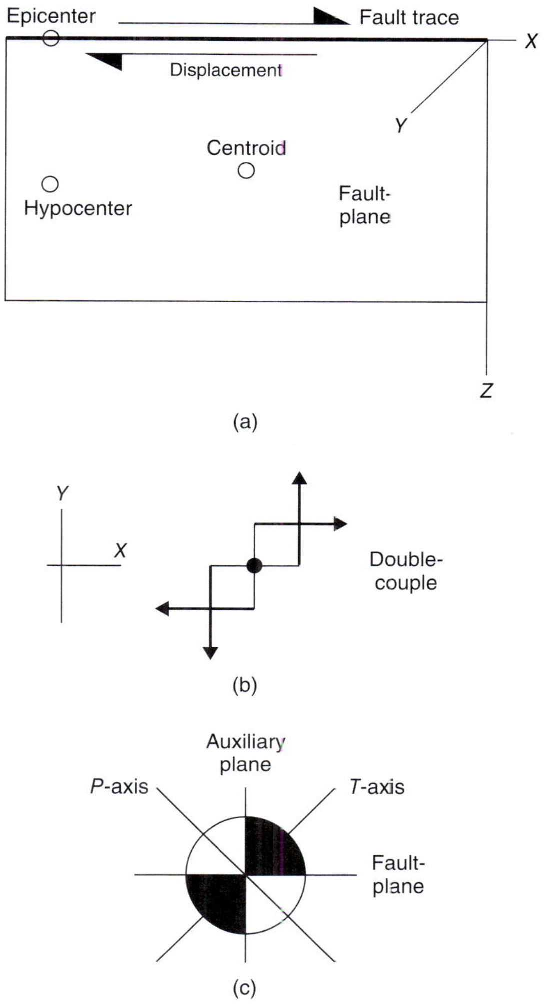

Fig. 2.1 Schematic diagrams of earthquake focal mechanism. (a) Fault-plane diagram – final rupture area (see text). (b) Double-couple source, equivalent forces yield the same displacement as the extended fault (see item *a*) rupture in a far-field. (c) Equal-area projection on the lower hemisphere (Aki and Richards 2002, p. 110) of quadrupole radiation patterns. The null (*N* or *B*) axis is orthogonal to the *T*- and *P*-axes, or it is located on the intersecting fault and auxiliary planes, that is, perpendicular to the paper sheet in this display. *Source:* Kagan (2006), Fig. 1.

During routine determination of the focal mechanisms, it is impossible to distinguish these planes. Their intersection is the null-axis (*N*-axis or *B*-axis), the *P*-axis is in the middle of the open lune, and the *T*-axis in the middle of the closed lune. These three axes are called the "principal axes of an earthquake focal mechanism," and their orientation defines the mechanism.

In the system of coordinates of *TPN* axes, shown in Fig. 2.1c, the second-rank seismic moment tensor matrix is

$$\mathbf{M} = M \times \mathrm{diag}[1, -1, 0], \qquad (2.1)$$

where M is a scalar seismic moment of an earthquake, measured in Newton-m (Nm). In an arbitrary system of coordinates all entries in 3×3 matrix (2.1) are non-zero. However, the tensor is always traceless, and for a *DC* it has a zero determinant. Hence for a *DC* source, it has only four degrees of freedom: one for the norm of the tensor (proportional to the scalar seismic moment) and three for orientation. The three representations of focal mechanism shown in Fig. 2.1 as well as in (2.1) are mathematically equivalent; Kagan (2005b) discusses interrelations between these parameterizations (see also Chapter 8).

In the geological and geophysical literature *DC* focal mechanisms usually are classified depending on their geometric relation to the Earth's surface: strike-slip, normal, and thrust mechanisms are distinguished (Frohlich 2001; Kagan 2005b). There are many methods for displaying sets of focal mechanisms in a 2-D plot (Frohlich 2001; Kagan 2005b).

2.2 Earthquake catalogs

Modern earthquake catalogs are collections of estimated earthquake origin times, hypocenter or centroid locations, measures of earthquake size (scalar seismic moment or appropriate empirical magnitude), and earthquake focal mechanisms or seismic moment tensors (Aki and Richards 2002). Such datasets in a certain sense fully describe each earthquake; for instance, one can compute far-field, low-frequency seismic radiation using the above information. Nevertheless, detailed study of earthquake occurrences shows that this description is far from complete, since each earthquake represents a process with seismic moment tensor or focal mechanism varying in extended time–space. Furthermore, because earthquakes have fractal features, even defining an individual earthquake is problematic: earthquake catalog records are the result of a complex interaction of fault ruptures, seismographic recordings, and their interpretations (see Sections 3.3 and 6.2).

In Figure 2.2 we display a map of earthquake centroids in the global CMT (Centroid-Moment-Tensor) catalog (Ekström *et al.* 2012, and its references). Earthquakes are mostly concentrated at tectonic plate boundaries. Each earthquake in this catalog is characterized by a centroid moment tensor solution.

Figure 2.3 displays a map of focal mechanisms for southern California. To create this map we combined historical, geologic, and instrumental information on seismicity. Details are provided by Kagan *et al.* (2006). The focal mechanisms

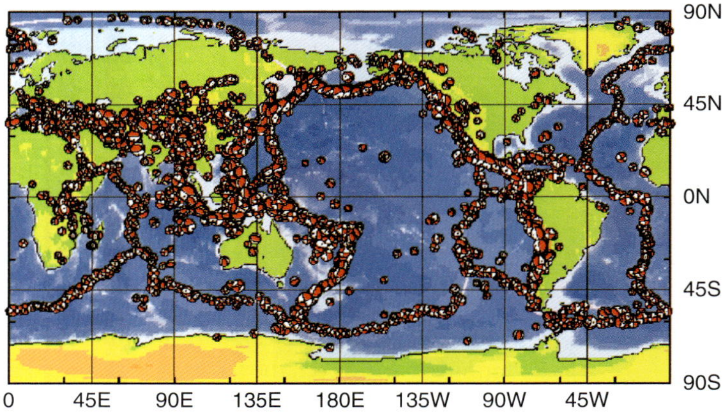

Fig. 2.2 Location of shallow (depth 0–70 km) earthquakes in the Global Centroid Moment Tensor (GCMT) catalog, 1976/1/1–2012/12/31. Earthquake focal mechanisms are shown by stereographic projecting the lower focal hemisphere (Aki and Richards 2002). The size of the symbol is proportional to the earthquake magnitude. (Courtesy of Göran Ekström and the GCMT project.) *Source:* http://www.globalgcmt.org/

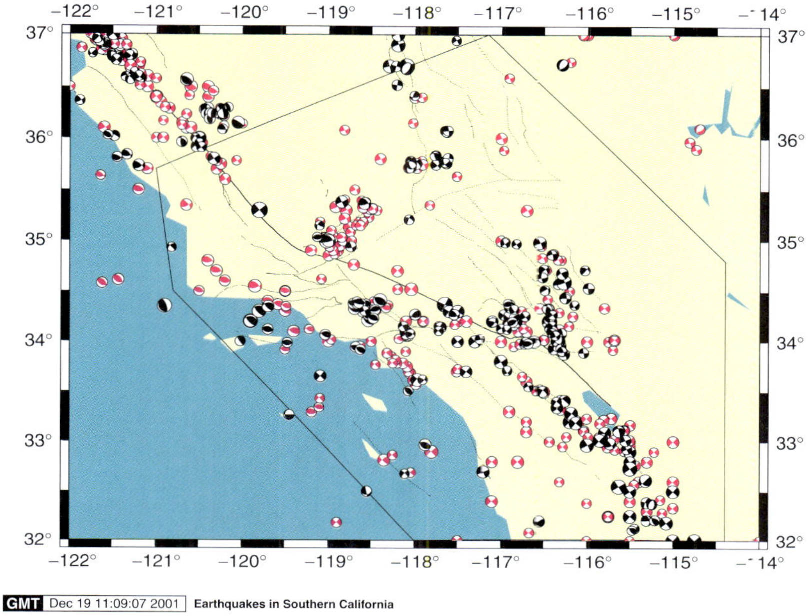

Fig. 2.3 The southern California catalog and polygon: a region wherein the catalog is believed to be accurate and complete (see also Fig. 2.5). Time period 1800–2002. Black beachballs – known solutions; orange beachballs – imputed solutions, obtained through interpolation from known focal mechanisms. Thin curved lines are active earthquake faults, including the San Andreas fault on which many earthquakes are concentrated. *Source:* Kagan *et al.* (2006), Fig. 1.

can be characterized by a 3-D rotation from a fixed position; an alternative, more compact way to represent each mechanism is a normalized quaternion (Kagan 1991, 2005, Chapter 8).

We often use an extended source representation for large earthquakes: we replace each epicenter point by a series of rectangular dislocations which cover the whole rupture area, see Figure 2.4. This figure displays focal mechanisms for California for earthquakes with a magnitude larger or equal to 6.5 (Kagan *et al.* 2006; Wang *et al.* 2009).

There are many other datasets which characterize earthquake processes, such as detailed investigation of the earthquake rupture for particular events, or earthquake fault maps and descriptions of certain faults. The unique advantages of an earthquake catalog include relative completeness, uniformity of coverage, and quantitative estimates of errors. These properties make catalogs especially suitable for statistical analysis and modeling. The catalogs can be roughly subdivided into two categories: global or regional catalogs covering large areas (continents or their large parts), and local catalogs for particular areas such as southern California (Fig. 2.3) and still smaller areas (Kagan 2003).

Mathematically, earthquake occurrences can be represented as a multidimensional stochastic process (Kagan 1992a, 2006):

$$T \times \mathbf{R}^3 (= \mathbf{R}^2 \times H) \times M \times \mathbf{SO}(3) (= \Phi \times S^2) \quad \text{(time-space-size-orientation)},$$

where $\mathbf{R}^3$ or $\mathbf{R}^2$ is the Euclidian space, H is the depth dimension, M is the scalar seismic moment, and $\mathbf{SO}(3)$ is the 3-D special orthogonal (rotation) group. The latter may be represented as a rotation by the angle Φ around a rotation pole

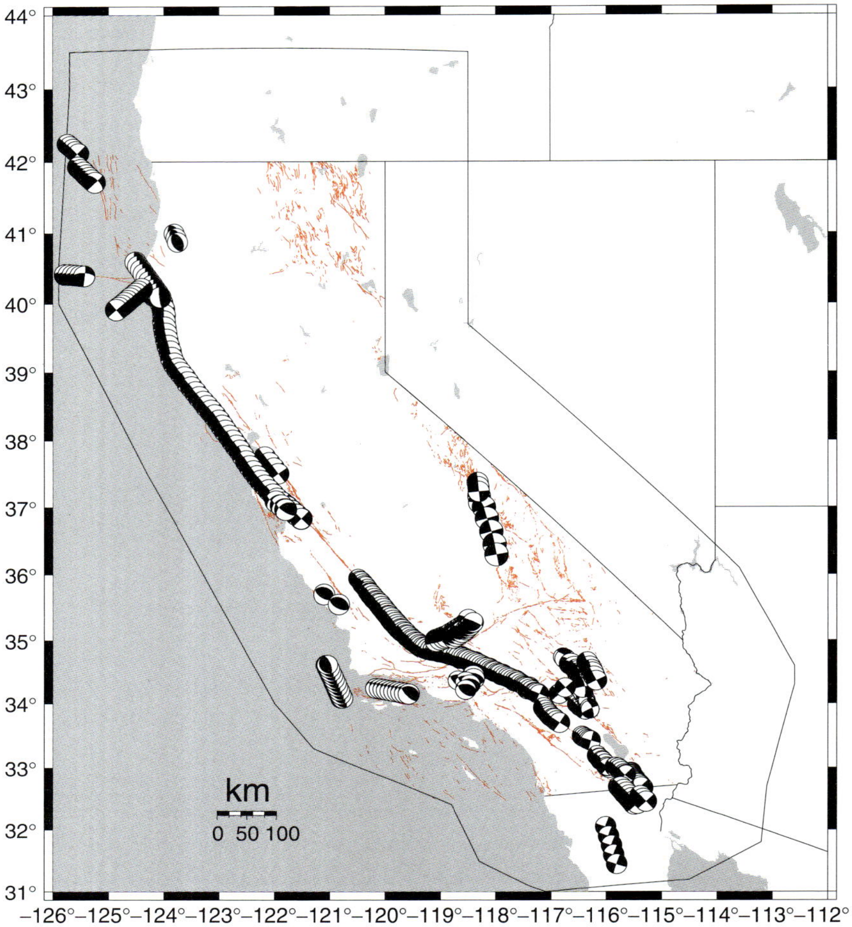

Fig. 2.4 Plot of the catalog of extended ruptures for earthquakes in California, 1800–2006. Beach balls are focal mechanisms. Thin red curved lines are active earthquake faults. The two longest chains of focal mechanisms along the San-Andreas fault are the 1906 (north) and 1857 (south) large earthquakes. Black solid lines are States' boundaries. The contour around California is an extension of the testing area specified by Schorlemmer and Gerstenberger (2007). *Source:* Wang *et al.* (2009), Fig. 3.

distributed over the 2-D sphere S^2 (Chapter 8). Multiple studies summarized in Kagan (2006) and Kagan and Vere-Jones (1996) indicate that marginal earthquake distributions are scale-invariant for all the above variables. The fractal pattern breaks down for large size, distance or time intervals. We discuss this in Chapters 4–8.

An important feature of the available earthquake catalogs is the range (X_r) of the above variables related to the average estimation error (A) in estimating them. The ratio of the range to an error describes roughly the information one can obtain from a catalog. These ratios are only approximate to one order of magnitude (see more in Kagan 2003): a summary is shown in Table 2.1.

From Table 2.1 we see that the temporal structure of earthquake occurrences can be detailed with great accuracy. The locations of earthquake foci are estimated relatively accurately in the horizontal plane, but vertical errors are often significantly larger. This effectively reduces the available spatial information.

Table 2.1
Information available in earthquake catalogs.

Notes – *items in the first column*: 2, 3 – global regular catalogs; 4–6 moment-tensor catalogs (Kagan 2003); 2′ – local catalogs; 2″, 3″ – wave correlation catalogs (e.g., Hauksson and Shearer 2005; Shearer *et al.* 2005; Hauksson *et al.* 2012). *Source:* Kagan (2006, Table 1).

#	Variable	Accuracy (A)	Range (X_r)	X_r/A
1	Origin time, T	0.01–1 s	5–25 y	$10^9 - 10^{11}$
2	Horiz. space, R^2	3–10 km	3000 km	10^3
2′	Horiz. space, R^2	0.5 km	200 km	$10^{2.5}$
2″	Horiz. space, R^2	0.02 km	2–20 km	$10^2 - 10^3$
3	Vert. space, R	5–15 km	50 km	10
3″	Vert. space, R	0.1 km	10 km	10^2
4	Moment magn., m	0.07	6.0	10^2
5	Rot. angle, Φ		120°	
6	Rot. pole, S^2	10°	360°	10^3

The influence of location errors and other nuisance variables often extends well above a catalog's reported accuracy values (Kagan and Knopoff 1980; Kagan 1991a, 2003). Similarly, boundary effects can be observed at distances substantially smaller than a region's total size. Therefore, the scale-invariant range of the spatial distribution is likely to be smaller than the 10^2–10^3 shown in Table 2.1. Focal mechanisms, which have been reliably obtained in only the last 35 years, also have large uncertainties (Kagan 2000; 2003).

Catalogs are a major source of information on earthquake occurrence. Since the late nineteenth century certain statistical features have been established: Omori (1894) studied temporal distribution; Gutenberg and Richter (1944) investigated size distribution; quantitative investigating spatial patterns started late (Kagan and Knopoff 1980).

Kostrov (1974) proposed that earthquake displacement can be described by a second-rank symmetric tensor. Gilbert and Dziewonski (1975) were the first to obtain a tensor solution from seismograms. However, statistical investigating catalogs remained largely restricted to time-size-space regularities. Tensor or focal mechanism analysis is difficult because we lack the appropriate statistical tools to analyze either second-rank tensors or quaternions which properly represent earthquake focal mechanisms (Kagan 1991c, 2005b, see Chapter 8). Kagan and Knopoff (1985a,b) and Kagan (1992b,c, 2000, 2005b) were the first to investigate the statistical properties of the seismic moment tensor (Chapter 8).

2.3 Description of modern earthquake catalogs

We studied earthquake distributions and clustering for the global CMT catalog of moment tensor inversions compiled by the GCMT group (Ekström *et al.* 2012). The present catalog contains more than 38,000 earthquake entries for the period 1976/1/1 to 2012/12/31. Earthquake size is characterized by a scalar seismic moment M. The magnitude threshold m_t for the full catalog is $m5.8$ (Kagan 2003), the earthquake record above the threshold is considered to be complete. Earthquake moment magnitude m_W is related to the scalar seismic moment M via

(Kanamori 1977; Hanks and Kanamori 1979)

$$m_W = \frac{2}{3}\log_{10} M - C, \tag{2.2}$$

where seismic moment M is measured in Newton-m, and C is usually taken to be between 6.0 and 6.1. Below we use $C = 6.0$ (Hanks 1992).

The PDE (Preliminary Determinations of Epicenters) worldwide catalog is published by the USGS (U.S. Geological Survey 2012). The catalog is distributed in preliminary form within a few minutes after each earthquake and in the final form with a few months latency. The catalog measures earthquake size, using several magnitude scales, and has provided the body-wave (m_b) and surface-wave (M_S) magnitudes (Lay and Wallace 1995) for most moderate and large events since 1965 and 1968, respectively. The moment magnitude (m_W) estimate has been added recently.

Determining one measure of earthquake size for the PDE catalog entails a certain difficulty. For example, Kagan (1991b) calculates a weighted average of several magnitudes to use in the likelihood search. Kagan (2003) also analyses systematic and random errors for various magnitudes in the PDE catalog. At various times different magnitudes have been listed in the PDE catalog, and establishing their relationships is challenging. Therefore, we chose a palliative solution: for each earthquake we use the maximum magnitude among those magnitudes shown (Kagan and Jackson 2012). This arrangement is easier to carry out and the results are easily reproducible. Because of the magnitude saturation effect (Kagan 2003) and in particular the magnitude availability, the maximum magnitude is usually m_b for moderate earthquakes and M_S for large ones. For larger recent earthquakes the maximum magnitude is most likely m_W. Depending on the time period and the region, the magnitude threshold of the PDE catalog is of the order 4.5 to 4.7 (Kagan 2003; Romashkova 2009). The catalog contains more than 50,000 shallow earthquakes with $m_b \geq 5$ since 1965.

The ANSS (Advanced National Seismic System 2008) composite catalog is a world-wide earthquake catalog that is created by merging the master earthquake catalogs from contributing ANSS institutions and then removing duplicate solutions for the same event.

The CalTech or California Institute of Technology (CIT) dataset (Hileman *et al.* 1973; Hutton and Jones 1993; Hutton *et al.* 2006; 2010) which begins in 1932, was the first instrumental local catalog to include small earthquakes ($m \geq 3$). The available catalog ends at 2001/12/31. The magnitude threshold of the 1932–2001 catalog is close to $m3$ (however, see Kagan (2004) and Section 6.2.1 for the threshold variations after strong earthquakes). In recent years, even smaller earthquakes have been included in the catalog, so for the 1989–2001 time period, a threshold of $m2$ is assumed. In many cases we selected earthquakes in a rectangular window (latitudes > 32.5°N and ≤ 36.5°N, longitudes > 114.0°W and ≤ 122.0°W) to analyse.

In Fig. 2.5 we show epicenters in the catalog compiled by Hauksson and Shearer (2005) for the period 1984–2002. They applied waveform

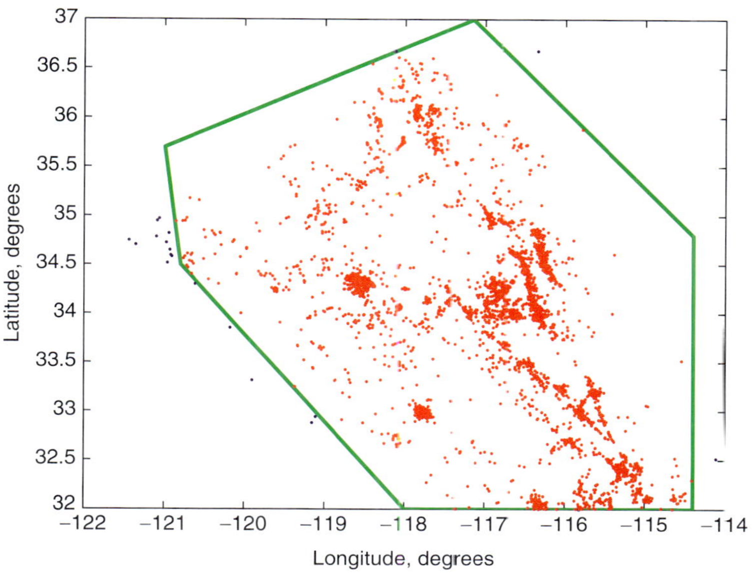

Fig. 2.5 Epicenter distribution of earthquakes in southern California in the Hauksson and Shearer (2005) catalog. Time interval is 1984–2002, magnitude threshold $m_t = 3$. A 6-point box with the following coordinates is used: North latitude −32.0°, 34.8°, 37.0°, 35.7°, 34.5°, 32.0°; West longitude −114.4°, 114.4°, 117.15°, 121.0°, 120.8°, 118.0°. Earthquake distribution is considered to be reasonably homogeneous and complete in this box for the CIT catalog. The area of the box is $S \approx 233,300$ km^2. *Source:* Kagan (2007a), Fig. 1.

cross-correlation to obtain precise differential times between nearby events. These times can then be used to greatly improve the relative location accuracy within clusters of similar events. The accuracy of the hypocenter location is very high; in many cases the horizontal errors do not exceed 0.1 km. This new catalog resolves individual faults in what previously appeared to be diffuse earthquake clouds. For the faults that are vertical or close to vertical, such as the 1992 Landers (epicenter coordinates 34.20°N, 116.44°W) or the 1999 Hector Mine (34.59°N, 116.27°W) earthquakes, the epicenters delineate the major faults as well as subsidiary features. However, for the 1994 Northridge earthquake (34.21°N, 118.54°W), epicenters form a cloud. This happens because the rupture plane of the earthquake was far from vertical (Thio and Kanamori 1996). Other relocated catalogs for southern California were produced by Shearer *et al.* (2005) and Hauksson *et al.* (2012).

The problem for almost all catalogs, especially local and regional, is their inhomogeneity: since any local seismographic network is bounded, the accuracy and catalog completeness vary considerably within the catalog area. This inhomogeneity is especially strong for the seismographic networks concentrated on island chains where stations represent a quasi-linear array; the location accuracy differs strongly along the direction of the array as compared to the orthogonal direction.

Another problem, especially for earlier catalogs, is the low precision of the hypocenter coordinates and magnitudes. Until recent years, the magnitude was specified up to one decimal place, and depth and epicentral coordinates were also rounded off. In our earlier work with earthquake catalogs, to save space, these variables were discretized, leading sometimes to loss of useful available information.

2.4 Earthquake temporal occurrence: quasi-periodic, Poisson, or clustered?

The periodic or quasi-periodic hypothesis of large earthquake occurrence has long been accepted by geoscientists. Gilbert (1884, p. 52) says:

> *[at the previous earthquake] locality the accumulated earthquake force is for the present spent, and many generations will probably pass before it again manifests itself.*

In fact, no rigorous demonstration of earthquake quasi-periodicity has been published since Gilbert's paper (see Kagan *et al.* (2012) and Section 6.3).

Similar hypotheses are called "seismic gap" or "seismic cycle" models. A seismic gap, according to such a model, is a fault or a plate segment for which the time since the previous large earthquake has not been long enough for the stress to build up.

Since earthquake occurrence is multidimensional and periodicity is a property of a one-dimensional process, the seismic record needs to be converted into a temporal series. The characteristic earthquake hypothesis (Schwartz and Coppersmith 1984) implies a sequence of recognizably similar events and provides the logical basis for discussing quasi-periodicity. Recurrence intervals and their statistics are meaningless without a clear definition of the characteristic earthquake (Davis *et al.* 1989; Jackson and Kagan 2006; Kagan *et al.* 2012). A characteristic earthquake is assumed to release most of the deviatoric stress and elastic strain around a segment. Other earthquakes are significantly smaller than the characteristic one and hence can be ignored when the total seismic moment release is calculated.

Implicitly assuming that recent large earthquakes were characteristic (although he did not use that term), Fedotov (1965, 1968, 2005) postulated that segments of plate boundaries that had not ruptured for some time were due to suffer large earthquakes. His hypothesis, if true, would have significantly advanced long-term forecasting.

McCann *et al.* (1979) adopted the gap model and produced a colored map of "earthquake potential" for close to a hundred circum-Pacific zones. They assumed that the seismic potential increases with the time since the last large earthquake. Nishenko (1991) refined the seismic gap model so that it could be more rigorously tested. He used a quasi-periodic, characteristic recurrence model to estimate conditional earthquake probabilities for 125 plate boundary segments around the Pacific Rim.

Kagan and Jackson (1991) compared the model predictions of McCann *et al.* (1979) against later earthquakes. They found that large earthquakes occurred more frequently in the very zones where McCann *et al.* had estimated low seismic potential. In other words, they found that large earthquakes are rather clustered in time. Kagan and Jackson (1999, see also Section 6.3) showed that large earthquakes have significantly higher probability of occurring in the neighbourhood of a recent strong event. Kagan and Jackson (1995) also found that earthquakes after 1989 did not support Nishenko's (1991) gap model. Rong *et al.* (2003) concurred: both predictions were inconsistent with the later earthquake record.

Bakun and Lindh (1985) proposed that a magnitude 6 earthquake would occur at the Parkfield, California segment of the San Andreas fault with a 95%

probability in the time window 1985–1993. The prediction model was based largely on the characteristic, quasi-periodic earthquake hypothesis. This was the only prediction reviewed and approved by the U.S. government. However, no such earthquake occurred untill 28 September 2004, when a magnitude 6.0 earthquake struck near Parkfield (Bakun *et al.* 2005).

Meanwhile, a complicated form of the seismic gap model was applied to estimate earthquakes probabilities in the San Francisco Bay region (Working Group 2003). The Working Group concluded that

> *there is a 0.62 [0.38–0.85] probability of a major, damaging [$m \geq 6.7$] earthquake striking the greater San Francisco Bay Region over the next 30 years (2002–2031).*

Stark and Freedman (2003) argue that the probabilities defined in such a prediction are meaningless because they cannot be validated. They point out that in weather predictions, 50% probability of rain can be tested by counting the ratio of the number of rainy days to the total having this predictive value. No such possibility exists for the predictions concerning San Francisco Bay (Working Group 2003) or Parkfield. Stark and Freedman (2003) finish their review of the San Francisco Bay earthquake prediction with the advice that readers "should largely ignore the USGS probability forecast."

This lack of falsifiability and the inability to construct an improved hypothesis both contradict the fundamental requirements of modern scientific method (Kuhn 1965; Popper 1980). After the gap model was formulated in its present form more than 30 years ago (McCann *et al.* 1979; Nishenko 1991; Schwartz and Coppersmith 1984), the model proponents did not attempt to verify its fundamental assumptions with a critical test. The apparent failure of the predictions (see above) was not extensively analyzed and explained (see, for example, debate, in *Nature* magazine, 1999 or Bakun *et al.* 2005). Jackson and Kagan (2006) review the implications of the Parkfield 2004 event for earthquake prediction, the characteristic earthquake hypothesis, and the earthquake occurrence in general. They argue that a simpler *null* hypothesis based on the Gutenberg-Richter (G-R) law (see Chapter 5) and Poisson time behavior better explains the Parkfield event sequence.

Clear examples of "uncharacteristic" earthquakes include the disastrous 2004 Sumatra and 2011 Tohoku megaquakes, each of which ripped through several previously hypothesized "segment boundaries." These tragic failures should have been the last straw (Kagan *et al.* 2012). What additional evidence could possibly be required to refute the characteristic earthquake hypothesis?

Despite this breakdown of scientific methodology, the potentially incorrect model continued to be used for seismic hazard assessment in the U.S. and many other countries (Jackson and Kagan 2006; Kagan *et al.* 2012). How could this happen? Geosciences are largely observational and descriptive disciplines. Earth scientists are not trained to formulate falsifiable hypotheses, critically test them, apply statistically valid proofs to validate models, systematically review possible sources of error, thoroughly rule out alternative explanations, and dismiss or improve the models thereafter. Retrospective analyses cannot provide a rigorous foundation for any model of earthquake occurrence (Kagan *et al.* 2012). Arbitrarily chosen data sets are fine for formulating hypotheses, but not for validating them. Kagan (2006) argues that the major challenge facing earthquake seismology is that new methods for hypothesis verification need to be developed. These methods should yield reproducible, objective results, and be as effective, for instance, as double-blind, placebo-controlled randomized testing in medical research (Kagan 1997b). In Chapters 10–11 it is proposed that only prospective testing earthquake forecasts with the prediction algorithm and a testing routine specified in advance, may satisfy these requirements. Section 6.3 discusses in more detail the quasi-periodic earthquake occurrence model and its relation to Omori's law.

2.5 Earthquake faults: one fault, several faults, or an infinite number of faults?

Figure 2.6 displays several earthquake fault models. In particular, Fig. 2.6a represents the most commonly used geometry of an earthquake fault: a planar boundary between two elastic blocks. Other block boundaries are usually considered to be free. When Burridge and Knopoff (1967) proposed this model, it was the first mathematical treatment of earthquake rupture and a very important development. Since then, hundreds of papers have been published using this model or its variants. We show below why seismology needs a much more complicated geometrical model to represent brittle shear earthquake fracture:

- The old model (Fig. 2.6a) is a closed, isolated system, whereas tectonic earthquakes occur in an open environment. This model justifies spurious quasi-periodicity, seismic gaps, and seismic cycle models (Section 2.4). No rigorous observational evidence exists for the presence of these features in earthquake records.
- An earthquake fault in the model (Fig 2.6a) is a well-defined simple geometrical object – a planar surface with the dimension 2.0. In nature, an earthquake fault system is a fractal set. This set is not a surface; its dimension is about 2.2 (Chapter 7).
- Two distinct scales are present in the diagram (Fig. 2.6a): irregularities of the planar fault surface and block size. Resistance to block motion due to breakage of microscopical surface inhomogeneities is described as a friction force. The friction is an appropriate model for man-made surfaces, where the scale of surface inhomogeneities is limited. In contrast, earthquakes are scale-invariant. The geometry and mechanical properties of an earthquake fault zone are the result of spatial self-organization. They are fractal (Kagan 2007a; Chapter 7).
- A displacement incompatibility problem is circumvented in the plot (Fig. 2.6a) because of the flat plate boundaries. Real earthquake faults always contain triple junctions (see, for example, Figs. 2.3 and 2.4; further deformation is impossible without creating new fractures and rotational defects (disclinations).

- Because the block boundary is planar, stress concentrations are practically absent after a major earthquake. Hence these models have few or no aftershocks.
- All earthquakes in the model have the same focal mechanism. Any variations in mechanisms obvious during even a cursory map inspection (as in Fig. 2.3) have not been taken into account.

King (1983), Turcotte (1986) and Andrews (1989) suggested that due to kinematic effects at fault junctions, the fault geometry of earthquakes may be represented as a scale-invariant matrix of faults. Gabrielov *et al.* (1996) developed a mathematical framework for calculating the kinematic and geometric incompatibility in a tectonic block system, both rigid and deformable. They concluded that due to geometric incompatibilities at fault junctions, new ruptures must be created to accommodate large plate tectonic deformations. Indeed, plate tectonic observations indicate that hundreds of km of deformation occur over the several million years of plate boundary existence (e.g., the San Andreas fault system).

Figs. 2.6b,c display two alternative models of earthquake faults: a smooth curved surface boundary and a fractal surface. Unless the smooth boundary is a surface of rotation, no large-scale block displacement is possible. Similarly, to move blocks along the fractal boundary, one needs to break the surface inhomogeneities. In contrast to the model of Fig. 2.6a, the largest inhomogeneities in Fig. 2.6c are comparable to the block size.

Obviously, if major faults comprising a plate boundary are not strictly parallel, fault junctions are unavoidable. The question is whether large deformations

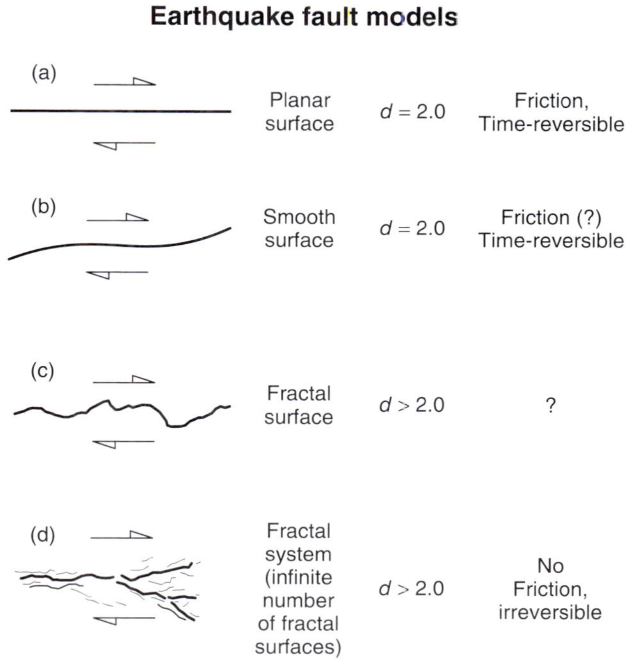

Fig. 2.6 Earthquake fault models. *Source:* Kagan (2006), Fig. 4.

can be accommodated by a few faults delineating a few tectonic blocks (see, for example, Ben-Zion and Sammis 2003), or whether an infinite number of faults must exist to account for such deformations.

The above considerations suggest again that the conventional models of tectonic block deformation need complete revision. If the number of faults and fault junctions is infinite, these junctions, as McKenzie and Morgan (1969) and Gabrielov *et al.* (1996) suggest, constitute "asperities" and 'barriers' for fault rupture propagation. These geometric complexities, not friction, should control the developing fault system and the stop-and-go feature of the earthquake rupture propagation. Kagan (1987) shows that when the earthquake rupture plane rotates, as in triple junctions, the third-rank seismic moment tensor, which can be identified with asperities or disclinations, becomes non-zero.

In Fig. 2.6d we show a fractal boundary zone. The number of micro-blocks in the zone is infinite, or is not well defined. The size of the blocks diminishes, up to the point where they constitute what is called an earthquake fault gouge (Kagan 1982). In this case, a complex fault pattern cannot be characterized as a surface: it is a fractal set of dislocations. In Chapters 7 and 8 we attempt to describe this pattern quantitatively.

2.6 Statistical and physical models of seismicity

As we mentioned above, hundreds of papers are scattered in geophysical and physical journals on the statistical properties of seismicity. They propose physical or phenomenological models of earthquake occurrence. We describe the papers and their results briefly.

Several authors, starting with Bak *et al.* (2002), have attempted to collapse the time–distance–size earthquake distribution into one plot (see also Corral 2005; Baiesi and Paczuski 2005). Such plots prominently demonstrate the scale-invariant structure of seismicity known previously from marginal distributions. However, as mentioned earlier, although temporal, spatial, and magnitude distributions are scale-invariant for small values of pertinent variables, for larger values the scale-invariant pattern is replaced by finite-scale effects (Chapters 5–7). Moreover, even for small variable values, the distributions are influenced by various random and systematic effects. Studying such errors is difficult in a collapsed multidimensional plot.

There are several groups of physical seismicity models. Most of them employ the geometrical and mechanical scheme illustrated in Fig. 2.6a as their major paradigm: two blocks separated by a planar surface (Dieterich 1994; Rice and Ben-Zion 1996; Langer *et al.* 1996, see also Kanamori and Brodsky 2004). Our earlier criticism of this model is also valid for these attempts: they ignore the spatial and mechanical complexity of the earthquake fault zone. Consequently, the deficiencies listed in the previous section are present in these models as well. Furthermore, since these paradigms describe only one boundary between blocks, they do not account for a complex interaction between other block boundaries and, in particular, triple junctions. Seismic maps (Figs. 2.3 and 2.4) convincingly demonstrate that earthquakes occur mostly at the boundaries of relatively rigid elastic blocks. However, if the blocks are rigid, stress concentrations at other block boundaries and a block's triple junctions should influence the earthquake patterns at any boundary. Thus, even after a large earthquake, the stress on a particular boundary can be restored almost immediately due to the influence of the block's other boundaries and their junctions.

Lyakhovsky *et al.* (2005) base their seismicity model on the damage rheology theory. In this case, where the mechanical properties of the rock medium are modeled, even elementary geometrical properties of a fault system are not considered. As a result, the fault geometry and earthquake focal mechanism distribution fall outside their work.

As we mentioned in Section 2.4, theoretical developments need to be critically tested against observational evidence. Otherwise, they remain in the realm of speculation. At the present time, numerical earthquake models have shown no predictive capability exceeding or comparable to empirical predictions based on earthquake statistics. Even if a theoretical or physical model exhibits some predictive skill, we should always question whether the predictive power comes from a deeper theoretical understanding, or from the earthquake statistics results embedded in the model.

The models described above have a large number of adjustable parameters, both obvious and hidden, to simulate a complicated pattern of seismic activity. Dyson (2004) says that Enrico Fermi advised him

> ... *My friend Johnny von Neumann used to say, with four parameters I can fit an elephant* ...

Thus, it is not clear whether a model fit to observations is the result of data selection or is due to over-fitting, that is, using excessive number of parameters. A common belief among geoscientists is that "the more detailed the model, the better" (Marzocchi and Zechar 2011). The number of degrees of freedom in such a model increases uncontrollably.

The observational evidence in support of these models generally consists of particular earthquakes in specific regions. In a random process there is always the possibility of using a large corpus of data to select a particular series of events which seem to agree with theoretical predictions (Shearer and Stark 2012). For the model confirmation to be credible, the criteria for data selection must be prescribed in advance (Jackson and Kagan 2006; Kagan *et al.* 2012).

Therefore, physical models may not have a theoretical predictive capability. As Sub-section 12.1.2 reports, although several phenomenological models have issued a quantitative prediction of future seismicity, no physical model has yet attempted to compete in such tests.

2.7 Laboratory and theoretical studies of fracture

In engineering science extensive work has been performed on the conditions of tensile crack initiation and propagation (e.g., Anderson 2005). However, these efforts have concentrated on the problem of a single crack: the most important problem for engineers. Tensile cracks create voids, ending the interaction of the two crack faces. However, earthquakes occur in the Earth's interior, where a considerable lithostatic pressure should prevent the voids appearing.

Moreover, engineering concepts as opposed to physical theories generally cannot be transferred into a new scientific field without major experimental work. This may explain why fracture mechanics did not significantly improve our understanding of earthquake rupture process (Kagan 2006). Similarly, another engineering science discipline, the study of friction, is mostly based on experiments with man-made surfaces. Though widely used in earthquake mechanics, its contribution to the theoretical foundations of this science is still uncertain (ibid). Vere-Jones (2010, p. 649) suggests

> *A descriptive or ad hoc model cannot be trusted outside the range of the data to which it has been fitted.*

The problem of crack propagation and branching, far more relevant to earthquakes, has recently been addressed in several papers. In laboratory experiments, a tensile crack develops instabilities which make its path and propagation velocity highly chaotic and unpredictable (Marder 1998; Sharon and Fineberg 1999; Bouchbinder *et al.* 2005). These instabilities and a sensitive dependence on the initial conditions are due to crack propagation, especially at a speed close to the elastic wave velocity. Stress and fracture conditions in laboratory specimens differ significantly from those in earthquake fault zones: in the laboratory the boundary effects are controlled by an experimenter. Therefore, fractures can self-organize only at spatial scales much smaller than those of the specimen. In fault zones, the stress, rock mechanical properties, and fault geometry are self-organized as large-scale self-similar patterns develop.

The calculations of molecular dynamics (Marder 1998; Sharon and Fineberg 1999; Buehler and Gao 2006) demonstrate that basic properties of tensile fracture can be effectively derived from simple laws. Similarly, precise laboratory measurements of fault propagation demonstrate multiple branching of fault surfaces. These simulations reproduce the fractal character of a fracture. Moreover, calculating the total energy balance in laboratory fracture experiments (Sharon and Fineberg 1996, 1999) demonstrates that almost all elastic energy goes into creating a new surface. Although the conditions during tensile fracture differ from those of the shear failure in earthquakes, the above result may be significant for the problem of the heat paradox for earthquake faults (Scholz 1996).

3 Stochastic processes and earthquake occurrence models

Kagan (2006) summarized the discussion in this chapter. Most statistical distributions considered so far have been one-dimensional marginal distributions of the earthquake point process. Two enhancements of this picture need to be presented: multidimensional distributions are to be constructed and the point structure of the process needs revision. In Section 5.5 we demonstrate that the focal zone of an earthquake, especially a large one, cannot be regarded as a point. Figure 6.1, for example, shows that earthquake rupture duration needs to be taken into account when very small time intervals are considered. In Fig. 7.7 we show the influence of inter-earthquake time intervals on the spatial structure of earthquake distribution.

A more basic way to study the multidimensional structure of earthquake occurrence is to apply the theory of stochastic point processes (Daley and Vere-Jones 2003), not ordinary statistical methods. The first applications of this theory to earthquake process were made by Vere-Jones (1970), Kagan (1973), and Ogata (1988). Many researchers (Console *et al.* 2003b; Helmstetter and Sornette 2003, 2004, and others) have recently applied the theory of stochastic point processes to analyze earthquake occurrence and clustering. The major impetus for these investigations is the application of statistical methods for earthquake prediction, both long- and short-term. Below we briefly review the available methods for earthquake occurrence analysis and their application in earthquake forecasting. We then discuss how these methods can be improved.

In Section 3.3 we discuss the continuum branching model of earthquake rupture: a multidimensional model based on random stress interactions. The model uses very few free parameters and appears to reproduce all the fundamental statistical properties of earthquake occurrence.

3.1 Earthquake clustering and branching processes

Almost any earthquake forecast requires proper accounting for earthquake clustering, mainly for aftershocks. If present, foreshocks may be used to calculate a mainshock probability. Even if we are mainly interested in a long-term earthquake forecast, the influence of short-term earthquake clustering on the results

Earthquakes: Models, Statistics, Testable Forecasts, First Edition. Yan Y. Kagan.

should be estimated. Moreover, a faithful modeling of the earthquake clustering is needed for any short-term forecast.

Clustering presents a special challenge since modern local catalogs have a magnitude range extending over several units: in California and Japan, the lower magnitude threshold is close to 1.0, whereas the largest earthquake may exceed *m*8. In such catalogs one should expect the aftershock numbers approaching or even exceeding millions after a very strong event. Handling these earthquakes and accounting for various systematic and random effects both present serious difficulties.

Figure 3.1 displays a sketch of earthquake catalog data and their models in the magnitude-time format. The left part of all the diagrams is the past for which no information is available, and similarly for the right or future part. Some earthquakes are detected below the magnitude threshold, shown as a dashed line.

Aftershock sequences have traditionally been taken into account by catalog declustering (Schorlemmer *et al.* 2007). Declustering can be used only as a preliminary step in seismicity analysis: it is subjective; and many different techniques are available but they are not optimized and have not been rigorously tested (see

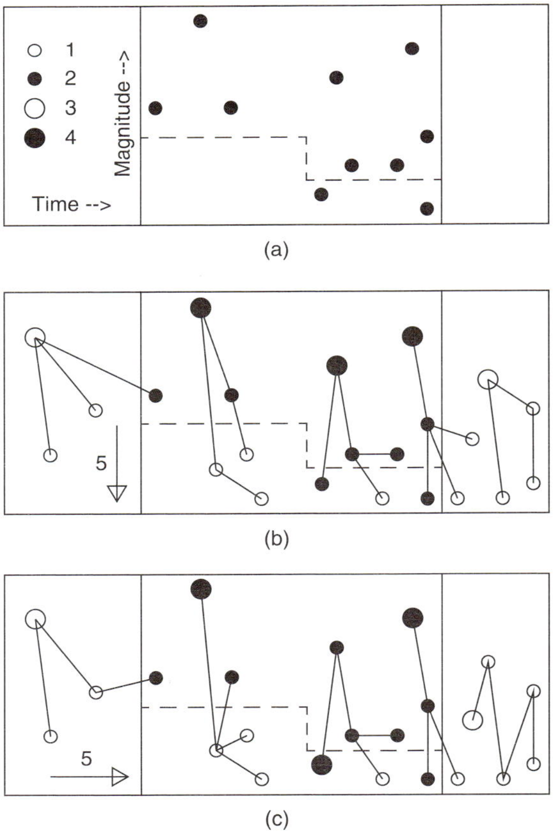

Fig. 3.1 Earthquake branching models. Open circles indicate (1) unobserved or modeled events; filled circles (2) stand for observed earthquakes. The dashed line represents observational magnitude threshold; the earthquake record above the threshold is complete. Many small events are not registered below this threshold. Large circles (3, 4) denote the initial (main) event of a cluster. The diagonal solid lines connecting the events represent hypothesized triggering influences. Arrows (5) indicate the direction of the branching process: down magnitude axis in (b) and along time axis in (c). (a) Observational data. (b) Branching-in-moment (magnitude) model. (c) Branching-in-time model. *Source:* Kagan (2006), Fig 17.

Chapter 4). We must use quantitative statistical methods to describe earthquake clustering. Only an application of stochastic point process theory can provide a robust approach to the problem (see below in this section).

However, the multidimensional nature of earthquake occurrence, fractal or power-law properties of earthquake statistical distributions, and inhomogeneities of earthquake distributions all make it difficult to create and statistically analyze stochastic models. Over the years several such models of earthquake occurrence have been proposed and almost all are based on the theory of branching processes (Harris 1963; Athreya and Ney 1972). Branching is expected to model the well-known property of primary and secondary clustering for aftershock sequences: a strong aftershock (or foreshock) tends to have its own sequence of dependent events (see also Fig. 3.2). The branching continues for each next generation of events. These multidimensional models are:

(A) Point process branching along the magnitude axis, introduced by Kagan (1973) and shown in Fig. 3.1b.

(B) Point process branching along the time axis (Hawkes 1971; Hawkes and Adamopoulos 1973; Hawkes and Oakes 1974) – called the Hawkes self-exciting process. In earthquake occurrence studies the process is called the CBM (Critical Branching Model), see Kagan and Knopoff (1987b) or the ETAS (Epidemic Type Aftershock Sequence) model (Ogata 1988), respectively (see Fig. 3.1c). Hawkes and Tukey (see discussion section in Kagan 1973) debate the difference between branching in earthquake size and in time.

(C) Continuum-state critical branching process which develops along the time axis (Kagan and Knopoff 1981; Kagan 1982; see Section 3.3).

The first two models (A and B) use the Poisson cluster process to approximate the earthquake occurrence. In these models, earthquake clusters are assumed to follow the Poisson occurrence on temporal or magnitude axes. Earthquakes within a cluster are modeled by a multidimensional branching process which reproduces a temporal-spatial pattern of dependent events (mostly aftershocks) around the initial event of a sequence (Kagan 1973; Kagan and Knopoff 1987b; Ogata 1988, 2004). Bremaud and Massoulie (2001) proposed a variant of Hawkes'

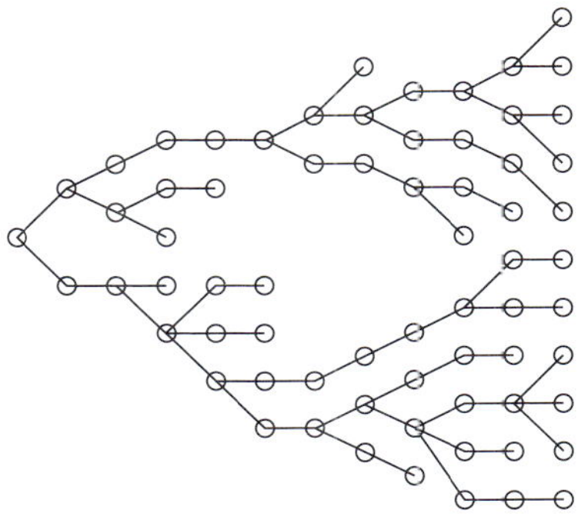

Fig. 3.2 An example of a "genealogical" tree of a critical branching process. The process starts with one "particle" of zero generation. Each particle produces the Poissonian number of descendants; for the critical branching process the mean number is equal to one. The development of any particle is independent of all the other particles in this or previous generations. Simulations of time, position, and orientation of descendant offspring are shown in Fig. 3.3. *Source:* Kagan (2006), Fig. 18.

process with no independent events (immigrants). However, in earthquake catalogs limited in time-span, we need to introduce independent events.

Models (A and B) employ in one form or another the classical statistical properties of earthquake occurrence: the G-R relation and Omori's law. Model (A) reproduces the G-R relation as the result of branching along the magnitude axis and uses Omori's law to describe earthquake clustering in time. Model (B) combines the G-R relation and Omori's law in a fairly empirical fashion to approximate seismicity. Math-physical model (C) yields the G-R law as the consequence of critical branching (Vere-Jones 1976). It applies a version of Omori's law to the temporal distribution of micro-dislocations and simulates the position and orientation of dislocations to reproduce the entire earthquake process (Section 3.3). As we discuss below, other models may have certain advantages in earthquake forecasting and the representation of seismicity. But the phenomenological model (B) is now almost exclusively used to statistically analyze and simulate earthquake occurrence (Kagan and Knopoff 1987b; Kagan and Jackson 2000; Ogata 2004).

Models (A) and (B) can be parameterized to analyze earthquake catalogs. The optimal parameter values can then be found by the maximum likelihood search (Kagan 1991b; Ogata 1988, 2004; Chapter 9). To account for earthquake clustering, one can put the obtained parameter values back into the model and find the probabilities for each event to be foreshock–mainshock–aftershock (Kagan and Knopoff 1976; Zhuang *et al.* 2004). If these probabilities are known, a catalog can be either declustered in an objective manner, or dependent events can be taken into account by a special procedure.

Most of the statistical models for earthquake occurrence (Kagan and Knopoff 1987b; Ogata 1988; Kagan 1991b) treat earthquake catalogs as a population set, with earthquakes considered as individual entities. As we discuss in Chapter 6, "an individual earthquake" is not a physical entity. Instead it is the result of the interpretation of seismograms and the selection by catalog compilers. Thus, extrapolations of observed features to smaller inter-earthquake time intervals, smaller size earthquakes, etc., may see a model breakdown. Such an approximation of deterioration is caused not by the physical properties of earthquake occurrence, but by the peculiarities of earthquake identification technique and catalogs. Why is this?

3.2 Several problems and challenges

1. Earthquake spatial distribution is very complex: the depth inhomogeneity, the fractal character of the spatial pattern, and various hypocenter location errors all make model parameterization difficult and create various biases in estimating parameters. Recent applications of stochastic point processes for seismicity analysis often yield results which are incompatible or unstable: slight variations in the data, assumptions, or processing techniques yield significantly different parameter values (Kagan 1991b, Chapter 9, in particular Section 9.5.2). It is difficult to see whether these contradictions are caused by biases of analysis, data defects, or differences in parametrization.

2. A critical and careful analysis of random and systematic errors in the earthquake catalogs needs to be performed before each statistical analysis. Otherwise, unless the effect being studied is very strong, the analysis results are almost surely artifacts. The problem is that most errors in the earthquake data are caused by systematic effects, so they are more difficult to identify and to correct (Kagan 2003).
3. There is no effective statistical tool to select proper models and check whether they fit the data. Likelihood methods and the "Akaike Information Criterion" (AIC) dependent on them (see Ogata 2004; Daley and Vere-Jones 2004) apparently work only for regular processes: quasi-Gaussian in a continuous case and quasi-Poisson for discrete (point) processes. However, an earthquake occurrence is controlled by scale-invariant, fractal distributions, diverging to infinity. Although these infinities can be regularized by using renormalization procedures similar to the techniques used in model (C), statistical tests applicable to such distributions have not been developed yet. Calculating the likelihood function for aftershock sequences illustrates this point: the rate of aftershock occurrence after a strong earthquake increases by a factor of thousands (see, for instance, Figs. 10.10, 10.11). Log(1000) = 6.9; hence, one close aftershock yields a contribution to the likelihood function analogous to about 7 free parameters.
4. What can be done in the present situation to obtain reliable statistical results? The model's number of degrees of freedom should be kept as small as possible: the new adjustable parameters are to be introduced only if they are critically tested against the data in various catalogs and against different tectonic environments.
5. Earthquake catalogs are incomplete in the wake of strong events (Chapter 6). They are also incomplete generally for small earthquakes (Section 5.2). Both of these effects need to be carefully accounted for (Kagan 2004).
6. Until now, only worldwide seismicity or seismicity in certain seismic zones has been analyzed. Several tectonic provinces have not been sufficiently investigated: deep earthquakes, oceanic earthquakes, earthquakes in stable continental areas, and volcanic earthquakes. The dependence of earthquake clustering on the rate of tectonic deformation should also be investigated: for example, in continental areas (and specifically in California) aftershock sequences occur in zones of fast and slow deformation rate. Are the clustering properties of earthquakes the same in these conditions? A study of earthquake occurrence in these tectonic environments should yield important information on the general properties of seismicity.
7. Apparently all the statistical models based on Omori's law fail to capture the properties of long-term earthquake clustering. Kagan and Jackson (1991a) argued that, in addition to short-term clustering which manifests in foreshock–mainshock–aftershock shallow event sequences, long-term clustering also occurs. The latter phenomenon is common both to shallow and deep earthquakes. They conjectured that short-term clustering results

from stress redistribution in a brittle crust; long-term clustering is most likely due to space-temporal irregularities of mantle convection.

8. Earthquake probabilities calculated using model (B) have a serious defect: if a strong event is preceded by a foreshock or a number of foreshocks, this large quake is considered dependent. Model (A) does not present this difficulty; the largest event in a cluster is always the mainshock.
9. Point models by definition provide only a point forecast. Each future earthquake is characterized by its location, magnitude, time, and possibly its focal mechanism. In reality, earthquakes are spatially extended and they are not instantaneous. This is especially important for large events. Therefore, to compute seismic occurrence maps, a point forecast needs to be supplemented by an extended source model. In contrast to models (A) and (B), model (C) is in principle defined in a continuum which can simulate realistic, complex rupture processes extended in time, space, and fault orientation.

3.3 Critical continuum-state branching model of earthquake rupture

3.3.1 Time-magnitude simulation

Kagan and Knopoff (1981) proposed the model of the earthquake occurrence based on the continuum-state critical branching random process. This model, which uses only a few parameters, reproduces the most important time-magnitude statistical properties of the earthquake occurrence. Kagan (1982) expanded the model to include a full geometrical description of earthquake occurrence so that the resulting design presents a complete kinematic picture of an earthquake process (Subsection 3.3.2). The model was based on the propagation (governed by a critical branching process) of infinitesimal dislocation loops.

The simulation proceeds in three stages. In the first stage the branching family trees are started from a number of initial ancestors as in Fig. 3.2. The second stage of simulation involves adding time delays between the appearance of the parent and the offspring. The delay is power-law distributed (Fig. 3.3a), with the probability density function (PDF) similar to Omori's law

$$X(t) \propto t^{-1-u}. \tag{3.1}$$

For shallow earthquakes Kagan and Knopoff (1981) find that $u \approx 1/2$.

Kagan and Knopoff (1987a) show that the distribution (3.1) may have a simple explanation: stresses at the end of an earthquake rupture are below the critical value and thereafter change randomly according to a one-dimensional Brownian motion (see more in Chapter 6 and Subsection 5.3.3). A new rupture starts when stress reaches a critical level. The level-set of this motion is a fractal set with a dimension $u = 0.5$ (Mandelbrot 1983). The distribution of time intervals is Lévy type which has the PDF (cf. Eq. 6.10 below)

$$f_{1/2}(t) = \frac{1}{t\sqrt{2\pi t}} \exp\left(-\frac{1}{2\,t}\right). \tag{3.2}$$

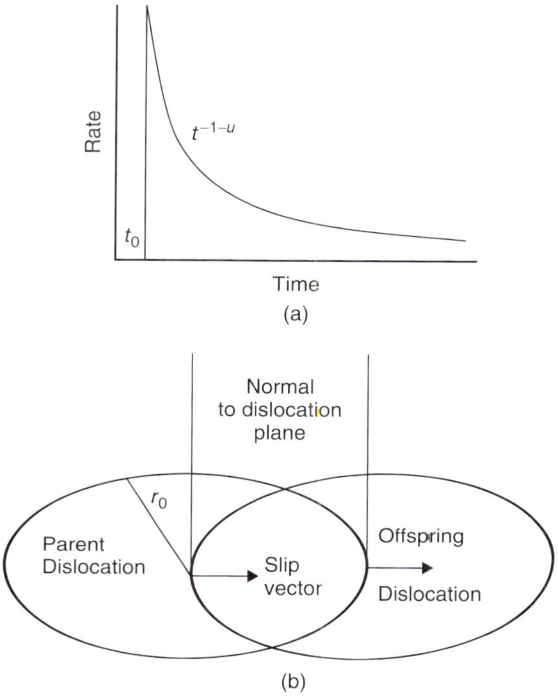

Fig. 3.3 Schematic diagram of fault propagation. (a) Temporal rate of occurrence of dependent shocks. For shallow earthquakes $u \approx 0.5$, t_0 corresponds to rupture time for a dislocation disk in (b). (b) Spatial propagation of a synthetic fault. The initial infinitesimal circular dislocation of radius r_0 gives rise to a secondary event. The center of this dislocation is situated on the boundary of the initial dislocation loop. Solid lines indicate the vector that is normal to the fault plane in both dislocations; arrows show slip vectors. The fault-plane and slip vector of the secondary dislocation rotate following the rotational (angular) Cauchy distribution (Eq. 8.54). The secondary dislocations can produce new events according to the same law (see Fig. 3.2). *Source:* Kagan (2006), Fig. 19.

With this information available, a cumulative plot of the number of elementary events against time can be obtained. In seismological terms, each elementary event is supposed to contribute a fixed amount to a scalar seismic moment release, so that cumulative plots can be interpreted as analogs to the cumulative moment-release plots used in discussing real earthquake seismograms (Fig. 3.4). However, in Section 5.3 we show that due to the varying orientation of earthquake sources, the cumulative moment may differ from a scalar sum of elementary events.

The intense clustering of the critical process results in this cumulative plot taking on a self-similar, quasi-step-function appearance. By convoluting the derivative of this cumulative function with a suitably shaped Green's function, a record can be obtained which may be compared with the trace of a seismograph or its envelope in reality. By applying similar criteria to those used to identify real seismic events, Kagan and Knopoff (1981) were able from the time series record to list simulated "events," each with its own seismic moment or magnitude.

More realistic synthetic seismograms can be obtained by applying the Green's functions that account for varying position and focal mechanism orientations of each elementary dislocation (see the next subsection). The individual events may then be selected from such seismograms, each event having occurrence time, hypocenter and centroid position and focal mechanism defined by averaging the locations and focal mechanisms of the elementary dislocations comprising the event. In this way a synthetic seismic catalog can be produced in which

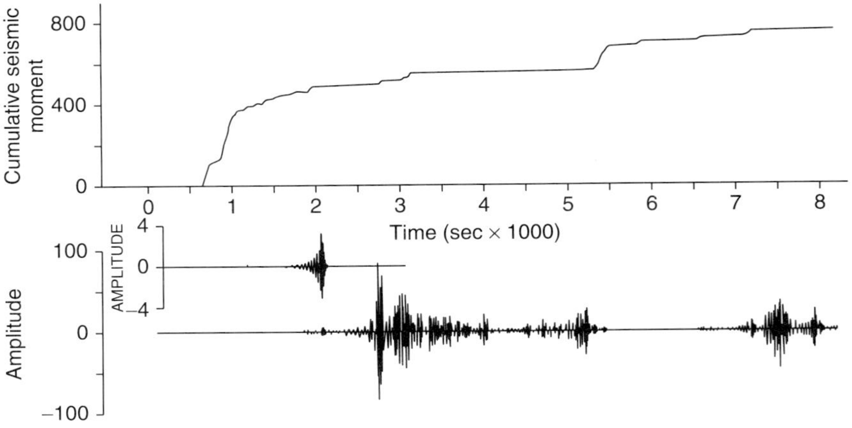

Fig. 3.4 Cumulative event curve interpreted here as "cumulative seismic moment" for a realization of the critical branching process model (upper curve). The middle curve is the theoretical seismogram for an impulse source (earthquake). The lower curve is an illustration of the filtered signal from which events and their "seismic moments" can be determined. *Source:* Kagan and Knopoff (1981), Fig. 3.

events are listed in time sequence and associated with a hypocenter, magnitude and a focal mechanism. Processing the synthetic catalog through a maximum likelihood procedure similar to that used for real catalogs (Section 3.1, Chapter 9) yields similar values of basic parameters describing an earthquake occurrence.

3.3.2 Space-focal mechanism simulation

In the third stage of modeling, the spatial coordinates (location of the dislocation disc center, its orientation, and slip direction) are simulated according to Fig. 3.3b. This model of earthquake rupture incorporated results of 2-, 3-, and 4-point spatial statistical moment studies (Kagan and Knopoff 1980; Kagan 1981a,b, see also Chapter 7) and reproduced the inferred geometrical properties of hypocenter distributions. Although in principle dislocations are infinitesimal, in practical simulations the dislocation loops are finite with a disc radius r_0. However, this radius can be taken to be as small as possible, implying infinitesimal seismic moment value for the loop. In such a case the critical branching process converts into a continuum-state process (Jirina 1958).

The rotation of the focal mechanisms follows the 3-D rotational Cauchy distribution (see Eq. 8.54). Most rotations are infinitesimal, though in rare cases large rotations give rise to fault branching seen by inspection of synthetic ruptures (Kagan 1982; see Fig. 3.5). As we explained above, the 3-D rotations are described by the group **SO**(3); hence, our model is a random branching walk on non-commutative groups. From the results of this stage, it is possible to obtain a visual picture of the resulting "fractures" by plotting the intersection of the elementary dislocation discs with a fixed plane (see Fig. 3.5).

It is partly from such pictures that the angular Cauchy distribution, rather than some other angular analogue of the stable distributions, has been chosen. As can be seen from Section 8.1, this distribution also has a simple, physical explanation:

theoretical arguments (Zolotarev 1986; Kagan 1990) and simulations demonstrate that the stress tensor in the medium with defects follows Cauchy distribution. This stress distribution should result in the rotational Cauchy distribution of earthquake focal mechanisms (ibid.).

The obtained distribution of fault traces (see Fig. 3.5) looks like actual earthquake fault maps. The spatial statistical moment functions are also qualitatively similar to those in Fig. 7.1: the 1/D and 1/V behaviors have been reproduced, although no rigorous tests have been attempted (Kagan and Vere-Jones 1996).

For extended rupture, like that shown in Fig. 3.5, we can calculate the seismic moment tensor of an earthquake or earthquake sequence

$$\mathbf{M} = \sum_{i=0}^{N} \mu \left(\prod_{\xi_i} q_i \times \cdots \times q_j \times \cdots \times q_0 \right), \qquad (3.3)$$

where q_0 is a quaternion (see Chapter 8) corresponding to the initial dislocation as in Fig. 3.2. The quaternion product, ending with q_i, describes a combination of

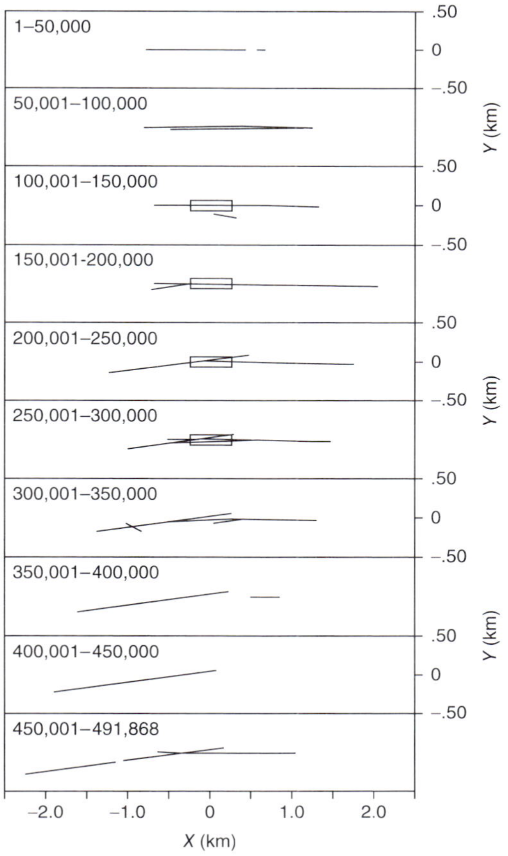

Fig. 3.5 Stages in the evolution of an episode in the branching simulation model: intersection with a fixed *XY* plane of the synthetic dislocation discs. The range of dislocation numbers for one random realization is shown in each frame. *Source:* Kagan (1982), Fig. 7.

3-D rotations at the path ξ_i in a branching process leading to the i-th dislocation. In a branching process such a path is unique. Each of the quaternion product q_j components follows the rotational Cauchy distribution Eq. (8.54). Thus, the quaternion product in the formula represents the orientation of the i-th dislocation. The operator $\mu(.)$ converts the orientation (quaternion) into the seismic moment tensor (Kagan and Jackson 1994, their Appendix; Eq. 8.37).

There has been renewed interest recently in the branching earthquake fault model. In a limited test, Libicki and Ben-Zion (2005) used a simplified procedure to reproduce some properties of Kagan's (1982) model.

Despite the interest in the model, and the many questions which it raises, it has not (to our knowledge) been the subject of any significant analytical investigations. In particular, it is not clear how key aspects, such as its evident spatial and temporal self-similarity, or the clustering of events subsequently identified as earthquakes, can be deduced from its defining elements. The main objectives of this section are to briefly summarize the model structure, and to indicate some possible directions for pursuing these questions (Kagan and Vere-Jones 1996).

Part II

Statistics

4 Statistical distributions of earthquake numbers: Consequence of branching process

The discussion in this chapter mainly follows Kagan (2010a). The standard assumption long used in earthquake hazard analysis (Cornell 1968; Allen 2007) is that earthquake numbers are reasonably well described by the Poisson distribution (see Eq. 4.14 below). However, it has also been known for a long time that earthquakes are clustered in time and space: their distribution is over-dispersed compared to the Poisson law. One conventional way to treat this problem is to "decluster" an earthquake catalog (Schorlemmer *et al.* 2007). But there are several declustering procedures, mostly based on *ad-hoc* rules. Hence traditional declustered catalogs are not unique and usually not fully reproducible. Therefore, it is important to derive and investigate earthquake number distribution in real earthquake catalogs.

Kagan (1996), Jackson and Kagan (1999), and Kagan and Jackson (2000) have all used the negative binomial distribution (NBD) to approximate earthquake numbers in catalogs. The NBD has a higher variance than the Poisson law and can be shown (Kagan 1973, see below) to result from the branching nature of earthquake occurrence.

In principle, several other over-dispersed discrete distributions, such as generalised Poisson distributions (Consul 1989) or generalised negative binomial distributions (Tripathi 2006; Hilbe 2007) can be used to approximate earthquake numbers. However, the NBD has the advantage of relative simplicity and is supported by theoretical arguments (Kagan 1973). As we discuss below, in addition to negative binomial and Poisson distributions, several other statistical discrete distributions can describe earthquake numbers. A general discussion of such integer-valued distributions can be found in Johnson *et al.* (2005) and Kotz *et al.* (2006).

Over the years many papers have analyzed various aspects of earthquake numbers distributions; for example, see Vere-Jones (1970), Shlien and Toksöz (1970), Dionysiou and Papadopoulos (1992). These publications investigated the distributions empirically by counting earthquake numbers in catalogs and trying to approximate them by various statistical laws. Here we explain these distributions as a consequence of the stochastic branching model.

Earthquakes: Models, Statistics, Testable Forecasts, First Edition. Yan Y. Kagan.

Therefore, in addition to the NBD and the Poisson distributions, in this chapter we investigate the geometric and logarithmic distributions in several earthquake catalogs and show their applicability in certain conditions. After presenting the theoretical derivation of these distributions, we explore the statistical parameter estimation for these laws. Then we apply these methods to several earthquake catalogs. Two global (GCMT and PDE, see Section 2.3) and one local Californian catalog are studied and the results are displayed in the tables and diagrams in this chapter. These results can be used in earthquake forecast testing (Schorlemmer *et al.* 2007; Kagan *et al.* 2010; Zechar *et al.* 2013).

4.1 Theoretical considerations

4.1.1 Generating function for the negative binomial distribution (NBD)

Several stochastic models of earthquake occurrence were proposed and almost all were based on the theory of branching processes (Kagan 2006; Chapter 3). These multidimensional models are:

(a) The supercritical point process branching along the magnitude axis, introduced by Kagan (1973) and shown in Fig. 3.1b. Earthquake occurrence constitutes a downward cascade in this model.

(b) Critical (or rather slightly subcritical) point process branching along the time axis (see Section 3.1 and Fig. 3.1c).

Both models shown in Figs. 3.1b,c use the Poisson cluster process to approximate earthquake occurrence. Earthquake clusters are assumed to follow the Poisson occurrence. Earthquakes within a cluster are modeled by a multidimensional branching process, which reproduces a temporal -spatial pattern of dependent events (mostly aftershocks) around the initial one in a sequence (Kagan 1973; Kagan and Knopoff 1987b; Ogata 1988, 1998). In Section 3.1 we describe how these models reproduce the statistical properties of earthquake occurrence: the Gutenberg-Richter (G-R) relation and Omori's law.

The distribution of earthquake numbers is a discrete law. The simplest way to describe discrete distributions is by using the probability generating function (Bartlett 1978; Evans *et al.* 2000). Given the generating function $\phi(z)$, the probability function (PF) for k events $f(k)$ can be obtained as

$$f(k) = \frac{1}{k!}\left[\frac{d^k\phi(z)}{dz^k}\right]_{z=0}. \tag{4.1}$$

Following the graph of inter-earthquake connections as shown in Fig. 3.1b, we investigate the earthquake numbers distributions for space–time intervals larger than the average dimensions of earthquake clusters. Thus, we neglect space–time differences between cluster members. We assume that the independent and dependent events are occurring along the $\Xi = \log M$ axis (M is the scalar seismic moment) with constant rates

$$\nu\,(d\,\Xi) = \nu \cdot d\,\Xi; \quad \beta\,(d\,\Xi) = \beta \cdot d\,\Xi \qquad \text{for} \quad \Xi < \Xi_{\mathrm{m}}, \tag{4.2}$$

where Ξ_{m} is the logarithm of the maximum seismic moment, M_m; ν and β are rates of independent and dependent events, respectively.

Events occurrence can be modeled as an "immigration and birth" process, where independent, spontaneous earthquakes (mainshocks) are treated as "immigrants." Any immigrant may spawn offspring, who may spawn their own offspring, etc., with the whole family making up a cluster. Figure 3.2 illustrates the branching process, in this figure the critical branching process is assumed with the average number of offsprings equal 1.0. In this model the offspring number could be significantly larger than 1.0.

Seismologists generally call the largest family member the "mainshock," the smaller events preceding in time are called the "foreshocks," and subsequent members the "aftershocks." The conditional generating function for the number of events with $\Xi \geq \Xi_t$ in a cluster, including the mainshock, is (Bartlett 1978, Eq. 3.4(7), p. 76)

$$\phi\,(z \mid \Xi_{\,t}) = \frac{z\,(M_t/M)^{\beta}}{1 - z\,[\,1 - (M_t/M)^{\beta}\,]}, \tag{4.3}$$

where $|$ is the sign for conditional function, β is the index of the seismic moment distribution, $\beta = \frac{2}{3}b$; b is the parameter of the G-R law (Chapter 5), and M_t is the seismic moment detection threshold of a seismographic network ($M \geq M_t$ and $\Xi_{\,t} = \log\ M_t$). In this formula we modify Bartlett's equation for the "birth and immigration" (but no "deaths") population process.

In numerical calculations in this chapter we take

$$\beta = 0.62, \tag{4.4}$$

the value suggested in Chapter 5.

Equation 4.3 characterizes the geometric distribution, with the PF

$$f\,(k) = (1-p)^{k-1}p \quad \text{for} \quad k = 1, 2, 3, \ldots, \tag{4.5}$$

with $p = (M_t/M)^{\beta}$. A more common form of the geometric distribution (Evans *et al.* 2000) is

$$f\,(k) = (1-p)^{k}p \quad \text{for} \quad k = 0, 1, 2, \ldots. \tag{4.6}$$

It would be appropriate for the number of dependent shocks in a cluster. Its conditional generating function is

$$\phi\,(z \mid \Xi_{\,t}) = \frac{(M_t/M)^{\beta}}{1 - z\,[\,1 - (M_t/M)^{\beta}\,]}. \tag{4.7}$$

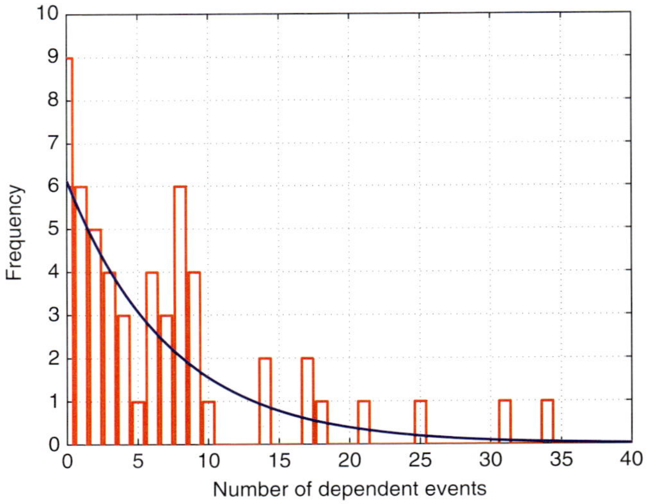

Fig. 4.1 Distribution of aftershock numbers: Red bars – distribution of aftershock numbers $m \geq 4.7$ in the PDE catalog 1977–2007, following all of $7.1 < m_W \leq 7.2$ earthquakes in the GCMT catalog. The blue line is an approximation of the distribution using the geometric law (Eq. 4.6) with the parameter $\hat{p} = 0.1279$, calculated using (4.26). *Source:* Kagan (2010a), Fig. 2.

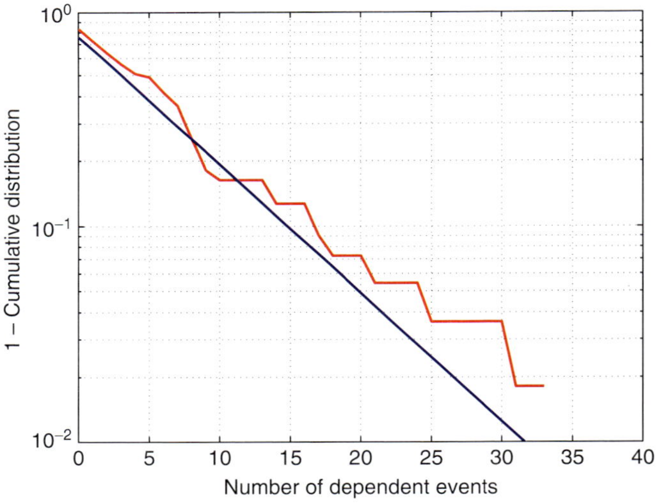

Fig. 4.2 Survival function (1 – Cumulative distribution) of aftershock numbers. Same catalogs and distributions as in Fig. 4.1 are used. Since $f(0) \neq 0$ in Fig. 4.1, the function is non-zero for zero value of abscissa axis. *Source:* Kagan (2010a), Fig. 3.

The geometric distribution is a discrete analogue of the exponential distribution for continuous variables. For a small p-value

$$f\ (k) \approx p\ \ \exp - (kp). \tag{4.8}$$

Figs. 4.1–4.3 illustrate the fit of the geometric law to aftershock number distribution.

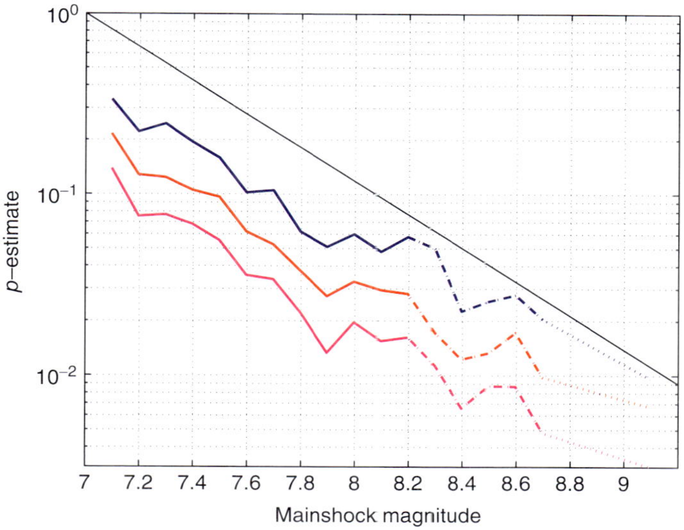

Fig. 4.3 Dependence of the $\hat{p}$-estimate (4.26) on mainshock magnitude. The same catalogs as in Fig. 4.1 are used. The blue line is for $m \geq 5.0$ aftershocks during 1-day interval after a mainshock, the red line is for $m \geq 4.7$ aftershocks during 1-day interval after a mainshock, and the magenta line is for $m \geq 4.7$ aftershocks during a 7-day interval after a mainshock. Solid lines connect data points with more than 4 mainshocks, dashed lines are for 4 or less mainshocks, and dotted lines connect the estimate for 2004 $m9.1$ Sumatra earthquake. The thin black line corresponds to $p \propto 10^{-1.5\ m\beta}$ (see Eqs. 4.4, 4.7). *Source:* Kagan (2010a), Fig. 4.

For the distribution of the total number of events in an earthquake catalog, we obtain (Bartlett 1978, Ch. 3.41, p. 83)

$$\begin{aligned}\phi(z) &= \exp \int_{\Xi_t}^{\Xi_m} [\phi(z|\Xi) - 1]\, \nu\, d\,\Xi \\ &= \exp \left\{ \left[\frac{\log\left(1 - z[1 - (M_t/M_m)^\beta]\right)}{-\beta\,(\Xi_m - \Xi_t)} - 1 \right] \nu(\Xi_m - \Xi_t) \right\} \\ &= \left[\frac{(M_t/M_m)^\beta}{1 - z[1 - (M_t/M_m)^\beta]} \right]^{\nu/\beta}. \end{aligned} \tag{4.9}$$

The above generating function is for the NBD.

In this derivation we assume that the "immigration and birth" process starts at the maximum seismic moment: we assume the moment distribution to be a truncated Pareto distribution with a "hard" limit on the maximum moment (Chapter 5). However, a more appropriate seismic moment distribution (a tapered Pareto distribution) uses a "soft" limit or a corner moment (ibid.). To consider this, Kagan and Jackson (2000, Eq. 29) propose some modifications of the NBD formulas.

The last line in Eq. (4.9) means that the distribution of events (counting mainshocks as well) in earthquake clusters would be logarithmic (or logarithmic series) with the PF

$$f(k) = \frac{[1 - (M_t/M_m)^\beta]^k}{k\ \log\ [(M_t/M_m)^\beta]} = \frac{[1 - (M_t/M_m)^\beta]^k}{k\ \beta\ (\Xi_m - \Xi_t)} \quad \text{for} \quad k = 1, 2, 3, \ldots. \tag{4.10}$$

The standard form of the logarithmic distribution (Evans *et al.* 2000) is

$$f(k) = \frac{-1}{\log(1-\eta)} \cdot \frac{\eta^k}{k} \quad \text{for} \quad k = 1, 2, 3, \ldots . \tag{4.11}$$

Figure 4.5 shows two approximations of empirical distribution by the logarithmic law.

For a small seismic moment threshold, the logarithmic distribution can be approximated by a tapered Pareto distribution (cf. Eq. 5.5)

$$f(k) \approx \{ k \log [(M_m/M_t)^{\beta}] \exp[k (M_t/M_m)^{\beta}] \}^{-1} \text{ for } k = 1, 2, 3, \ldots, \tag{4.12}$$

that is, the clusters with small and intermediate mainshocks follow the Pareto style distribution with a heavy-tail at the largest earthquakes. Clauset *et al.* (2009) discuss the difference between continuous (Pareto) and discrete power-law distributions and their properties. The exponential tail should be observed only for clusters with large mainshocks having a magnitude close to maximum.

The average size for a group of dependent events is

$$S(k) = \frac{[(M_m/M_t)^{\beta} - 1]}{\beta(\Xi_m - \Xi_t)} . \tag{4.13}$$

It is clear from Eqs. 4.10 and 4.13 that the number of dependent events decreases as $M_t \rightarrow M_m$: the largest earthquakes are distributed according to the Poisson law as is assumed in model of Fig. 3.1b.

The Poisson distribution has the PF

$$f(k) = \frac{\lambda^k \exp(-\lambda)}{k!} . \tag{4.14}$$

For this distribution its mean and variance are equal to its rate λ.

What are the distributions of earthquake numbers for the branching-in-time model (see Fig. 3.1c)? Can we calculate these distributions in the same way as model Fig. 3.1b? Indeed, when the first event in a sequence of earthquakes is the largest one, as is often the case, there is little ambiguity in calling it a mainshock, and the other dependent events may be called aftershocks, though some of them are actually aftershocks of aftershocks. In such a case there should be little difference between Fig. 3.1b and Fig. 3.1c models. If, however, the first event in a sequence is smaller than the subsequent events, it is typically called a foreshock. Retrospectively, it is relatively easy to subdivide an earthquake catalog into fore-, main-, and aftershocks. However, in real-time forecasting it is uncertain whether the most recent large event registered by a seismographic network is a foreshock or a mainshock. Although the subsequent triggered events are likely to be smaller and would thus be called aftershocks, there is a significant chance that some succeeding earthquakes may be bigger than the predecessors that triggered them (Chapter 6; Kagan 1991b).

As we discussed in Section 3.2, the difficulty in the model shown in Fig. 3.1c is that some connections between events may not be observable. Suppose that there is a potential foreshock–mainshock pair: a larger earthquake is preceded by a smaller one. However, the first event is below the magnitude threshold but the larger earthquake is above the threshold. Then this second event would be treated as independent (immigrant); that is, the branching-in-time calculations would miss this connection (Kagan 1991, 2006; Sornette and Werner 2005). It is possible that the number of such missed connections is small, and the event number distributions are similar to those derived above. Nevertheless, such distribution estimates for this model would be approximate. For the branching-in-magnitude model (Fig. 3.1b), the derived expressions are exact.

The equations obtained in this sub-section are based on the theory of population processes. In simple models of such processes (Bartlett 1978), it is assumed that individuals start to reproduce immediately after their birth. This is not the case for an earthquake occurrence: after a large event, the aftershocks, even large ones, cannot be observed for a period of time (Section 6.2.1). Furthermore earthquakes are not instantaneous, point events; and their rupture takes time. Hence, the derived formulas would over-estimate the number of dependent events. Nevertheless, as we see below, during analysis of the earthquake catalogs the theoretical models provide some valuable insight into the quantitative behavior of earthquake occurrence.

4.1.2 NBD distribution expressions

There are many different representations of the NBD (see several examples in Anscombe 1950; Shenton and Myers 1963). The most frequently used (we call it *standard*) form of the PF for the NBD generalizes the Pascal distribution (Feller 1968, Eq. VI.8.1; Hilbe 2007, his Eq. 5.19):

$$\begin{aligned} f\,(k) = {} & \frac{\tau\,(\tau+1)\,\ldots\,(\tau+k-2)\,(\tau+k-1)}{k!} \\ & \times\ \theta^{\tau}(1-\theta)^k = \binom{\tau+k-1}{\tau-1} \times\ \theta^{\tau}(1-\theta)^k \\ = {} & \binom{\tau+k-1}{k} \times\ \theta^{\tau}(1-\theta)^k = \frac{\Gamma(\tau+k)}{\Gamma(\tau)\ k!} \times\ \theta^{\tau}(1-\theta)^k, \end{aligned} \tag{4.15}$$

where $k = 0, 1, 2, \ldots$, Γ is the gamma function, $0 \le \theta \le 1$, and $\tau > 0$. If the parameter τ is integer, then this formula (Pascal distribution) is the probability distribution of a certain number of failures and successes in a series of independent and identically distributed Bernoulli trials. For $k + \tau$ Bernoulli trials with success probability θ, the negative binomial gives the probability for k failures and τ successes, with success on the last trial. If $\tau = 1$, this equation corresponds to (4.5): the geometric distribution. Therefore, the latter distribution can be considered a special case of the NBD.

The average of k for the NBD is

$$E(k) = m_1 = \tau\,\frac{1-\theta}{\theta}. \tag{4.16}$$

and its variance

$$D(k) = m_2 = \tau \frac{1-\theta}{\theta^2}. \tag{4.17}$$

The negative binomial distribution generally has a larger standard deviation than the Poisson law. Thus, it is often called an 'over-dispersed Poisson distribution' (Hilbe 2007). For $\theta \to 1$ and $\tau(1-\theta) \to \lambda$ expression (4.15) tends to (4.14) (Feller 1968, p. 281); the negative binomial distribution becomes the Poisson one; the latter distribution is a special case of the former.

Anraku and Yanagimoto (1990, their Eq. 1.1) and Hilbe (2007) propose the following NBD density form, which they obtain as a mixture of the Poisson distributions with the gamma distributed rate parameter λ (see, for example, Hilbe 2007, his Eqs. 5.1–5.9 or 7.21–7.33)

$$\begin{aligned} f(k) &= \frac{\Gamma(1/\alpha + k)}{\Gamma(1/\alpha)\, k!} \cdot \frac{\lambda^k}{(\lambda + 1/\alpha)^k\, (1+\alpha\lambda)^{1/\alpha}} \\ &= \frac{\lambda^k}{k!} \cdot \frac{\Gamma(1/\alpha + k)}{\Gamma(1/\alpha)\, (\lambda + 1/\alpha)^k} \cdot \frac{1}{(1+\alpha\lambda)^{1/\alpha}}, \end{aligned} \tag{4.18}$$

which converges to the geometric distribution if $\alpha = 1$, but to the Poisson distribution (4.14) when $\alpha \to 0$,

$$f\ (k) = \frac{\lambda^k}{k!} \cdot 1 \cdot \frac{1}{\exp(\lambda)}. \tag{4.19}$$

Comparison with (4.15) shows that $\alpha = 1/\tau$ and $\lambda = \tau\ (1-\theta)/\theta$. Distribution (4.18) is called an *alternative* form of the NBD in Wikipedia and we accept this term for our purposes.

In addition to the above representations of the NBD, we use Evans' (1953) expressions. Evans provides the parameter uncertainty equations for the estimates by the statistical moment method. His formula for the PF is

$$f\ (k) = \frac{\Gamma(\lambda/a + k)}{\Gamma(\lambda/a)\ k!} \times \frac{[a/(1+a)]^k}{(1+a)^{\lambda/a}}. \tag{4.20}$$

If we make $a = \alpha\lambda$, this equation converts to (4.18).

The probability generating function (Bartlett 1978) for the NBD standard form is

$$\phi\ (z) = \left[\frac{\theta}{1 - z(1-\theta)}\right]^{\tau}. \tag{4.21}$$

For the alternative form (4.18) of the NBD the generating function is

$$\phi\ (z) = \frac{1}{[1 + \alpha\lambda - z\alpha\lambda]^{1/\alpha}}. \tag{4.22}$$

Comparing (4.21) with (4.9), it is clear that if the seismic moment threshold is close to M_m, the NBD approaches the Poisson distribution for earthquake counts.

This is confirmed by extensive seismological practice (see also Figs. 4.11 and 4.12).

After comparing (4.21) with (4.9), we propose the following relations between the parameters

$$\theta = \left(\frac{M_t}{M_m}\right)^{\beta}, \tag{4.23}$$

and

$$\tau = \nu/\beta. \tag{4.24}$$

However, as mentioned at the end of Section 4.1.1, such relations are valid only when larger shocks start producing dependent events immediately after their rupture. In earthquake catalogs there is a significant delay of aftershock generation (Section 6.2.1). Thus, these expressions are over-estimated. Moreover, in all available earthquake catalogs there are very few or no mainshocks with a size approaching the maximum magnitude. Therefore, we would not expect the observational θ-estimates to be close to that of (4.23). However, as we will see later, the dependence of θ on the magnitude threshold, the maximum magnitude, and β are visible in earthquake records.

4.1.3 Statistical parameter estimation

For the Poisson distribution (4.14) the estimate of its parameter is the average earthquake rate per time interval ΔT

$$\hat{\lambda} = \frac{n\ \Delta T}{T}, \tag{4.25}$$

where T is the time-span and n is the total number of events in a catalog. The estimate of p for the geometric distribution in (4.5) is

$$\hat{p} = \frac{1}{1 + m_1}, \tag{4.26}$$

where $m_1 = \hat{\lambda}$ is the average event number (the first statistical moment).

For the logarithmic distribution (4.11) there is no simple expression to evaluate its parameter η. Patil (1962, Table 1) and Patil and Wani (1965, Table 2) propose tables for calculating the maximum likelihood estimate (MLE) of the parameter after determining the average number of events.

For the standard form (4.15) of the NBD, we use (4.16) and (4.17) to obtain the estimates of the NBD parameters by the statistical moment method

$$\hat{\theta} = \frac{m_1}{m_2}, \tag{4.27}$$

and

$$\hat{\tau} = \frac{m_1^2}{m_2 - m_1}, \tag{4.28}$$

where m_1 and m_2 are the average and variance of the empirical number distribution. Below we sometimes use the term *moment estimate* for the estimate by the statistical moment method. For the Poisson process, $m_2 = m_1$. Hence the estimate (4.28) would be unstable if the NBD process approaches the Poisson process, and the estimate uncertainty would be high.

For the alternative form of the NBD (4.18), we obtain the following moment estimates

$$\hat{\lambda} = m_1, \tag{4.29}$$

and

$$\hat{\alpha} = \frac{m_2 - m_1}{m_1^2}. \tag{4.30}$$

Evans' (1953) parameter a is estimated similarly to (4.30)

$$\hat{a} = \frac{m_2 - m_1}{m_1} = \frac{m_2}{m_1} - 1. \tag{4.31}$$

Evans (1953, p. 203, see '*Estimation by Method 1*') derives approximate estimates of the parameters' uncertainties

$$\sigma_\lambda = \sqrt{\lambda(a+1)/N}. \tag{4.32}$$

and

$$\sigma_a \approx \sqrt{2(a+1)/N + a(a+1)(3a+2)/(\lambda N)}. \tag{4.33}$$

as well as the covariance between these two estimates

$$\mathrm{Cov}(\hat{\lambda},\ \hat{a}) \approx a(a+1)/N. \tag{4.34}$$

In these equations N is the number of time intervals with earthquake counts.

The maximum likelihood estimate (MLE) of any parameters for the discussed distributions can be obtained by maximizing the log-likelihood function

$$\ell = \log \prod_{j=0}^{\infty} [f(k_j)]^{P(k_j)} = \sum_{j=0}^{\infty} P(k_j) \log f(k_j), \tag{4.35}$$

where $P(k_j)$ is the observational frequency of earthquake numbers in interval j. Function $f(k)$ is defined by the expression (4.14), (4.15), (4.18), or (4.20) for the Poisson, the standard NBD, the alternative NBD, and Evans' formula, respectively. To evaluate parameter uncertainties of the MLE method we need to obtain the Hessian matrix (the second partial derivatives of the likelihood function) of the parameter estimates (Wilks 1962; Hilbe 2007).

4.2 Observed earthquake numbers distribution

4.2.1 Statistical analysis of earthquake catalogs

Several problems arise when the theoretical considerations of Section 4.1 are applied to earthquake catalogs. Due to the limited sensitivity of a seismographic network and its sparse spatial coverage, catalogs are incomplete in magnitude, and inhomogeneous in time and space (Kagan 2003). In particular, the magnitude threshold of completeness varies in time, usually decreasing during the catalog time-span. Furthermore, after strong earthquakes due to mainshock coda waves and interference by other stronger events, small aftershocks are usually absent for a period of a few hours to a few days (Section 6.2.1; Kagan 2004).

An additional problem in comparing theoretical calculations to observations is identifying earthquake clusters. Due to delays in dependent events, temporal distribution described by Omori's law, clusters usually overlap in time and space (see Chapter 6). Only in zones of slow tectonic deformation can aftershock clusters of large earthquakes be easily distinguished. Sometimes this aftershock sequence decay takes centuries (Ebel *et al.* 2000; Ebel 2009). In more active regions, many earthquake clusters overlap.

It is possible, in principle, to use stochastic declustering in defining statistical interrelations between various events. For the branching-in-magnitude (Fig. 3.1b) such a procedure was first applied by Kagan and Knopoff (1976, see Table XVIII); for the branching-in-time model, Kagan and Jackson (1991a) and Zhuang *et al.* (2004) proposed the identification procedure (see also Chapter 9 and Tables 9.2–9.5). For branching-in-time models, connections of small vs. large events (such as foreshock–mainshock ties) create an additional difficulty in stochastic reconstruction. See more discussion on this topic at the end of Section 4.1.1. Handling these connections unambiguously is difficult.

Moreover, the equations derived in Section 4.1 ignore the temporal and spatial parameters of earthquake occurrence. Hence, they are valid for large space-time windows exceeding the size of the largest earthquake cluster. The time-span of available earthquake catalogs is very limited. For the largest earthquakes, the duration of the aftershock sequences is comparable or exceeds a typical catalog length. Additionally, when studying earthquake number distribution, we often need to subdivide the catalog into several sub-units, thus reducing again the temporal or spatial window. Furthermore, the theoretical model neglects long-term modulations of seismicity (Kagan and Jackson 1991; Lombardi and Marzocchi 2007). Therefore, we should not expect close agreement between the theoretical formula and empirical results. Only general regularities in distribution behavior can be seen.

4.2.2 Observed earthquake numbers distributions

Figure 4.1 shows the distribution of shallow (depth 0–70 km) aftershock numbers in the PDE catalog, for all $m7.1 - 7.2$ events in the GCMT catalog (see

Section 2.3). The bar height at zero means, for example, that there are 9 such GCMT earthquakes with no aftershocks. We count the total aftershock number during the first day within a circle of radius R (see Kagan 2002b and Fig. 5.15)

$$R(m) = 20 \cdot 10^{m-6} \text{ km}. \tag{4.36}$$

Even taking into account location errors in both catalogs, the radius of 200 km for the $m7$ event guarantees that almost all the first day aftershocks would be counted. The geometric distribution curve seems to approximate the histogram satisfactorily. Comparing the observed cumulative distribution with its approximation in Fig. 4.2 also confirms that the geometric law appropriately describes the aftershock number distribution for mainshocks of given size.

For the geometric distribution, Fig. 4.3 shows the dependence in the p-parameter on the mainshock magnitude. The $\hat{p}$-values decay approximately by a factor of $10^{-1.5\beta}$ with a magnitude increase by one unit. This behavior is predicted by Eqs. 4.4 and 4.7.

Figure 4.4 displays an example of earthquake numbers in equal time intervals (annual in this case) for the CIT catalog (see also Fig. 5 by Kagan and Jackson 2000). Even a casual inspection suggests that the numbers are over-dispersed compared to the Poisson process: the standard deviation is larger than the average. The number of peaks are easily associated with large earthquakes in southern California: the $m7.5$ 1952 Kern County, the $m7.3$ 1992 Landers, and the $m7.1$ 1999 Hector Mine earthquakes. Other peaks can usually be traced back to strong mainshocks with extensive aftershock sequences.

In large time intervals, one would expect a mix of several clusters, and according to Eq. 4.9, the numbers would be distributed as the NBD. The small time intervals may still have several clusters due to weaker mainshocks. However, one

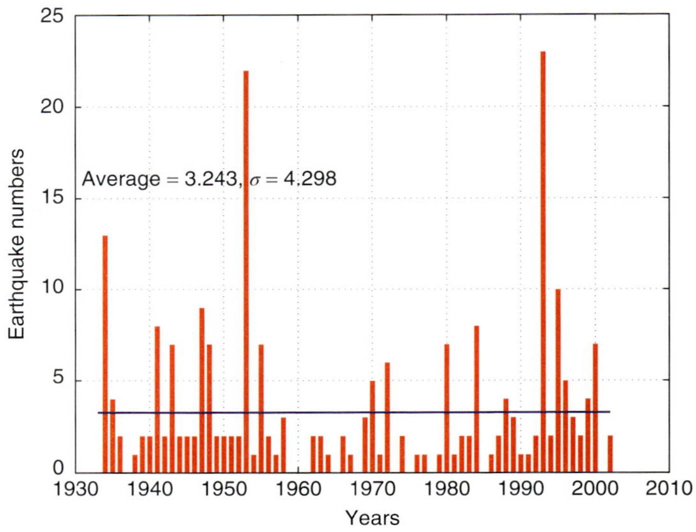

Fig. 4.4 Annual numbers of earthquakes $m \geq 5.0$ in southern California, 1932–2001. *Source:* Kagan (2010a), Fig. 5.

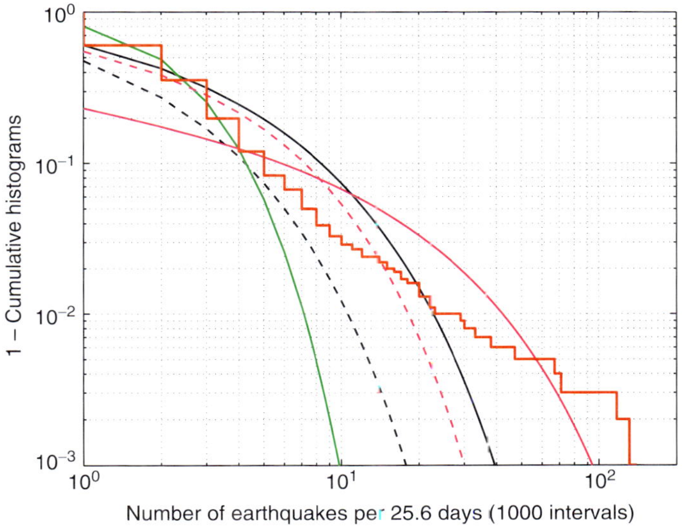

Fig. 4.5 Survival function (1 − Cumulative distribution) of earthquake numbers for the CIT catalog 1932–2001, $m \geq 4.0$. The step-function shows the observed distribution in 25.6 days time intervals (1000 intervals for the whole time period). The green solid curve is the approximation by the Poisson distribution (4.14 , 4.37); the magenta solid line is the NBD approximation (Eqs. 4.15 , 4.18 , and 4.38), with parameters estimated by the MLE method; the magenta dashed curve is the NBD approximation with parameters estimated by the statistical moment method (see Section 4.1.3); the black dashed line is the approximation by the logarithmic distribution (4.10, 4.11); the black solid line is the approximation by the logarithmic law for the zero-truncated distribution (see text). See also a comment at the end of Fig. 4.2 caption. *Source:* Kagan (2010a), Fig. 6.

small time interval would likely contain only one large cluster. Therefore, their distribution would likely be approximated by the logarithmic law (4.10 –4.11).

The observed distribution in Fig. 4.5 is compared to several theoretical curves (cf. Kagan 1996). The Poisson cumulative distribution is calculated using the following formula

$$F(k) = P\,(N < k) = \frac{1}{k!} \int_{\lambda}^{\infty} y^k\, e^{-y}\, dy = 1 - \Gamma(k+1,\ \lambda), \qquad (4.37)$$

where $\Gamma(k+1,\ \lambda)$ is an incomplete gamma function. For the NBD

$$F(k) = P\,(N < k) = \frac{1}{B(\tau,\ k+1)} \int_0^{\theta} y^{\tau-1}\, (1-y)^k\, dy, \qquad (4.38)$$

where θ and τ are defined in (4.15) and $B(\tau,\ k+1)$ is a beta function. The right-hand part of the equation corresponds to an incomplete beta function, $B(\tau,\ k+1,\ x)$ (Gradshteyn and Ryzhik 1980).

For the logarithmic distribution (4.10) two parameter evaluations are made: one based on the naïve average number counts and the other on the number counts for a "zero-truncated" empirical distribution (Patil 1962, p. 70; Patil and Wani 1965, p. 286). The truncation is made because the logarithmic distribution is not defined for a zero number of events (see Eqs. 4.10, 4.11). Thus, we calculate the average event count for only 60% of intervals having a non-zero number of earthquakes.

These theoretical approximations produce an inadequate fit to the observation. The Poisson law fails because there are strong clusters in the catalog. The NBD fails for two reasons: the clusters are truncated in time and the cluster mix is insufficient, especially at the higher end of the distribution at the right-hand part of the plot. Moreover, as we mentioned above, the long-term seismicity variations may explain the poor fit. The logarithmic distribution fails at the lower end, since several clusters, not a single one, as expected by the distribution, are frequently present in an interval. The quasi-exponential tail (see Eq. 4.12) is not observed in the plot, since in the CIT catalog there are no events with a magnitude approaching the maximum (corner) magnitude. For California the corner magnitude should be of the order of $m8$ (Bird and Kagan 2004; see Fig. 5.3).

We produced similar plots for other global catalogs (PDE and GCMT), and the diagrams also display a power-law tail for small time intervals. However, this part of a distribution is usually smaller. This finding is connected to the smaller magnitude range of these catalogs; fewer large clusters of dependent events are present in these datasets.

Figure 4.6 shows a cumulative distribution for annual earthquake numbers. Again the fit by the Poisson law (4.37) is poor, whereas the NBD (4.38) is clearly a better approximation.

4.2.3 Likelihood analysis

Figure 4.7 displays a two-dimensional plot of the log-likelihood function (4.35) for the NBD standard version (4.15). Such plots work well for parameter estimates: if the relation is non-linear or the parameter value needs to be restricted (if, for example, it goes into the negative domain or to infinity, etc.), such plots provide more accurate information than does the second-derivative matrix. (Similar maps to determine earthquake size distribution parameters are shown in Figs. 5.1, 9.2 and 9.3.)

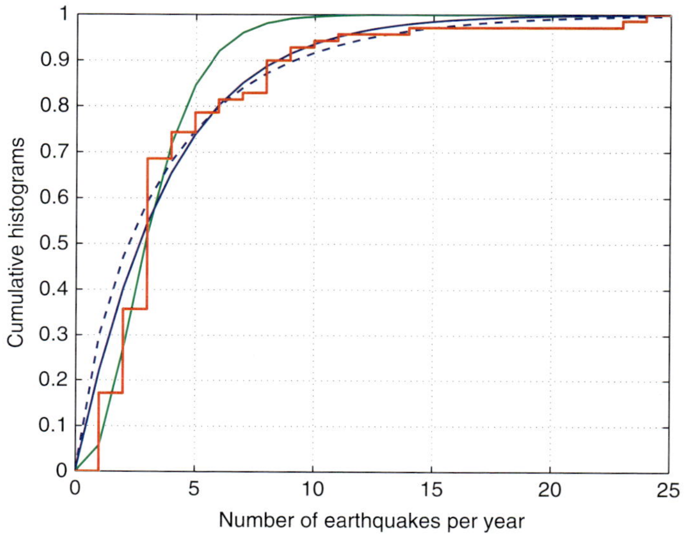

Fig. 4.6 Cumulative distribution of annual earthquake numbers for the CIT catalog 1932–2001, $m \geq 5.0$. The step-function shows the observed distribution, and the solid green curve is the theoretical Poisson distribution (4.37). Two negative binomial curves (4.38) are also displayed: for the dashed blue curve the parameters θ and τ are evaluated by the statistical moment method, for the solid blue curve MLEs are used. The negative binomial curves fit the tails much better than the Poisson does. *Source:* Kagan (2010a), Fig. 7.

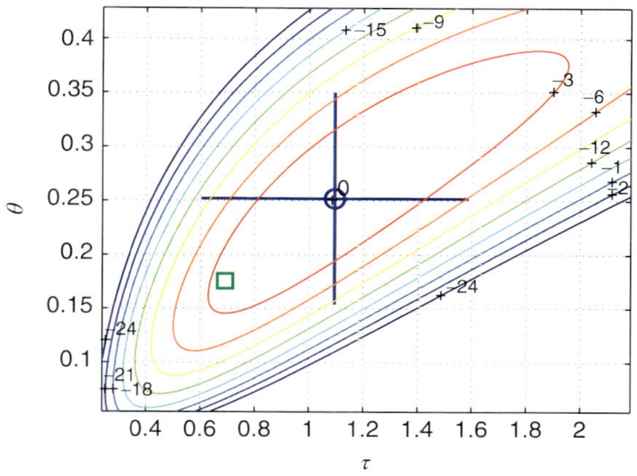

Fig. 4.7 The log-likelihood function map for the CIT earthquake catalog 1932–2001, $m \geq 5.0$, annual event numbers are analysed. The standard representation of the NBD (4.15) is used. Natural logs are used in the map. The green square is the estimate of θ and τ by the moment method, whereas the blue circle shows the MLE parameter values. An approximate 95%-confidence area, based on asymptotic relations, corresponds to the red contour labeled "−3". Two orthogonal line intervals centered at the circle are 95% confidence limits for both parameters, obtained by MATLAB (Statistics Toolbox). The correlation coefficient ρ between these estimates (also evaluated by MATLAB) is 0.867. *Source:* Kagan (2010a), Fig. 8.

Figure 4.7 shows that the parameter estimates are highly correlated. Furthermore, isolines are not elliptical, as required by the usual asymptotic assumption; thus, uncertainty estimates based on the Hessian matrix may be misleading. The 95% confidence limits obtained by the MATLAB (Statistics Toolbox) procedure testify to this. Wilks (1962, Chap. 13.8) shows that the log-likelihood difference is asymptotically distributed as $\frac{1}{2}\chi^2(2)$ (chi-square distribution with two degrees of freedom, corresponding to two parameters of the NBD model). Thus, the isoline [−3.0] at the log-likelihood map should approximately equal 95% confidence limits. The moment estimates (Eqs. 4.27 and 4.28) are within the 95% limits.

For the PDE catalog, where $\rho = 0.993$, the effect of the high correlation coefficient (ρ) on parameter estimates of the standard NBD is demonstrated more strongly in Fig. 4.8. In addition to the parameters for the empirical frequency numbers, estimated by the MLE and the moment method, parameter estimates for 100 simulations, produced applying the MATLAB package, are also shown here. These simulations used the MLEs as their input. The parameters scatter widely over the plot, showing that regular independent uncertainty estimates cannot fully describe their statistical errors. Similar simulations with other catalogs and other magnitude thresholds show that for the standard NBD representation, simulation estimates are widely distributed over the τ, θ plane. Depending on the original estimation method, the moment or the MLE, the estimates are concentrated around the parameter values used as the simulation input.

Figure 4.9 displays the likelihood map for the *alternative* form (4.18) of the NBD. It shows a different behavior from that of Fig. 4.7. Isolines are not inclined in this case, indicating that the correlation between the parameter estimates should be slight.

Figure 4.10 shows the likelihood map for Evans' distribution. Again the isolines are less inclined with regard to axes, showing a relatively low correlation between the estimates. Using formulas (4.32 −4.34), we calculate 95% uncertainties shown in the plot. The correlation coefficient (ρ) between the estimates is ~ 0.15.

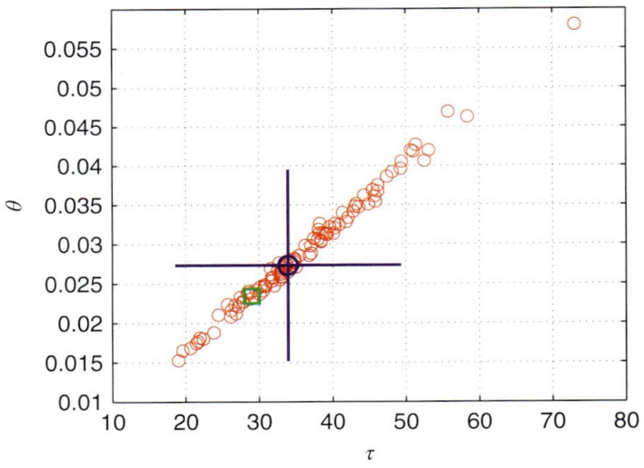

Fig. 4.8 The standard NBD parameter calculations for PDE earthquake catalog 1969–2007, $m \geq 5.0$, annual event numbers are analysed. The large green square is the estimate of θ and τ by the moment method, whereas the large blue circle shows the MLE parameter values: $\tau = 33.98 \pm 15.42$ and $\theta = 0.0274 \pm 0.0122$. Two orthogonal line intervals centered at the circle are 95% confidence limits for both parameters. Small circles are simulated parameter estimates, using MLEs. In simulations the parameter estimates for θ and τ are also MLEs (see above). *Source:* Kagan (2010a), Fig. 9.

Figure 4.11 shows the dependence of the log-likelihood difference $\ell - \ell_0$ on the magnitude threshold (ℓ_0 is the log-likelihood for the Poisson distribution, see Eq. 4.35). The difference increases as the threshold decreases, testifying again that large earthquakes are more Poissonian. The Poisson distribution is a special case of the NBD. Therefore, we can estimate at what log-likelihood difference level we should reject the former hypothesis as a model of earthquake occurrence. Wilks (1962; see also Hilbe 2007) shows that the log-likelihood difference is distributed for a large number of events as $\frac{1}{2}\chi^2(1)$ (chi-square distribution with one degree of freedom). This corresponds to one parameter difference between the Poisson and NBD models. The 95% confidence level corresponds to the $\ell - \ell_0$ value of 1.92. Extrapolating the curve suggests that earthquakes smaller than $m6.5$ in southern California cannot be approximated by a Poisson model. If larger earthquakes were present in a catalog, these events ($m \geq 6.5$) might also be considered as clustered, so this threshold would need to be set even higher.

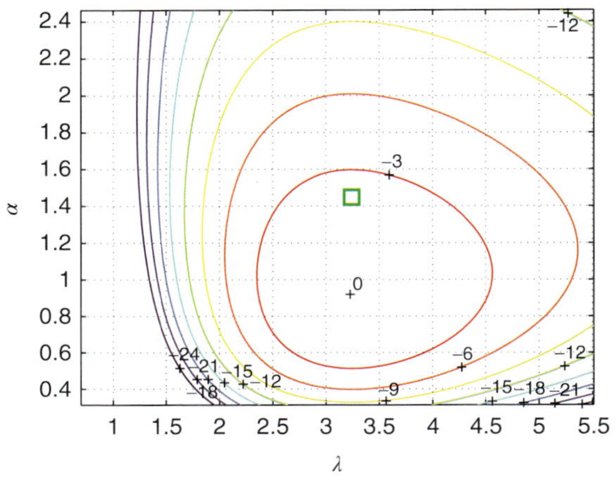

Fig. 4.9 The log-likelihood function map for CIT earthquake catalog 1932–2001, $m \geq 5.0$, annual event numbers are analysed. The alternative representation of the NBD (4.18) is used. The green square is the estimate of α and λ by the moment method. MLE corresponds to the plus sign near '0'. An approximate 95%-confidence area, based on asymptotic relations, corresponds to the red contour labeled −3. *Source:* Kagan (2010a), Fig. 10.

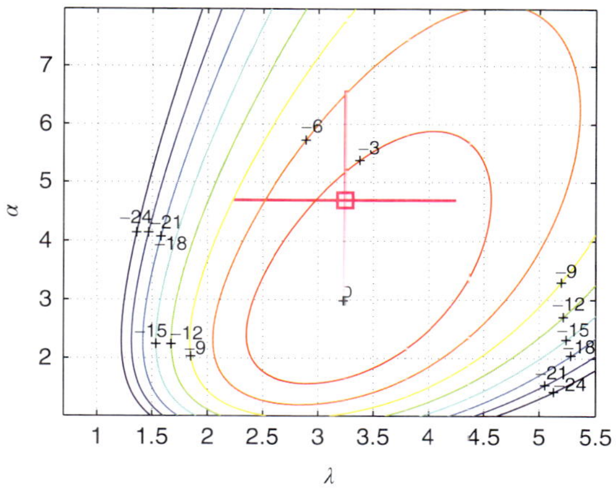

Fig. 4.10 The likelihood function map for CIT earthquake catalog 1932–2001, $m \geq 5.0$, annual event numbers are analysed. Evans' (1953) representation of the NBD (4.20) is used. The red square is the moment estimate of a and λ. An approximate 95%-confidence area, based on asymptotic relations, corresponds to the red contour labeled −3. Two orthogonal line intervals centered at the square are 95% confidence limits for both parameters, based on Evans' (1953) variance formula for moment estimates. *Source:* Kagan (2010a), Fig. 12.

4.2.4 Tables of parameters

Tables 4.1–4.3 show brief results of statistical analysis for three earthquake catalogs. These tables display three dependencies of NBD parameter estimates: (a) on the magnitude threshold (m_t); (b) on time intervals (ΔT) into which a catalog time-span is subdivided; and (c) on a subdivision of a catalog space window. Three NBD representations are investigated: the standard, the alternative, and Evans' formula. Since the parametrization of the last two representations is similar, we discuss below only the standard and the alternative set of parameters. We used the moment method, which is more convenient in applications, to determine the parameter values.

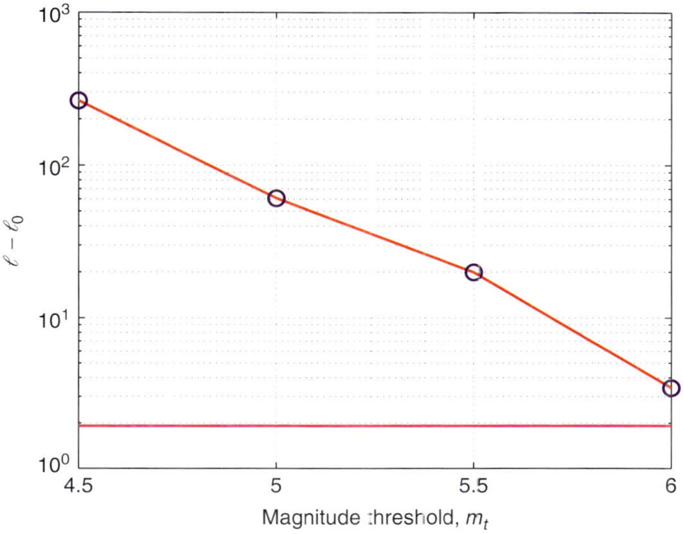

Fig. 4.11 Dependence of the log-likelihood difference ($\ell - \ell_0$) for the NBD and Poisson models 4color2 Fig. 4.11 of earthquake occurrence on the threshold magnitude. The CIT catalog 1932–2001 is used, annual event numbers are analysed. The magenta line corresponds to $\ell - \ell_0 = 1.92$; for a higher log-likelihood difference level the Poisson hypothesis should be rejected at the 95% confidence limit. *Source:* Kagan (2010a), Fig. 13.

Table 4.1
Values of NBD parameters for various subdivisions of the 1969– 2007 PDE catalog.

Subd	m_t	n	N	λ	α	a	θ	τ	ΔT
1	2	3	4	5	6	7	8	9	10
G	7.0	459	39	11.8	0.0089	0.1044	0.9055	112.7	365.2
G	6.5	1359	39	34.9	0.0207	0.7212	0.5810	48.32	365.2
G	6.0	3900	39	100.0	0.0284	2.835	0.2607	35.27	365.2
G	5.5	13553	39	347.5	0.0395	13.72	0.0679	25.33	365.2
G	5.0	47107	39	1207.9	0.0335	40.504	0.0241	29.82	365.2
G	5.0	47107	5	9421.4	0.0118	111.06	0.0089	84.83	2848.8
G	5.0	47107	10	4710.7	0.0269	126.52	0.0078	37.24	1424.4
G	5.0	47107	20	2355.4	0.0309	72.72	0.0136	32.39	712.2
G	5.0	47107	39	1207.9	0.0335	40.50	0.0241	29.82	365.2
G	5.0	47107	50	942.1	0.0422	39.78	0.0245	23.68	284.9
G	5.0	47107	100	471.1	0.0541	25.50	0.0377	18.47	142.4
G	5.0	47107	200	235.5	0.0670	15.78	0.0596	14.92	71.2
G	5.0	47107	500	94.2	0.1137	10.72	0.0854	8.792	28.5
G	5.0	47107	1000	47.1	0.1620	7.632	0.1158	6.172	14.2
0	5.0	2225	39	57.1	0.1686	9.621	0.0942	5.930	365.2
1	5.0	7740	39	198.5	0.0645	12.80	0.0725	15.50	365.2
2	5.0	3457	39	88.6	0.0722	6.400	0.1351	13.85	365.2
3	5.0	3010	39	77.2	0.1475	11.38	0.0808	6.780	365.2
4	5.0	30675	39	786.5	0.0329	25.85	0.0372	30.43	365.2

Notes: In column 1: G means that the whole global catalog is used, 0–plate interior, 1 – Active continent, 2 – Slow ridge, 3 –Fast ridge, 4 – Trench (subduction zones), see Chapter 9; n is the number of earthquakes, N is the number of time intervals, ΔT – subdivision interval duration in days. *Source:* Kagan (2010a), Table 1.

The parameter variations in all subdivisions of Tables 4.1–4.3 exhibit similar features:

(a) In the whole PDE catalog the parameter α decreases as the threshold, m_t, increases ($\alpha \to 0$ corresponds to the Poisson distribution). The θ-value displays the opposite behavior: when $\theta \to 1.0$, the NBD approaches the Poisson law. The θ parameter shows a similar behavior in the GCMT and CIT catalogs (the negative parameter values for $m_t = 7.0$ in the GCMT catalog reveal that the NBD is inappropriate fit for our observations). The α parameter displays no characteristic features for the GCMT and CIT catalogs. The $m_t = 2.0$ results for the CIT catalog are obtained for the period 1989–2001, so they cannot readily be compared to results for other magnitude thresholds.
Figure 4.12 displays the dependence of θ on the magnitude threshold for all catalogs. The decay of the parameter values can be approximated by a power-law function; such behavior can be explained by comparing Eqs. (4.9) and (4.21). This diagram is similar to Fig. 6 in Kagan and Jackson (2000), where the parameter θ is called Υ, and the magnitude/seismic moment transition is as shown in (2.2).

Table 4.2
Values of NBD parameters for various subdivisions of the 1977–2007 GCMT catalog.

Notes: See notes for Table 4.1. *Source:* Kagan (2010a), Table 2.

Subd	m_t	n	N	λ	α	a	θ	τ	ΔT
1	2	3	4	5	6	7	8	9	10
G	7.0	307	31	9.903	−0.0211	−0.2094	1.2649	−47.29	365.2
G	6.5	1015	31	32.742	0.0159	0.5211	0.6574	62.83	365.2
G	6.0	3343	31	107.839	0.0181	1.956	0.3384	55.15	365.2
G	5.8	5276	31	170.194	0.0138	2.356	0.2980	72.25	365.2
G	5.8	5276	5	1055.200	0.0026	2.706	0.2698	389.9	2264.4
G	5.8	5276	10	527.600	0.0076	3.996	0.2001	132.0	1132.2
G	5.8	5276	20	263.800	0.0107	2.821	0.2617	93.52	566.1
G	5.8	5276	31	170.194	0.0138	2.356	0.2980	72.25	365.2
G	5.8	5276	50	105.520	0.0220	2.320	0.3012	45.48	226.4
G	5.8	5276	100	52.760	0.0268	1.411	0.4147	37.38	113.2
G	5.8	5276	200	26.380	0.0417	1.100	0.4763	23.99	56.6
G	5.8	5276	500	10.552	0.0795	0.8392	0.5437	12.57	22.6
G	5.8	5276	1000	5.276	0.1079	0.5693	0.6372	9.267	11.3
0	5.8	172	31	5.548	0.0604	0.3353	0.7489	16.55	365.2
1	5.8	723	31	23.323	0.0357	0.8323	0.5458	28.02	365.2
2	5.8	336	31	10.839	0.0528	0.5720	0.6361	18.95	365.2
3	5.8	537	31	17.323	0.0234	0.4055	0.7115	42.72	365.2
4	5.8	3508	31	113.161	0.0268	3.033	0.2480	37.32	365.2

(b) The θ-values are relatively stable for both global and the CIT catalogs, showing that the event number distribution does not change drastically as the time interval varies. It is not clear why the α significantly changes as the time intervals decrease.
When the ΔT changes, the behavior of the α and the θ, is generally contradictory: both parameters increase with the decrease of the time interval. This trend is contrary to the dependence of these variables on the magnitude threshold [see item (a)]. The θ increase may suggest that the distribution approaches the Poisson law as $\Delta T \to 0$, whereas the α trend implies an increase in clustering. Such anomalous behavior is likely to be caused by a change of the earthquake number distribution for small time intervals. Fig. 4.5 demonstrates that the upper tail of the distribution is controlled by the logarithmic series law (4.10). Although the logarithmic distribution is a special case of the NBD, it requires one parameter versus NBD's two degrees of freedom. Therefore, it is possible that when the logarithmic distribution is approximated by the NBD, the parameters α and θ of two different NBD representations behave differently. The dependence of the distribution parameters on the time sampling interval needs to be studied from both theoretical and observational points of view.

Table 4.3 Values of NBD parameters for various subdivisions of the 1932–2001 CIT catalog.

Notes: In column 1: G means that the whole CIT catalog is used, SE–south-east part of southern California, NE – north-east, SW–south-west, NW – north-west; n is the number of earthquakes, N is the number of time intervals, ΔT–subdivision interval duration in days. *Source:* Kagan (2010a), Table 3.

Subd	m_t	n	N	λ	α	a	θ	τ	ΔT
1	2	3	4	5	6	7	8	9	10
G	6.0	25	70	0.357	2.1360	0.7629	0.56726	0.4682	365.2
G	5.0	226	70	3.229	1.4422	4.6564	0.17679	0.6934	365.2
G	4.0	2274	70	32.49	1.2976	42.154	0.02317	0.7706	365.2
G	3.0	17393	70	248.5	0.9184	228.20	0.00436	1.0888	365.2
G	2.0	5391	13	414.7	1.2958	537.36	0.00186	0.7717	365.2
G	3.0	17393	5	3478.6	0.1062	369.46	0.00270	9.4155	5113.6
G	3.0	17393	10	1739.3	0.1569	272.87	0.00365	6.3740	2556.8
G	3.0	17393	20	869.7	0.3389	294.73	0.00338	2.9507	1278.4
G	3.0	17393	50	347.9	0.6514	226.58	0.00439	1.5353	511.4
G	3.0	17393	70	248.5	0.9184	228.20	0.00436	1.0888	365.2
G	3.0	17393	100	173.9	1.1803	205.28	0.00485	0.8473	255.7
G	3.0	17393	200	86.97	1.7691	153.85	0.00646	0.5652	127.8
G	3.0	17393	500	34.79	3.9832	138.56	0.00717	0.2511	51.1
G	3.0	17393	1000	17.39	6.5226	113.45	0.00874	0.1533	25.6
SE	3.0	8281	70	118.3	1.7967	212.55	0.00468	0.5566	365.2
NE	3.0	2839	70	40.56	4.2550	172.57	0.00576	0.2350	365.2
SW	3.0	2622	70	37.46	3.7734	141.34	0.00703	0.2650	365.2
NW	3.0	3651	70	52.16	1.7630	91.953	0.01076	0.5672	365.2

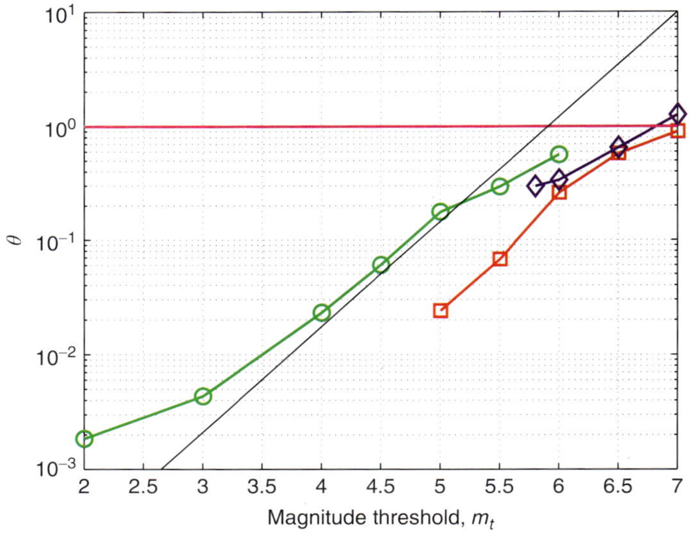

Fig. 4.12 Dependence of the θ-value for the NBD model (4.15) of earthquake occurrence on the threshold magnitude. Three catalogs are used: the green curve is for the CIT catalog 1932–2001, the red curve is for the PDE catalog 1969–2007, and the blue curve is for the GCMT catalog 1977–2007. The horizontal magenta line is $\theta = 1.0$, corresponding to the Poisson occurrence. The thin black line corresponds to $\theta \propto 10^{1.5\ m\ \beta}$ (see Eqs. 4.4 and 4.7). *Source:* Kagan (2010a), Fig. 14.

(c) Two types of spatial subdivision are shown in the tables. For global catalogs we subdivide seismicity into five zones according to the tectonic deformation which prevails in each zone (Chapter 9 and Kagan *et al.* 2010). The CIT catalog was subdivided into four geographic zones. We also made a similar subdivision for the global catalogs (not shown). In all these subdivisions both α and θ are relatively stable, whereas a sum of parameter τ-values for the subdivided areas approximately equals the parameter's value for the whole area. For example, the τ-values for the CIT catalog are: $1.0888 \approx 0.5566 + 0.235 + 0.265 + 0.5672(1.6238)$. By definition, the λ parameter equals the sum for the sub-areas.

5 Earthquake size distribution

This chapter continues to develop ideas presented in several of our publications (Kagan 2002a,c, 2010b; Bird and Kagan 2004; Kagan and Jackson 2013). The distribution of earthquake sizes is usually invoked as a first confirmation for virtually any model of seismicity. Moreover, this distribution is by far the most studied feature of statistical seismology. Starting with its first discussion by Ishimoto and Iida (1939) and then Gutenberg and Richter (1944), it has been established that earthquakes increase in number as a power-law as their sizes decrease. This dependence is usually referred to as the magnitude-frequency or the G-R relation, and its parameter (see Eq. 5.2) is commonly known as the "b-value". A very large body of literature exists concerning the earthquake size distribution, its interpretation and possible correlation with geotectonics, stress, rock properties, etc. For example, a search of the ISI database (Thomson Reuters Scientific 2012) (isiknowledge.com/) for keywords like "**earthquake* and b-value**" yields about 200 publications in the last four years (2009–2012). However, that proliferation has not led to a deeper understanding of earthquake generation.

5.1 Magnitude versus seismic moment

In this book M denotes the scalar seismic moment, and m denotes the magnitude of an earthquake, b is the parameter for the magnitude distribution. If the earthquake size follows the pure power-law distribution (like Eqs. 5.2, 5.3), the parameter β for the seismic moment distribution is $\beta = b/1.5$. Equation 2.2 allows us to use the moment magnitude as a proxy for a physical quantity: seismic moment. All moment magnitude values quoted in this book use the GCMT catalog (Section 2.3) M-estimates as input to (2.2).

Magnitude (m) is an empirical measure of earthquake size and many different magnitude scales are currently used (see Kagan 2003; Castellaro *et al.* 2006). The magnitude is used here as an auxiliary variable mostly for illustration and comparison with old data and results. We consider the scalar seismic moment M-values as a primary, proper physical measure of earthquake size (see also

Earthquakes: Models, Statistics, Testable Forecasts, First Edition. Yan Y. Kagan.

Chapter 8). Several types of errors need to be investigated in earthquake size measurements. Some of them are known to be connected with earthquake *magnitude* determination: saturation of all magnitude scales (Kanamori 1977), which is explained by the finite seismogram frequency for a seismographic network. Other types of errors are common to both magnitude and seismic moment determination (Kagan 2003).

Relatively high-frequency seismic waves are used to determine magnitude; the effects of scattering, multipathing, focussing and unfocussing are stronger as the wave periods decrease. These effects cause great variations of wave amplitude which lead to larger uncertainties and biases in magnitude measurements (cf. Ekström and Dziewonski 1988).

Seismographic networks are limited in detecting weak earthquakes and their essential parameters such as hypocenter location, origin time, and magnitude. This results in another limitation of magnitude distributions: at the lower magnitude end progressively larger number of events are missing from catalogs. Unfortunately, this lower magnitude cutoff is neither sharp nor uniform over time and space, especially for local catalogs.

Seismic moment is proportional to the amplitude of seismic waves at zero or close to zero frequency; hence its accuracy is higher than that using magnitudes. Kagan (2003) estimates that uncertainty in moment magnitude is on the order of 0.1–0.15 and is by a factor of 2 to 3 smaller than regular magnitude uncertainties.

Because of saturation and other systematic effects, each magnitude can only be evaluated over a limited range of earthquake size. Different magnitude scales are then compared by using a regression relation

$$m_1 = C_1 + C_2 m_2, \tag{5.1}$$

where m_i are magnitudes and C_j are coefficients for a linear regression. Although both magnitudes in (5.1) usually have errors of similar size, regular, not orthogonal regression, is commonly used (Castellaro *et al.* 2006). This should cause a significant bias in converting one magnitude into another. Most earthquake catalogs initially use several magnitudes. To obtain a common magnitude value, catalog compilers transform various magnitudes, using variants of (5.1). Errors and systematic effects of such calculations should significantly shape the estimates of earthquake size and features of the size distribution.

Inspecting the value of the C_2 coefficient in various publications, one can see the degree of the problem in determining magnitudes. C_2 should be close to 1.0 at the range of earthquake size where both magnitudes are well defined. This is rarely the case: the C_2-value often reaches 0.7 or 1.3 (Kagan 2003, Fig. 14). These fluctuations of the conversion coefficient may cause spurious variations of the *b*-value. In contrast, when the moment magnitude in different catalogs is compared, the C_2 coefficient is close to 1.0 (ibid., Fig. 12).

Regular magnitude measurements are subject to many random and systematic errors (Kagan 2003). Kagan (1999, pp. 557–558) studied the correlation between the *b*-value estimates based on m_b and M_S magnitudes and found that the correlation coefficient is low (0.1–0.2). These coefficient values seem to indicate that b-value variations are not caused by regional tectonic or physical factors. Therefore, we investigate earthquake size distribution in this chapter, using only seismic moment tensor measurements. The most complete, extensive, and accurate catalog of tensor solutions is the GCMT dataset (Kagan 2003; see also Section 2.3).

5.2 Seismic moment distribution

The distribution of earthquake size is usually described by the classical G-R (Gutenberg and Richter 1944) magnitude-frequency relation

$$\log_{10} N(m) = a_t - b(m - m_t) \qquad \text{for} \quad m_t \leq m, \tag{5.2}$$

where $N(m)$ is the number of earthquakes with magnitude $\geq m$, m_t is the magnitude threshold – that is, the smallest magnitude/moment above which the catalog can be considered to be complete, and a_t and b are parameters: a_t is the logarithm of the number of earthquakes with $m \geq m_t$ (the seismic activity or earthquake productivity level) and b describes the relation between numbers of small and large earthquakes, $b \approx 1$. The G-R laws seems to apply to earthquakes as small as $m < -4.0$ (Boettcher *et al.* 2009; Kwiatek *et al.* 2011), with a rupture length on the order of a few centimeters, and as large as the 2004 $m9.1$ Sumatra, with about a 1000 km rupture.

Vere-Jones (2010, p. 648) points out that in the original formulation the G-R law was usually interpreted not as a statistical distribution but erroneously as a regression between earthquake numbers and magnitude. This leads to many errors and misunderstandings, still plaguing earthquake science.

The expression similar to (5.2) has been proposed in the above functional form by Vilfredo Pareto (1897, p. 305, his Eq. 1) for the financial income distribution. The G-R relation (5.2) can be transformed into the Pareto (power-law) distribution for the scalar seismic moment M. The distribution in a PDF form is (Kagan 2002a, 2010b)

$$f(M) = \beta \, M_t^{\beta} \, M^{-1-\beta} \qquad \text{for} \quad M_t \leq M; \tag{5.3}$$

in a survivor function form (1− Cumulative distribution) it is

$$F(M) = (M_t/M)^{\beta} \qquad \text{for} \quad M_t \leq M, \tag{5.4}$$

where β is the index parameter of the distribution. The magnitude threshold for the 1977–2008 GCMT catalog is $m_t = 5.8$; for the 1982–2008 catalog it is $m_t = 5.6$ (Kagan 2003). Ekström *et al.* (2012) provide threshold estimates for the whole 1976–2010 GCMT catalog.

Simple consideration of the finiteness of seismic moment flux or the deformational energy, available for earthquake generation, requires that the Pareto relation (5.3) be modified at the large size end of the moment scale. The distribution density tail must have a decay stronger than $M^{-1-\beta}$ with $\beta > 1$. This problem is generally solved by introducing into the distribution an additional parameter called the *maximum* or *corner* moment (M_x or M_c).

The tapered G-R (TGR) or tapered Pareto relation has an exponential taper applied to the cumulative number of events (cf. 5.4) with the seismic moment larger than M (Vere-Jones *et al.* 2001; Kagan 2002a)

$$F(M) = (M_t/M)^\beta \exp\left(\frac{M_t - M}{M_c}\right) \qquad \text{for} \quad M_t \le M < \infty, \tag{5.5}$$

here M_c is the parameter that controls the distribution in the upper ranges of M ("the corner moment"). Kagan (2002a) introduces three additional laws that can be used to approximate earthquake size distribution and provides equations for their interrelations. The b-value is not properly defined for such complicated magnitude-frequency relations. Therefore, we argue that b is an observed slope in some accessible magnitude range, whereas β is an asymptotic slope at small magnitudes (not necessarily observed); thus, they may differ by more than a factor of 1.5; they differ in the basic concept (see also Subsection 9.5.1).

The corner moment (M_c) can be thought of as the earthquake size which is rarely exceeded (e.g., about once per century for subduction earthquakes), and in maximum-likelihood fitting of (5.5) to actual large (sub)catalogs it is typically not very different from the second-largest earthquake moment.

Figure 5.1 displays the seismic moment or moment magnitude distribution and its approximation by (5.5) with a choice of several corner magnitudes (m_c) for the subduction zones (Bird and Kagan 2004). Log-likelihood 2-D map for evaluation of β and m_c is also shown. The β-value can be easily evaluated from this map. However, for M_c only the lower limit can be established, the upper limit

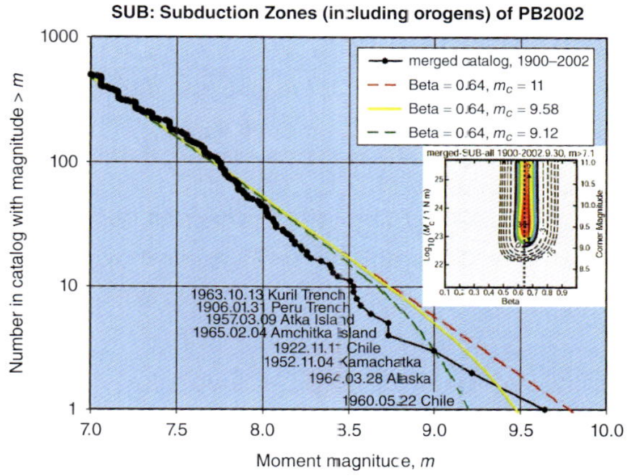

Fig. 5.1 The size distribution for subduction earthquakes (Bird and Kagan 2004). The inset shows the 2-D likelihood map which is used to estimate both the β and the corner moment (M_c) values. This map shows that for subduction zones only the lower M_c limit can be determined, the upper limit is ∞. *Source:* Kagan (2006), Fig. 5.

is infinity, and the MLE is poorly constrained. (Similar 2-D maps for earthquake number distributions are constructed in Chapter 4: Figs. 4.7, 4.9, 4.10.) Additional moment-frequency relations for other tectonic regions and likelihood maps are displayed in Figs. 9.2 and 9.3. Figure 9.3 shows the log-likelihood β / M_c maps in a more accessible format. Similar maps and methods for their construction are discussed, for instance, by Kagan (1997a) and by Kagan and Jackson (2000).

The corresponding PF for the TGR law is

$$f(M) = \left[\frac{\beta}{M} + \frac{1}{M_c}\right] (M_t/M)^{\beta} \ \exp\left(\frac{M_t - M}{M_c}\right). \tag{5.6}$$

The above distribution in both expressions (Eqs. 5.5 –5.6) was proposed by Pareto (1897, pp. 305–306, Eqs. 2, 2*bis*, and 5). Equations 5.6 and 5.5 are equivalent to (5.3) and (5.4), respectively, if $M_c \to \infty$.

Figure 5.2 shows the result of the maximum likelihood determination of the β-values for eight worldwide tectonic provinces (Bird and Kagan 2004). All 95% confidence limits include $\beta \approx 2/3$ value. This can be considered a universal parameter of earthquake size distribution (see more in Section 5.3).

Figure 5.3 displays the corner moment values evaluated for the same eight provinces. For convenience they are shown on the map of central America, where all the provinces are represented. In contrast to the β-value result, Bird and Kagan (2004, Table 5) find that at least four distinct values of the corner magnitude seem to be required, based on the 95% confidence limits. These values include Oceanic Spreading Ridges (normal faulting, corner magnitude range, $m_c = 5.7 - 6.0$); Oceanic Transform Faults (medium and fast velocities, range, $m_c = 6.4 - 7.0$); all the continental zones, Oceanic Transform Faults and slow velocity Oceanic Convergent Boundary (range, 7.4–8.7); and Subduction zone boundaries (range from 9.1 to $+\infty$).

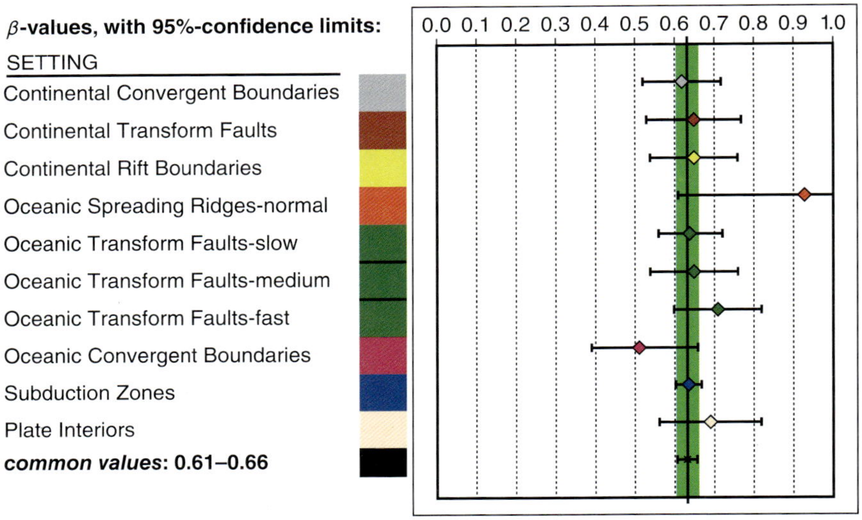

Fig. 5.2 The β-value global distribution. The β-values, determined by the maximum likelihood method for eight tectonic provinces (Bird and Kagan 2004), are shown with their 95% confidence limits. *Source:* Kagan (2006), Fig. 6.

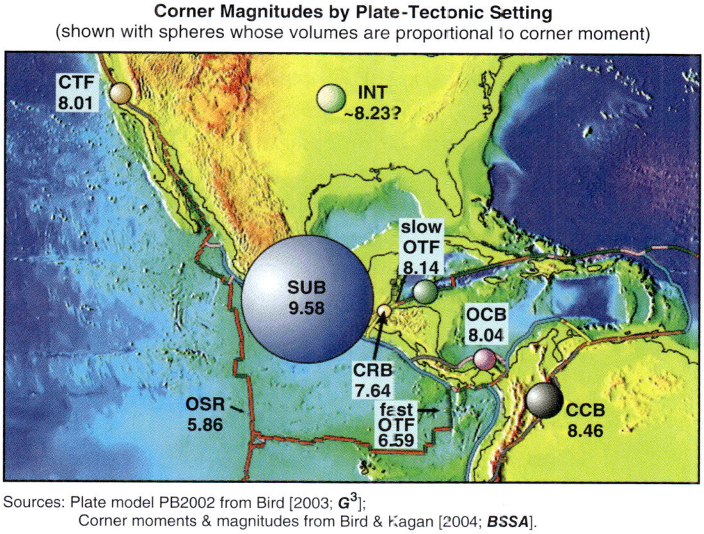

Fig. 5.3 Corner moment (M_c) distribution for eight tectonic provinces (Bird and Kagan 2004). In this figure the moment magnitude is calculated using $C = 6.033$ in (2.2). For plate boundary names and abbreviations, see Fig. 5.2. *Source:* Kagan (2006), Fig. 7.

Integrating the earthquake size distribution (5.5, 5.6), we can calculate the seismic moment rate (Kagan 2002c)

$$\dot{M}_s = \frac{\alpha_0 M_0^{\beta}}{1-\beta} M_c^{1-\beta}\Gamma(2-\beta), \tag{5.7}$$

where α_0 is the annual rate of earthquakes with moment M_0 or greater and Γ is the gamma function. We compare the seismic moment rate (Eq. 5.7) with the tectonic moment rate ($\dot{M}_T$):

$$\dot{M}_T = \mu\, W \int_A\!\!\int |\dot{\epsilon}| dA \propto \dot{M}_s, \tag{5.8}$$

where μ is the elastic shear modulus, W is the seismogenic width of the lithosphere, $\dot{\epsilon}$ is the average long-term horizontal strain rate, and A is the area under consideration (see also Section 5.6.2). When comparing the tectonic and seismic rate, the former should be multiplied by χ (the seismic coupling or seismic efficiency coefficient, $\chi \leq 1.0$). Bird and Kagan (2004, Eq. 11) propose another, more exact formula for calculating the tectonic moment rate appropriate to a plate boundary zone.

Subsequently we compare $\dot{M}_s$ (5.7) to the tectonic rate (5.8) (Kagan 2002c; Bird and Kagan 2004). Thus, plate tectonic predictions can be quantitatively related to seismic activity and the corner moment M_c calculated from such a comparison. Below we show (see Section 5.4) that because of the power-law property of earthquake size distribution, any naïve comparison of the cumulative seismic moment and tectonic deformation yields unreliable results in most cases.

In regions of high seismicity, instead of Eq. 5.7 we can use measured long-term seismic activity to infer the earthquake rate by extrapolating (5.5) to any moment level (see Bird and Liu 2007, Eqs. 4 and 5). Bird *et al.* (2010) presented an algorithm and tables for a long-term world-wide forecast of shallow seismicity based on the Global Strain Rate Map (GSRM) by Kreemer *et al.* (2003). Because GSRM does not estimate tectonic strain-rates in stable plate interiors, a simple empirical-averaging method has been used. Thus, the seismicity in plate interiors is represented by a uniform rate.

Since the seismicity level in plate interiors may vary by orders of magnitude, the uniform rate may strongly under- or over-estimate the seismicity rate. Therefore, we apply Eqs. 5.7 and 5.8 to evaluate first the tectonic moment rate and then a long-term forecast for these regions

$$\alpha_0 = \frac{\dot{M}_T \ \chi(1-\beta)\ \exp(-M_0/M_c)}{M_0^{\beta}\ M_c^{1-\beta}\ \Gamma(2-\beta)}. \tag{5.9}$$

By calculating α_0 for a particular choice of M_0, we may re-normalize Eq. 5.5 and obtain earthquake size distribution for any region with a known strain rate and corner moment. In Section 6.6 we use this relation to calculate the earthquake rate for the New Madrid (USA) area.

5.3 Is β ≡ 1/2?

5.3.1 Preamble

Theoretical analysis of earthquake occurrence (Vere-Jones 1976, 1977) suggests that, given its branching nature, the exponent β of earthquake size distribution should be identical to $1/2$. Properties of the critical branching process explain this result: the total number of events (individuals) in such a process is distributed asymptotically according to a power-law with exponent 0.5 (Otter 1949; Harris 1963, Ch. I.13). Such distributions, obtained by simulations, are shown as upper curves in Figs. 5.7 and 5.8.

The same values of power-law exponents are derived for percolation and self-organized criticality (SOC) processes in a high-dimensional space (see discussion by Kagan 1991b, p. 132). Similar values of exponents are obtained by theoretical arguments and simulations for dislocation avalanches in crystalline materials (Zaiser 2006, and its references, see also Section 5.3.4).

However, almost all the β or b measurements in earthquake catalogs result in estimates larger than $\frac{1}{2}$ (or 0.75 for the b-value). These estimates exhibit large fluctuations in different regions and tectonic zones. In almost all publications the variation of the b-value is attributed to different tectonics, rock stress, etc.

The following reasons for variability in the measured b- and β-values can be proposed:

1. Inappropriate use of magnitude scales other than the moment magnitude: only the moment magnitude distribution should be studied. Regular earthquake magnitudes have significant systematic and random errors

(Kagan 1999, 2003, 2010b; Section 5.1), making them inappropriate for rigorous statistical, quantitative investigation.

2. The maximum or corner moment M_c needs careful consideration (Kagan 1991a, 2002a; Bird and Kagan 2004). If M_c is close to the moment threshold and the earthquake size distribution is approximated by a plain Pareto (or G-R) distribution (Eqs. 5.3 and 5.4), the magnitude-frequency curve would shift downwards, and the b- or β-estimates would be strongly biased upwards. In such a case to avoid the bias, it is necessary to apply a two-parameter relation, which includes the maximum or corner magnitude/moment (Eqs. 5.5 and 5.6). In the latter 2-parameter approximation the β-value is defined by the part of the distribution close to the moment threshold, hence the corrected β would be smaller.
3. Mixing populations of earthquakes from tectonic settings that have different corner magnitudes: when earthquake populations with a varying corner moment M_c are placed in the study samples, a false increase in the β-values results. Mixing populations with different M_c may even yield a seemingly linear curve in a log–log plot, in such a case a two-parameter approximation of the magnitude-frequency relation would fail to avoid bias. For example, due to a significant corner moment variability, Kagan (2002a, Table 5, Section 5.2.3) determined the β-values on the order 0.8–1.1 for mid-ocean earthquakes. Similarly and apparently for the same reason, excessively large b-values for oceanic earthquakes were obtained by Okal and Romanowicz (1994) and Schorlemmer *et al.* (2005). Bird and Kagan (2004) gave correct evaluation.
4. Relative seismic moment errors increase with decreases in earthquake size, resulting in a spurious β increase (see Fig. 5.5).
5. The object of study should be earthquake sequences, not individual earthquakes: theoretical estimates, discussed above, are relevant for earthquake *sequences*, not individual events. Hence β-value estimates need to be corrected for this effect.
6. An earthquake is a tensor; its size, as given in moment tensor solutions, is a tensor sum of earthquake subevents. If these subevents have different focal mechanisms, their tensor sum would be smaller than that for scalar moments of subevents. Thus, even if the number of elementary earthquake events were distributed according to the power-law with the exponent $\beta \equiv \frac{1}{2}$, the distribution of earthquake size, as represented by the tensor sum, would have a larger exponent value. This follows from stochastic complexity in the source. If, for example, a source consists of positive and negative random n subsources, its total size would be proportional to $\sqrt{n}$: the size would be similar to that at the end of a Brownian random walk.
7. The corner moment M_c is likely to change significantly with the depth for shallow earthquakes (Kagan 1999, Section 1; Kagan 2002a). For shallow seismicity, earthquake populations with various depths are usually added up, thus β determinations are biased as discussed in item 3 above. Earthquake catalogs with only the hypocenter information cannot be used to investigate

this effect because hypocenters are often located at a lower or upper depth boundary of a fault rupture zone. On the other hand, the degree of accuracy of centroid depths in moment tensor catalogs is presently too low to study rigorously the M_c depth variation.

On the basis of statistical analysis of several earthquake catalogs and some theoretical observations, Kagan (1991b, pp. 129, 132–133) conjectured that the β-value is a universal *constant* (1/2). Three additional arguments can now be added to strengthen this hypothesis: (a) more careful measurements of the β parameter in modern earthquake catalogs (mainly the GCMT catalog) suggest (Kagan 1999; Bird and Kagan 2004; Boettcher and Jordan 2004) that the β is universal (see, for example, Fig. 5.2); (b) recent advances in space geodesy and quantitative plate tectonics allow for detailed calculation of tectonic deformation rate. By comparing tectonic and seismic moment rates, we can calculate the upper bound of M_c for earthquake moment distribution (see Eqs. 5.7 and 5.8); and (c) investigation of dislocation avalanches in crystals indicates that their size distribution is a power-law whose exponent has a universal value.

However, the β universality model can be challenged. One can argue that more careful measurements may reveal statistically significant variations in the β exponent. The aim here is to show that the β-value is a universal *constant*. To corroborate this argument we consider in detail various systematic and random effects on β evaluation that are listed in this subsection.

5.3.2 Catalog analysis and earthquake size distribution

Only shallow earthquakes (depth 0–70 km) are investigated in this section, because more data is available on them. Additionally, the seismic efficiency coefficient, or proportion of tectonic deformation released by such events, is close to 1.0 in continental and/or subduction settings (Bird and Kagan 2004). Geometry of deep earthquake faults is much less known than that for shallow seismicity, and may be more complex. For deeper earth layers (depth > 70 km) only a small part (less than 5%) of tectonic motion is released by seismicity (Kagan 1999; Frohlich and Wetzel 2007). Therefore, earthquake rupture in a brittle crust would be better modeled by the critical branching process theory.

To demonstrate the quality of approximating the observed magnitude-frequency relation (as in Fig. 5.1) by the tapered Pareto or the TGR relation in Fig. 5.4 we show the normalized difference between these variables. We assume that the earthquake numbers in a frequency plot follow the Poisson distribution (see Eq. 4.14). Kagan (2010a) demonstrates that the temporal distribution of large earthquakes approaches the Poisson law. Kagan *et al.* (2010, Table 2) show that in the GCMT catalog, due to its high moment threshold, the aftershock numbers are less than 25% of the total. For large number of events its standard deviation (σ) is a square root of the number. To normalize, we divide the difference by $1.96 \times \sigma$; hence the ± 1 ordinate value would correspond to the 95% confidence level. For most of the diagram, the difference between the

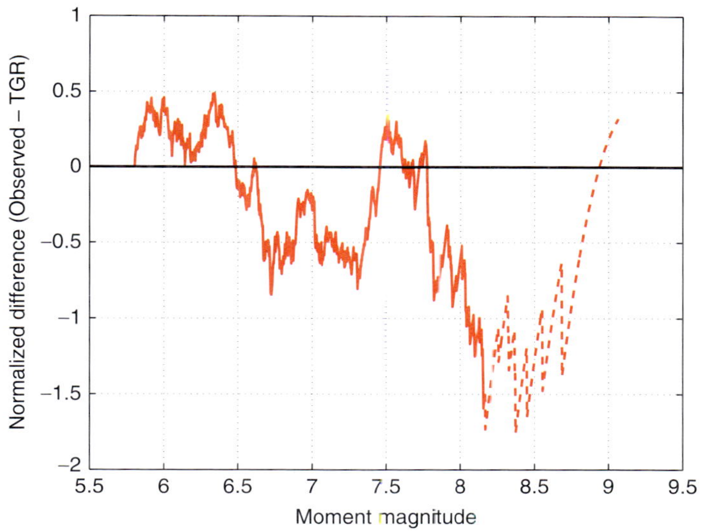

Fig. 5.4 Difference between observed magnitude-frequency relation and approximation by TGR law. GCMT catalog 1977–2008, magnitude threshold $m_t = 5.8$. The dashed line indicates where the difference is based on fewer than 10 events. *Source:* Kagan (2010b), Fig. 2.

observed and the TGR distributions is less than $1.96 \times \sigma$, thus the approximation is reasonably good.

5.3.3 Systematic and random effects in determining earthquake size

This subsection considers in detail four sources of magnitude / moment bias mentioned in the Preamble (items 4–7).

5.3.3.1 Scalar seismic moment errors

There is a bias in evaluating the G-R parameters (Eq. 5.2) due to random errors in determining magnitude (Molchan and Podgaetskaya 1973, p. 47, Eq. 7; Tinti and Mulargia 1985, p. 1690; Castellaro *et al.* 2006). The magnitude random error is assumed to be symmetric and Gaussian, the error analysis by Kagan (2003, see Figs. 12–15) seems to confirm this. Generally, as long as the moment or amplitude errors are relatively small, one should expect them to be Gaussian as they are the result of summing up many independent random uncertainties. Even if the errors are more heavy-tailed than those for the Gaussian distribution as Werner and Sornette (2008) suggest, this would only strengthen our conclusions about the magnitude uncertainty influence on the β-bias (see discussion at the end of this section).

The magnitude error causes a shift of the a estimate in (5.2) toward larger values: given the approximate symmetry of the error distribution, more weak earthquakes have their magnitude increased than vice versa:

$$a_{\text{estim}} = a_{\text{corr}} + \frac{b^2 \, \sigma_m^2 \log 10}{2}, \tag{5.10}$$

where σ_m is a standard error in the magnitude estimate, a_{corr} is the corrected (true) a-value, a_{estim} is the measured a-value. Otherwise, the bias may manifest as a right-hand (horizontal) magnitude shift in the G-R curve

$$m_{corr} = m_{estim} - \frac{b\,\sigma_m^2 \log 10}{2}, \tag{5.11}$$

where m_{corr} is the corrected magnitude, m_{estim} is the measured magnitude. For $b = 1$ the shift is of the same amplitude in (5.10) and (5.11).

If the magnitude errors do not depend on the magnitude, this error does not practically influence the estimated b-value (Tinti and Mulargia 1985). However, if σ_m is a function of m, the b estimates would be affected by magnitude errors. Rhoades (1996, 1997) derived the theoretical estimates of a bias in the b-value in such a case. Rhoades (1996) and Rhoades and Dowrick (2000) studied the influence of magnitude errors on b-value estimates and provided some approximate appraisals for the b-bias from such errors. In these evaluations they assumed that magnitude errors increase as the magnitude itself increases, and they found that the "measured" b-value decreases by about 4% compared to its true value.

The GCMT catalog supplies inversion errors for tensor moment components (Dziewonski *et al.* 1981; Dziewonski and Woodhouse 1983a,b; Ekström *et al.* 2012; see Section 2.3). It is important to distinguish magnitude and moment random errors. Magnitude is defined as a logarithm of appropriately scaled and averaged amplitude (A) of specific seismic waves: $m = \log_{10} A + C$, where C is a coefficient. Therefore, for small amplitude uncertainties the standard error in the amplitude measurements (σ_A) is related to the magnitude error (σ_m) as

$$\sigma_m \propto \frac{\partial m}{\partial A} \propto \frac{\sigma_A}{A}, \tag{5.12}$$

that is, the magnitude error is proportional to a *relative* amplitude error. A similar relation is valid for the scalar seismic moment (see Eqs. 5.13 and 5.16).

Kagan (2002a) measured the scalar seismic moment errors in the GCMT catalog and found that the relative moment errors actually *decrease* with the earthquake size, implying that the *magnitude* errors should also decrease. We define the relative moment error, ϵ, as the ratio of the error tensor norm to the moment tensor norm

$$\epsilon = \sqrt{\sum_{i,j} E_{ij}^2 \Big/ \sum_{i,j} M_{ij}^2}\,, \tag{5.13}$$

where E_{ij} and M_{ij} are standard error and moment tensor components, respectively. The distribution of ϵ for the GCMT catalog 1977–2008 is shown in Fig. 5.5 (compare Fig. 5 in Kagan 2002a). Since the influence of the magnitude threshold value is insignificant when determining the relative moment error, we use $m_t = 5.6$ in this plot. We calculate two regression lines approximating the dependence of the errors on the magnitude: the linear and quadratic curves

$$\log_{10}\epsilon = c_0 + c_1(m - 6) + c_2(m - 6)^2. \tag{5.14}$$

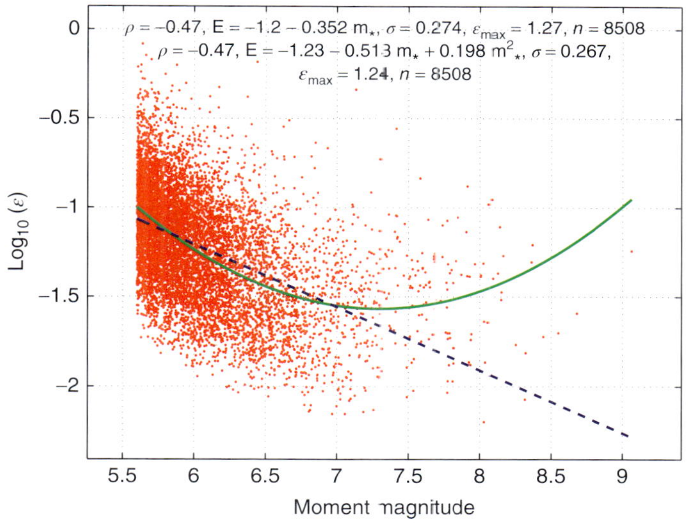

Fig. 5.5 Relative error ϵ versus moment magnitude for shallow earthquakes $m_t = 5.6$ in the 1977–2008 GCMT catalog. The curves show two approximations: linear and quadratic fits. *Source:* Kagan (2010b), Fig. 4.

Table 5.1 Parameter values for relative errors.

Approximation	n	c_0	c_1	c_2	ρ	σ
GCMT 1977–2008, $m_t = 5.6$						
Linear	8508	−1.20	−0.352	–	−0.47	0.274
Quadratic	8508	−1.23	−0.513	0.198	−0.47	0.267
GCMT 1982–2008, $m_t = 5.4$						
Linear	11600	−1.21	−0.447	–	−0.57	0.274
Quadratic	11600	−1.26	−0.537	0.208	−0.57	0.266

Notes: n, the number of $m \geq m_t$ events; for c_0, c_1, and c_2, see (5.14); ρ, correlation coefficient; σ, standard deviation of fit. For the linear approximation, the c_2 parameter is shown as a dash. *Source:* Kagan (2010b), Table 1.

We use $(m - 6)$ instead of m as an argument, so that the c_0-value would have a clear intuitive meaning – an ordinate of a regression curve at $m = 6$.

In Figure 5.5, the coefficient of correlation between ϵ and the magnitude is −0.47, indicating that relative moment errors decrease with the increase of m. Residual regression errors are close for both linear and quadratic cases: $\sigma = 0.274$ and $\sigma = 0.267$, respectively. Errors for earthquakes with $m > 6.5$ deviate significantly from a linear trend. However, since the number of strong earthquakes is small, the residuals of the linear and quadratic cases do not differ significantly.

The parameter values in (5.14) for two subsets of the GCMT catalog are listed in Table 5.1. The ϵ-values for earthquakes in the magnitude range 5.4–6.4 can be represented as

$$\epsilon \approx 0.056 \times 10^{-0.54(m-6)}, \tag{5.15}$$

for shallow events in the 1982–2008 GCMT catalog (see Table 5.1). For small ϵ the magnitude error σ_m is calculated as

$$\sigma_m \approx \frac{\epsilon}{1.5 \log 10}. \tag{5.16}$$

Modifying (5.11) or Eq. 10 in Rhoades and Dowrick (2000), we obtain the following magnitude correction for the magnitude estimates perturbed by random errors:

$$m_{\mathrm{corr}} = m_{\mathrm{estim}} - \frac{3}{4}\,\sigma_m^2\,\beta\,\log 10. \qquad (5.17)$$

To apply this formula to b or β correction, we estimate σ_m at two magnitude values (5.4 and 6.4) and use (5.17) to compute $\delta_c m = m_{\mathrm{corr}} - m_{\mathrm{estim}}$. Performing such calculations for the relative moment error ϵ in (5.16), we obtain the correction for β of shallow earthquakes 0.0013: the β-value is decreased by about 0.2%.

However, our calculations could not consider one important source of error. For many weak shallow earthquakes in the GCMT catalog, no solution can be obtained for the tensor components $M_{r\theta}$ and $M_{r\phi}$ (Dziewonski *et al.* 1981, p. 2829; Dziewonski and Woodhouse 1983a,b; Ekström *et al.* 2012; Frohlich and Davis 1999). In such a case, $E_{r\theta}$ and $E_{r\phi}$ as well as $M_{r\theta}$ and $M_{r\phi}$ are set to zero. About 4% of shallow earthquakes have this problem (Kagan 2003, pp. 195–196). For strike-slip events which predominate in this group, the tensor components $M_{r\theta}$ and $M_{r\phi}$ are close to zero. This means that if the true values of E_{ij} and M_{ij} were available for these events in (5.13), the numerator value should be much greater, but the denominator would be essentially the same. This would significantly increase the resulting ϵ-value.

Moreover, apparently the relative moment error, ϵ, is only part of the total seismic moment uncertainty. Dziewonski *et al.* (1981) and Dziewonski and Woodhouse (1983a,b) suggested that the standard errors obtained in the GCMT solutions are likely to be underestimated. Kagan (2000, 2002a) estimated that the reported errors are possibly one third to one half of the total. Werner and Sornette (2008) indicate that the magnitude uncertainty is more heavy-tailed than a Gaussian distribution. These error outliers can be expected to increase the magnitude random scatter and have a substantial effect on the β-bias. Because the bias in estimating β depends on the square of the magnitude estimation error (see Eq. 5.17), a systematic bias as high as 1–3% may be caused by the decrease of the relative magnitude uncertainty with the increasing earthquake size.

5.3.3.2 Earthquake sequences and their influence

As mentioned in the Preamble, theoretical β estimates are relevant only for earthquake *sequences*, not individual events. We posit that an earthquake belongs to a sequence and sequences are the theoretical entity of interest, rather than individual earthquakes. Registration of aftershock sequences (Kagan 2004; Kagan and Houston 2005; Enescu *et al.* 2009, see also Chapter 6 and Fig. 6.1) shows that immediate aftershocks observed in high-frequency seismograms are included in a mainshock or in large aftershocks in catalogs based on long period registration. Thus, for example, the GCMT catalog earthquakes include some close aftershocks.

In our continuum model (Section 3.3) earthquake sequences are produced by the critical branching process. Later aftershocks are separated into individual events due to temporal delays controlled by Omori's law (Kagan and Knopoff 1981). Occasionally, the first event in a sequence is weaker than the following events, in which case it is commonly called a "foreshock". Therefore, we could consider an earthquake cascade or entire foreshock–mainshock–aftershock sequence as one entity.

Here we attempt to study the seismic moment distribution for earthquake sequences. To define the sequences, we use the results of the likelihood analysis of earthquake catalogs (Kagan 1991b, see also Section 9.5). We approximate an earthquake occurrence by a multidimensional Poisson branching cluster process (see Fig. 3.1c). The parameters of the model are estimated through a maximum likelihood search. A similar scheme has been proposed recently by Zhuang *et al.* (2002, 2004).

As the result of the likelihood optimization, we evaluate probabilities (p_{ij}) of any i-th earthquake belonging to a j-th cluster or sequence ($\sum_j p_{ij} = 1$); p_{ii} corresponds to the probability that an earthquake is considered independent. We use these probabilities (p_{ij}) to assign a part of the seismic moment of the i-th event to the j-th earthquake; the j-th earthquake might again belong to some k-th group, etc. This process continues until all earthquakes and their interconnections in a catalog have been counted. In the end, some of the aftershock moments are transferred to their mainshocks if the aftershock has a non-zero probability of belonging to a particular mainshock cluster. As a result of this seismic moment reassignment, some earthquakes (former aftershocks) may have a seismic moment below the magnitude threshold, m_t. We remove these earthquakes from a new catalog.

As the number of earthquake sequences is always smaller than the number of earthquakes in a catalog, while the total moment in a catalog is constant, we should expect the β-value for sequences to be smaller than that for individual earthquakes. For deep and intermediate earthquakes, the difference in the β-values in calculations which use sequences and those using individual earthquakes is negligible. This small difference is due to a small number of aftershocks for these sequences in the GCMT catalog (Kagan 1999, Table 4).

Generally, we can treat the probabilities of being independent (p_{ii}) as corresponding to the weight of an earthquake as it is included in calculations. However, to make our computations similar to those used for real catalogs, we simulate new catalogs leaving in only earthquakes whose p_{ii} exceeds a random number distributed uniformly in the [0–1] interval. Thus, we obtain a "declustered" or residual catalog, in which we delete an earthquake according to its probability of being dependent.

Table 5.2 shows several β measurements for two GCMT shallow earthquake subcatalogs, 1977–2008 and 1982–2008, with the magnitude thresholds $m_t = 5.8$ and $m_t = 5.6$, respectively. For global datasets three types of computation were performed: (a) in the original list (row #1); (b) in a declustered catalog, where the seismic moment has been preserved for each earthquake (#2); and (c) in

Table 5.2
The β-values for various subdivisions of the GCMT catalog.

#	Earthquakes	$m_t = 5.6$		$m_t = 5.8$	
		β	n	β	n
1	2	3	4	5	6
1.	Global	0.6773	7369	0.6820	5450
2.	Global declustered	0.6480	5841	0.6568	4498
3.	Global aftershocks included	0.6229	5605	0.6366	4358
4.	Trenches (subduction zones)	0.6463	4805	0.6507	3223

Note: Shallow earthquakes in 1977–2008 ($m_t = 5.8$) and 1982–2008 ($m_t = 5.6$) GCMT catalog, n is earthquake number. *Source:* Kagan (2010b), Table 2.

a declustered catalog with aftershock moment transferred to an appropriate mainshock (#3) according to probabilities p_{ij} (see above). We performed similar measurements for earthquakes in subduction zones (#4, trenches) (Table 9.1, Kagan *et al.* 2010). Trench earthquakes have not been declustered, because some may have connections to outside events. Therefore the dependence probabilities can be biased.

In Table 5.2 the β-values are smaller by about 1–3% for the 1982–2008 dataset compared to the 1977–2008 catalog. A higher average accuracy of these solutions (Kagan 2003) and a larger magnitude range may explain this reduction. Bird and Kagan (2004; see also Fig. 5.3) showed that for global seismicity the minimum value of m_c for some tectonic zones is of the order 5.9–6.6. Extending the magnitude threshold to $m_t = 5.6$ expands the power-law part of the plot, and the influence of the corner magnitude on the β-estimate is smaller.

The β bias in Table 5.2 is also caused by the mix of different earthquake populations with various corner magnitudes, m_c (item 3 in the Preamble). This effect could explain why the β-values for trenches are significantly lower (by about 5%) than the global ones. A global earthquake set consists of many populations, of which oceanic rift zones have the smallest m_c-values (Bird and Kagan 2004, see Fig. 5.3). These oceanic events are excluded from the subduction zone (trench) dataset. Hence, the estimate of β for trench earthquakes is closer to the theoretical value (1/2) than for earthquakes in any other tectonic province.

As expected, the β-values decrease for the declustered catalogs, since excluded aftershocks have smaller moment values. This reduction is even stronger for catalogs where the aftershock moment is assigned to their potential mainshocks. The β-value decreases are about 4.5% and 8.5%, respectively.

These bias estimates depend on the correctness of the calculations used to estimate the earthquake probabilities. The likelihood procedure used to assign the probability of an event independence is influenced by the catalog quality, its duration and magnitude threshold. Given the presence of temporal boundaries, many relations between earthquakes are missed: some events at the beginning of the catalog may be aftershocks of previous strong quakes. Thus, instead of having a probability closer to zero, as expected, these aftershocks would have an independence probability equal to 1.0.

Due to the magnitude threshold, some connections between events are not observable. Suppose there is a potential foreshock–mainshock pair: a larger earthquake is preceded by a smaller one. However, the first event is below the magnitude threshold and the larger quake is above it. Then this second event would be treated as independent; our calculations would not include this connection (Kagan 1991b, see also discussion at the end of Subsection 4.1.1).

Moreover, the likelihood model used in our inversion is not perfect (Kagan 1991b, see Section 9.5). As a result, the independence probability values of earthquakes may not be fully counted, leading to a bias in the β computations. Therefore, the reduction of the β-values due to the influence of aftershock sequences is likely to be greater.

If we add up all the influences of aftershocks, we should see the β-value decrease to about 0.59–0.6. Kagan (1991b, p. 129) and Kagan (1999, Table 5) obtained a similar result for declustered shallow earthquakes.

5.3.3.3 Seismic moment tensor and its complexity

The previous discussion assumed that the scalar seismic moment is a fair measure of an earthquake size. In reality, seismograms are caused by excitation from many subevents during the main phase of an earthquake. Thus, the seismic moment tensor of an earthquake is a compound tensor sum of subevents. If all these subevents were identically oriented, the tensor sum would be proportional to a scalar sum of all the subevent scalar moments. However, detailed studies of earthquakes clearly indicate that subevent orientation sometimes changes significantly during ruptures.

The Bulletin of the Seismological Society of America (BSSA) published several special issues dedicated to a thorough analysis of a few large earthquakes like the 1992 Landers, the 1999 Hector Mine, the 1999 Chi-Chi, Taiwan, the 2002 Denali, the 2004 Sumatra, and so on. These studies detail a very complex geometrical picture of the quake rupture process. The focal planes and slip vectors of earthquake subevents often rotate several degrees and even tens of degrees (see Tsai *et al.* 2005 and Chapter 8). Therefore, the seismic moment tensor solution and the resulting estimate of an earthquake scalar moment are subject to random fluctuations from the stochastic misalignment of earthquake focal mechanism components.

In principle, we could avoid the systematic effect caused by source complexity if we used an earthquake's energy as a measure of its size. Energy is a positive scalar; thus, no bias due to source complexity would appear in the energy estimate. Unfortunately, estimates of the radiated seismic energy are highly uncertain and often differ by up to an order of magnitude (Perez-Campos *et al.* 2003). In contrast, the relative accuracy of evaluating seismic moment tensor is on the order of $10^{0.15}$ (Kagan 2003; Section 5.1).

Thus, because of random fluctuations during earthquake fracture, scalar moment earthquake estimates should always be lower than the sum of the subevents' scalar moments. This effect would also bias upwards the estimated β-values. Because we lack a comprehensive model of the earthquake rupture process which would enable us to estimate rigorously the resulting bias, we proceed by applying several approximations. These will give insight into the problem and provide an order of the magnitude estimate of possible systematic effects due to source complexity.

We can evaluate the influence of source complexity on the resulting estimate of earthquake size by initially assuming that small elementary subevents have their sign selected randomly. If the sign changes with equal probability ($p = 0.5$), the resulting sum of the subevents would be an ordinary random walk. The walk would converge to Brownian motion if the number of subevents, n, is large. The sum would be distributed according to the Gaussian distribution with the standard deviation proportional to $\sqrt{n}$. The final value of the sum would be proportional to its standard error. (In a count of the Brownian sum, we use an absolute value of the final walk position. Therefore, the total "moment" estimate is positive.) Consequently, in a critical branching process in which arbitrary sign descendants are added with the random probability $p = 0.5$, the distribution power-law index would increase by a factor of two: $\beta = 1.0$.

If we change the probability value from 0.5 to a higher level, $p = 0.5 + \delta$, this would produce Brownian motion with a drift (Feller 1971). For small n values, the walk behavior would resemble a regular Brownian motion, and later the sum would have a steady component $n\ \delta$. Thus, its behavior would be similar to the cumulative number increase with $p = 1$.

Figure 5.6 shows three simulated source-time functions with the time delay controlled by the Omori-type function. The cumulative functions for each curve are the sums of elementary subevents of the unit size and only the signs are different. The first curve with $p = 1$ (deterministic addition of events) is similar to Fig. 3 by Kagan and Knopoff (1981, see also Fig. 3.4). The initial step-like increase of this function would likely be interpreted as a mainshock, whereas a few steps at a later time would be labeled aftershocks. The $p = 0.75$ function increases the

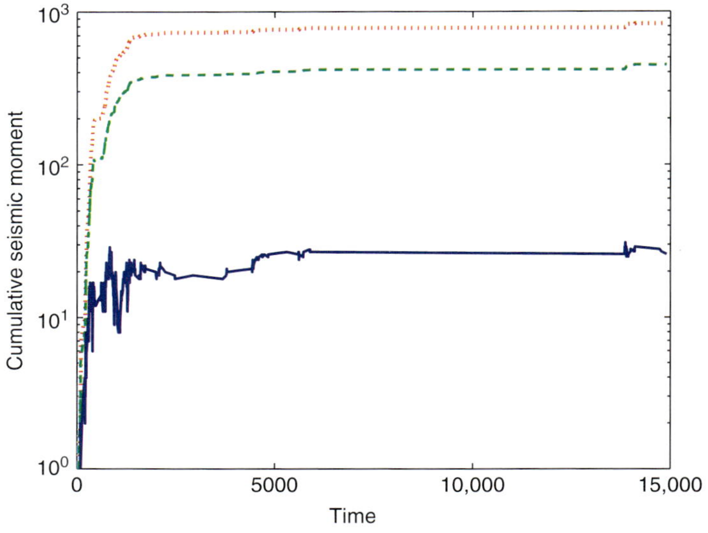

Fig. 5.6 Simulated source-time functions in a critical branching process. Dotted line – positive number addition ($p = 1.0$). Dashed line – unequal positive/negative number addition (random walk with a drift, $p = 0.75$). Solid line – equal positive/negative number addition $p = 0.5$ (random walk). *Source:* Kagan (2010b), Fig. 5.

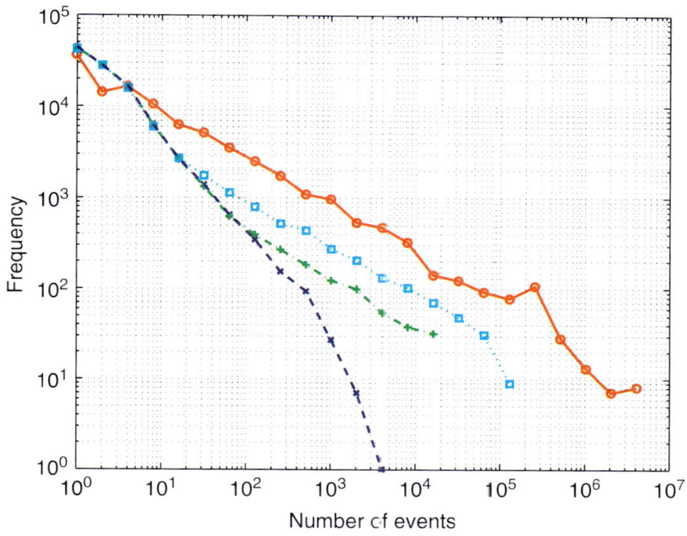

Fig. 5.7 Distribution of event numbers in a critical branching process. Red line with circles – positive deterministic number addition ($p = 1.0$). Blue line with x-marks – equal positive/negative number addition $p = 0.5$ (random walk). Green line with pluses – unequal positive/negative number addition (random walk with a drift, $p = 0.51$). Cyan line with squares – unequal positive/negative number addition (random walk with a drift, $p = 0.55$). *Source:* Kagan (2010b), Fig. 6.

same way as the first curve but with smaller amplitude; random fluctuations are not easily observable. The random walk function ($p = 0.5$) behaves more erratically and its total final ordinate is much smaller than that of the other two curves. Only the values of the curves at the end of a branching simulation, corresponding to the total moment of a sequence, are counted in our calculations of the moment for an earthquake sequence. For the p-values close to 0.5, the random branching walk could end up as a negative cumulative sum; as we explained above, we take the absolute value of the final ordinate of a simulation run. These values are assumed to correspond to the total seismic moment of an earthquake sequence.

Figure 5.7 illustrates the above considerations. We simulated a critical branching process and counted the sum of events at the end of each simulation run, such as the extreme right-hand points of three curves in Fig. 5.6. These numbers are shown in the diagram in the log–log format. While the event numbers are small (less than 10) in the left-hand part of the plot, the discretization effects are noticeable. For the largest sequences, random fluctuations are observable, because there are few of these sequences. In the mid-number range, the deterministic number addition ($p = 1$) distribution (solid curve) has an index $\beta = 0.5$. As expected, the Brownian walk addition (dashed curve) has $\beta = 1.0$. As explained above, the curves for the motion with a drift first follow the Brownian curve, then they are parallel to the solid curve. Thus, in the beginning their index is $\beta = 1.0$, and for larger numbers it changes to $\beta = 0.5$, confirming our predictions. The randomness in the number addition significantly increases the power-law distribution index.

In Fig. 5.8 we show a more complicated test. We sum up seismic moment tensors, instead of scalar quantities, in a critical branching process simulation. In each simulation run, we determine a norm of the total tensor sum which for the seismic moment tensor is equivalent to the scalar moment. Again the solid

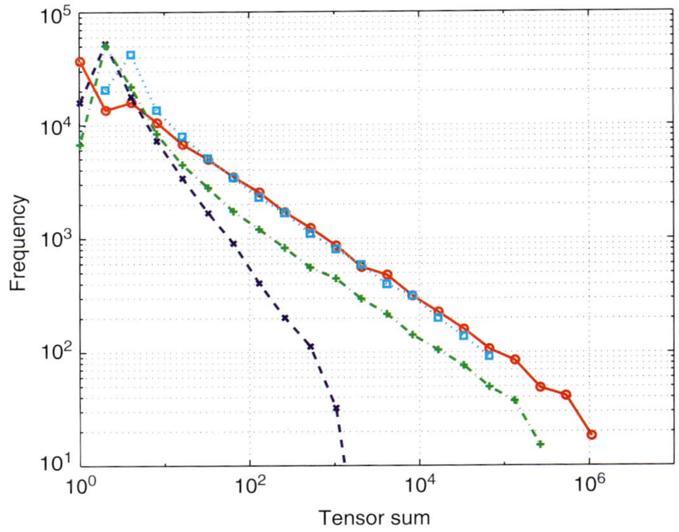

Fig. 5.8 Frequency plot of tensor sum norm in a critical branching process. The red line with circles – tensor sum with no rotation, the 3-D rotation angle, $\Phi = 0°$. The blue dashed line – random rotation ($\Phi \leq 120°$). The green line with pluses – limited random rotation, $\Phi \leq 80°$. The cyan line with squares – limited random rotation, $\Phi \leq 30°$. *Source:* Kagan (2010b), Fig. 7.

curve shows the distribution when the tensors are identical. For other curves the tensors are independently randomly rotated through the 3-D rotation angle, Φ (Kagan 2003, 2009, see also Subsection 8.4.2). The maximum rotation angle for a double-couple focal mechanism is 120°; therefore, these tensors are rotated in a uniformly random manner. The angle $\Phi = 0°$ corresponds to the zero rotation (solid curve). If $\Phi < 120°$, then the rotation is restricted, being uniformly random only for angles smaller than Φ.

Figure 5.8 appears to be similar to the previous plot (Fig. 5.7). If the tensors' orientation is identical, the power-law exponent $\beta = 0.5$. For a completely random orientation, $\beta = 1.0$, and for a restricted focal mechanism misalignment, the curves follow the latter distribution first and are then parallel to the former line.

As discussed earlier, many earthquake ruptures exhibit significant variations in focal mechanisms (cf. Tsai *et al.* 2005). However, detailed analyses of individual earthquakes are still rare and insufficient for rigorous statistical study. Therefore, for more comprehensive study we analyze the degree of misalignment in several mainshock/aftershock sequences. Kagan (2000, see also Fig. 8.12) investigated the correlation of earthquake focal mechanisms and showed that the degree of mechanism 3-D rotation increases between earthquakes with temporal and spatial differences. Hence, we hope that the immediate aftershocks of strong earthquakes would characterize the geometric complexity of mainshock rupture process.

To this end we studied all shallow (depth 0–70 km) earthquakes in the 1977–2008 GCMT catalog with a magnitude $m_1 = 7.5$ and higher. All earthquakes ($m \geq 5.6$) are considered aftershocks within the first 7 days of $m7.5$ earthquake occurrence and closer than

$$r = 75 \times 10^{(m_1 - 7.5)/2} \text{ km}, \tag{5.18}$$

(Kagan 2002b, see also Section 5.5). There are 105 such mainshocks in the catalog and 81 of them have one or more aftershocks.

To investigate the orientation differences between a mainshock and its aftershocks, we calculate the correlation invariant or seismic moment tensor dot-product J_3 (Kagan 1992b, 2009a, see also Section 8.8.2)

$$J_3 = \sum_{i,j} m_{ij}\, n_{ij}, \tag{5.19}$$

for the main event (m_{ij}) and the sum of normalized tensors for the whole 7-days aftershock sequence (n_{ij}). Summation of repeating indices is assumed. Both m_{ij} and n_{ij} are normalized. In (5.19) $J_3 = 2.0$ means that focal mechanisms are identical; $J_3 = -2.0$ corresponds to components of both tensors having the opposite sign.

A J_3 histogram in Fig. 5.9 displays the correlation between tensors. Most correlation invariant values are close to 1.5–2.0. Thus, the aftershock focal mechanisms are similar to that of their mainshock. However, some J_3-values are close to zero, and one is negative, testifying to a significant variation in the rupture process. The smallest J_3-value is due to the November 2000 New Ireland earthquake sequence. The sequence started with a $m = 8.0$ left-lateral main shock on 16 November and was followed by a series of aftershocks with thrust mechanisms primarily (Geist and Parsons 2005; Park and Mori 2007). The negative J_3-value signifies that the aftershocks have on average a slightly opposite orientation compared to their mainshock.

Figure 5.10 displays two distributions of the ratio for the tensor sum of the mainshock and its aftershocks to the sum of their scalar moments

$$R = |\sum m_{ij}| \;/\; \sum M, \tag{5.20}$$

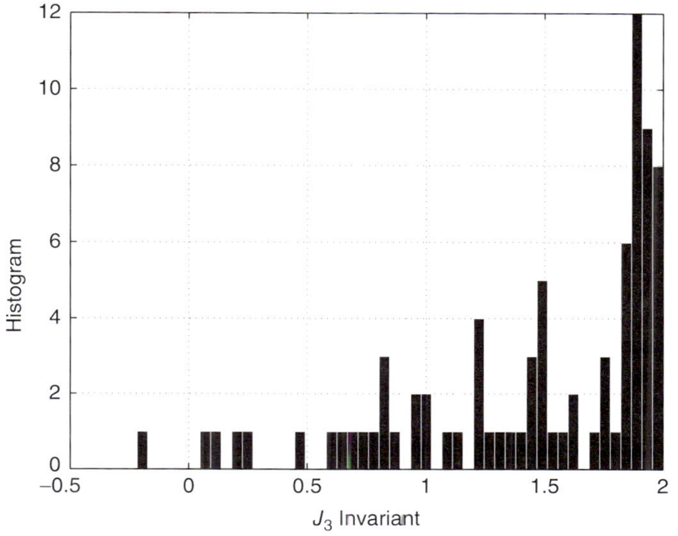

Fig. 5.9 Frequency plot of correlation tensor invariant (tensor dot-product) for $m7.5$ mainshocks and sum of their immediate aftershocks in 1977–2008 GCMT catalog. Average $\overline{J_3} = 1.458$, its standard error $\sigma_J = 0.5535$. *Source:* Kagan (2010b), Fig. 8.

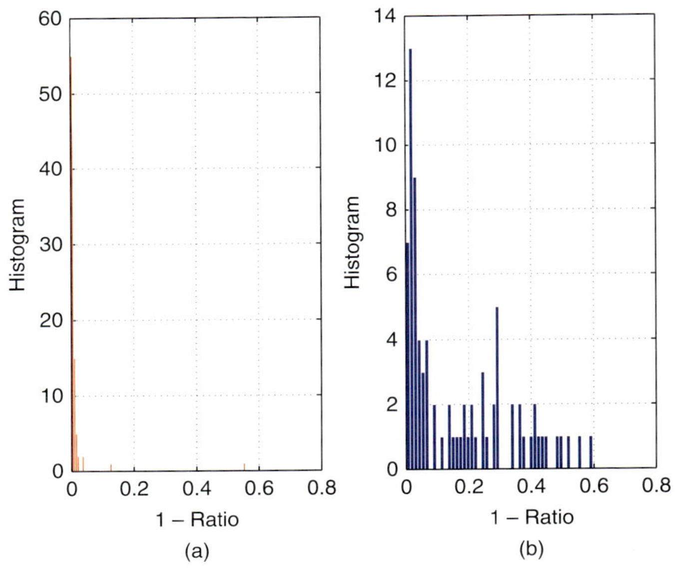

Fig. 5.10 Frequency plot of tensor/scalar sum ratio for $m7.5$ mainshocks and immediate aftershocks in 1977–2008 GCMT catalog: (a) Unnormalized sum –average $\overline{(1-R)} = 0.0128$, its standard error $\sigma = 0.0627$. (b) Normalized sum–average $\overline{(1-R)} = 0.1642$, its standard error $\sigma = 0.1639$. *Source:* Kagan (2010b), Fig. 9.

where $|m_{ij}|$ means the norm of the tensor. The aftershocks are selected according to the same criteria as in Fig. 5.9. If the aftershocks have the same focal mechanism as the mainshock, the ratio would be 1.0. A similar sum ratio was considered by Apperson (1991, Eq. 4) who called it "Seismic consistency" (C_s).

In the left-hand diagram of Fig. 5.10 the moment tensors are not normalized; in the right-hand plot they are. In the inversions of the earthquake rupture process (see the BSSA special issues mentioned above), several subevents of approximately equal size but significantly different orientation in focal mechanism are often observed. This is the reason we investigate the normalized sums.

Figure. 5.10 (a) and (b) show significantly varied focal mechanisms in aftershock sequences. Large fluctuations are seen in the normalized sums especially. This result suggests again that a conspicuous randomness occurs in the focal mechanism orientation of earthquake sequences. By implication this should also occur during an earthquake rupture process. Such random fluctuations may noticeably decrease the measured earthquake size and influence the β measurement.

What is the size of the β measurement bias due to source complexity? All estimates shown above are indirect. Earthquakes do not consist of an identically oriented or purely random collection of elementary sources. Various observations suggest that a rupture occurs over quasi-planar fault patches, so there should be a strong correlation between neighboring fault segments. This correlation is sometimes broken by significant fault branching. Kagan (1982, see Section 3.3.2) proposed a geometrical model of such stochastic rupture. Unfortunately, the degree of geometrical branching (ϕ_0) in this model is not well known. Therefore, we cannot easily simulate and study such branching sequences.

What β-value change can be proposed as result of randomness in the fault rupture orientation? Pure randomness yields a β-value increase by a factor of two (see Figs. 5.7 and 5.8). Unfortunately, we cannot yet quantitatively study the complex geometry of earthquake rupture. We need to extrapolate from the measured misalignments of close aftershocks. These measurements indicate that complexity, though far from being completely random, is nevertheless quite significant. For example, in Fig. 5.9 the correlation invariant is $J_3 = 1.46 \pm 0.55$. These values can be compared to purely random arrangements of double-couple focal mechanisms (Kagan 1992b), where $J_3 = 0 \pm 0.89$ has been obtained.

Similar conclusions could be inferred from the results of the 3-D rotation angle distribution. The average angle between the mainshock focal mechanism and mechanisms of immediate aftershocks is on the order of 10° (Kagan 2000, see also Chapter 8 and especially Section 8.7). For a completely random rotation, the maximum angle is 120° and the average angle is 75.2° (Grimmer 1979, Table 5; Kagan 2003; see also Section 8.7.3). Given the source complexity, as demonstrated by the rotation angles, the β bias should be around 8–12%.

Analogous conclusions could be drawn from Fig. 5.10. If immediate aftershocks had the same moment tensor solutions as their mainshock, the tensor/scalar sum ratio R in (5.20) should be 1.0, and $(1 - R)$ equal zero. Both average and standard deviation for $(1 - R)$ in the plots display significant non-zero values. We infer that the source is complex and the β bias may be on the order of a few percent.

To summarize the results of this subsection, we hypothesize that as a consequence of the random geometrical misalignment of a fault rupture, the measured β-value may be increased by at least a few percent (3–7%) from its true size. The estimate above is conservative. More work needs to be done to obtain a more reliable value which could lead to an even greater β-estimate decrease.

5.3.3.4 Centroid depth influence

Kagan (2002a, pp. 538–539, see also Preamble) proposed that the observed b-value differences in volcanic areas, at creeping faults, and at the boundary between brittle crust and plastic deformation in the upper mantle may be due to significantly varied corner moments. If earthquake populations with different m_c are mixed, the resulting statistical distribution could be interpreted as belonging to a power-law with the exponent β significantly exceeding 0.6–0.7, the value normally observed in tectonic earthquakes (see item 3 in Preamble).

The GCMT catalog supplies coordinates for the seismic moment centroid, the centroid is in the center of the moment release volume. The centroid distance from the fault edge cannot be smaller than a half of that particular earthquake rupture zone width. Closer to the surface or to the fault boundary the corner moment would approach zero. An inspection of seismic maps suggests that hypocenters of larger events are on average deeper than those for small earthquakes. Thus, the moment-frequency law (see Eqs. 5.5 and 5.6) would change due to an increase in the maximum earthquake size with depth.

As we explained in the Preamble, the depth accuracy for shallow earthquakes is presently insufficient to investigate observationally the dependence of the corner moment on depth. Therefore, we study a possible influence of the finite fault size by calculating a new distribution of earthquake size for a few simple models of the earthquake rupture pattern. These theoretical guesses would help evaluate the depth effect up to the order of magnitude.

We assume that the earthquakes are distributed over an infinite planar fault surface extending either vertically for 20 km (imitating conditions for strike-slip faults in California), or distributed over an inclined fault with a width of 200 km (as in some subduction zones, see, for instance, Bird and Kagan 2004). The variable corner moment for such faults is

$$M_c' = C\,\zeta^3, \tag{5.21}$$

where ζ is distance from a centroid to a fault edge and C is an appropriate coefficient. Here we assume that the moment is proportional to the cube of earthquake rupture size (see Section 5.5).

In one model we assume that earthquake centroids are distributed uniformly over the fault surface. Then, using the algebraic and numerical facilities of MATHEMATICA (Wolfram 1999), we calculate the new survivor function

$$\begin{aligned}\Psi_1(M) \propto \int_0^L F(M)\,d\zeta &= \left(\frac{M_t}{M}\right)^\beta \int_0^L \exp\left(\frac{M_t - M}{C\,\zeta^3}\right)\,d\zeta \\ &= \left(\frac{M_t}{M}\right)^\beta \frac{\Gamma\left(-\frac{1}{3}, \frac{M-M_t}{C\,L^3}\right)}{3\left(\frac{C}{M-M_t}\right)^{1/3}},\end{aligned} \tag{5.22}$$

where $F(M)$ is defined by (5.5), L is a half-width of a fault plane, and $\Gamma(.,.)$ is an incomplete gamma function (Abramowitz and Stegun 1972, Eq. 6.5.3).

The other possibility is to assume that the earthquake centroid density increases linearly with increasing depth up to the middle of the fault width. The density decreases to zero thereafter. Fig. 7.6 shows that such a feature is a common occurrence. Then the survivor function would be

$$\begin{aligned}\Psi_2(M) \propto \int_0^L F(M)\,\zeta\,d\zeta &= \left(\frac{M_t}{M}\right)^\beta \int_0^L \exp\left(\frac{M_t - M}{C\,\zeta^3}\right)\,\zeta\,d\zeta \\ &= \left(\frac{M_t}{M}\right)^\beta \frac{\Gamma\left(-\frac{2}{3}, \frac{M-M_t}{C\,L^3}\right)}{3\left(\frac{C}{M-M_t}\right)^{2/3}}.\end{aligned} \tag{5.23}$$

Figure 5.11 displays two survivor functions corresponding to Eqs. 5.5 and 5.22 . In a log–log plot the former function has a linear part for the moment M values that are significantly smaller than the corner moment M_c. We take a slope β to be 0.5. The curve has an exponential taper for M close to M_c. On the other hand, the latter function (5.22) is slightly convex even for small moment values. It is formed by a sum of distributions similar to (5.5) but with the corner moment increasing from zero to the maximum, M_c. Therefore, we can only calculate an

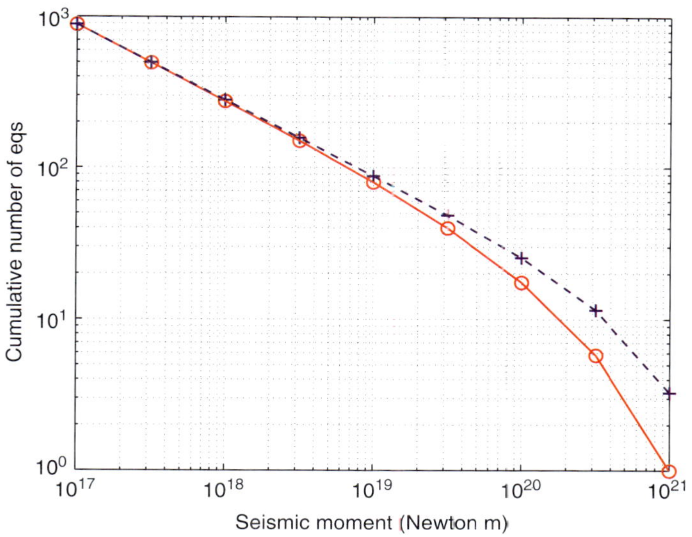

Fig. 5.11 Two theoretical moment-frequency curves: the number of earthquakes with moment (M) larger than or equal to M as a function of M, moment threshold $M_t = 10^{17.0}$ Nm ($m_t = 5.33$). Blue dashed line shows the TGR: the G-R law restricted at large seismic moments by an exponential taper with the corner moment $10^{21.0}$ Nm ($m_c = 8.0$). The slope of the linear part of the curve corresponds to $\beta = 0.50$. The red solid line is a plot for Eq. 5.22, with a half-width of a fault, $L = 10$ km, and $C = 1.0$. The curves are normalized so that the solid line has an ordinate 1.0 at the right-hand end. *Source:* Kagan (2010b), Fig. 11.

effective slope (β') of the curve; for the moment range $10^{17}-10^{19}$ Nm, the slope β' in the plot is 0.523.

We calculate three theoretical curves for both equations (5.22 and 5.23): (a) fault width $2L = 200$ km and $M_c = 10^{23}$ Nm ($m_c = 9.33$); (b) $2L = 200$ km and $M_c = 10^{22}$ Nm ($m_c = 8.67$); and (c) $2L = 20$ km and $M_c = 10^{21}$ Nm ($m_c = 8.0$). For the formula (5.22) the β'-values are 1.3, 2.8, and 6.6% higher than the original β-value, for the second formula these exponent increases are reduced by a factor of 1.4.

5.3.4 Dislocation avalanche statistics

Recent experimental and theoretical investigations have demonstrated that crystal plasticity is characterized by large intrinsic spatio-temporal fluctuations with scale-invariant characteristics. In other words, deformation proceeds through intermittent bursts with power-law size distributions (Zaiser 2006; Dahmen *et al.* 2009). In particular, Csikor *et al.* (2007), Dimiduk *et al.* (2006), Weiss and Marsan (2003) studied dislocation avalanches (micro-earthquakes) in ice and other crystals. They consistently obtain the power-law size distribution. The power-law size distribution in a single ice crystal extends over six decades of magnitude (Miguel *et al.* 2001).

Condensed-matter physicists explain new results on the scale-invariant distribution of the dislocation avalanche size by suggesting a new theoretical approach to crystal plasticity (Zaiser 2006). According to this interpretation, at a micro-scale the crystal deformation proceeds through intermittent bursts similar to earthquakes. Only at a larger, meso-scale does plastic deformation have the appearance of a smooth, homogeneous, quasi-laminar flow process. Crystal boundaries seem to influence this transition. Many regularities of dislocation avalanches behavior, observed in crystalline materials, seem to be caused by the interaction of dislocations with crystal boundaries and the orientation of the crystal lattice. Such interactions are extensively studied in the crystallographic (material) texture analysis (see Section 8.5). These regularities are of less interest for earthquake size distribution, since the earthquake rupture usually exceeds the average crystal size by many orders of magnitude. In polycrystal materials, the power-law distribution of bursts is also observed, but its size is limited by an upper cutoff.

Therefore, plastic, ductile deformation proceeds by two very different mechanisms: (a) intermittent displacement at the micro-scale with scale-invariant distribution of strain steps and the universal value of the power-law exponent (1.5); and (b) a smooth flow at larger scales. Because detailed quantitative observation at small sub-grain scales was not possible until recently, the first mechanism had been largely ignored (Zaiser 2006, p. 241).

Zaiser and Moretti (2005), Csikor *et al.* (2007) and Zaiser and Nikitas (2007) propose the following probability density function for the dislocation avalanche energy or strain

$$P(s) = C\, s^{-\tau} \exp[-(s/s_0)^2], \tag{5.24}$$

where C is a normalization constant, τ is a scaling exponent ($\tau \simeq 1.5$), and s_0 is the strain of the largest avalanches. This formula is similar to our (5.6) with exponent $\tau = 1 + \beta$, but the decay taper at large strain values is Gaussian-like. We use instead the exponential decay. Because statistics on the largest events are insufficient in both cases, we cannot distinguish by observation between these formulas. Zaiser and Nikitas (2007) found that the maximum strain s_0 is proportional to the specimen size.

Comparing these dislocation avalanche measurements with seismological observations leads to some problems. The most accurate measurements of earthquake size are for the seismic moment. Experimental laboratory observations provide the energy of acoustic emission bursts or strain step measurements. Zaiser (2006, pp. 212, 223) argues that both these measurements are approximately equivalent and yield similar values for the distribution density power-law exponent ($\tau = 1.5-1.6$).

Earthquake energy has the same distribution as the seismic moment. Kanamori (1977) cites the following relation between energy released by earthquakes, E, and their magnitude, m:

$$\log_{10} E = 1.5\, m + 11.8. \tag{5.25}$$

Because the moment and magnitudes have a similar dependence (see Eq. 2.2), the exponents for energy and moment power-law distributions should be identical.

Another problem of comparison is that experimental statistical distributions of dislocation avalanches are not processed by appropriate statistical techniques. Thus, the obtained values of the exponents may be biased. For example, Zaiser *et al.* (2008) use the least-square fit to calculate the exponent. Richeton *et al.* (2005) apply the Levenberg-Marquardt algorithm for this purpose. These methods are appropriate for fitting regression curves, but they are not the statistically efficient techniques for a parameter evaluation in a statistical distribution. They may yield biased estimates of an exponent parameter and especially its uncertainty (Vere-Jones 1988, 2010; Clauset *et al.* 2009, see also Section 5 2).

5.3.5 What are $\beta \equiv 1/2$ consequences?

In the previous subsections we have analyzed the index of the power-law distribution for earthquake size (the β-value) to argue that its true value is $1/2$, the value suggested by theoretical arguments. The direct β measurements for the scalar seismic moment based on catalog analyses (Kagan 1999; Bird and Kagan 2004) usually yield a value in the range 0.63–0.67, equivalent to the commonly known G-R b-value of 0.95–1.0. Four systematic and random factors that bias the β-value estimate upwards are investigated: dependence of errors on the magnitude, earthquake sequences, complexity of earthquake source, and a finite size of earthquake faults (items 4–7 in the Preamble). We found that these factors would increase the observational β-estimate by about 1–3%, 5–10%, 3–7%, and 2–6%, respectively. Of these values the third is most uncertain, because it is based on extrapolating the immediate aftershock focal mechanism distribution to the mainshock rupture time.

If we combine the above biases and apply them to the most accurately determined β-value, that is, β equals from 0.63 to 0.64 for subduction zones (Bird and Kagan 2004), the corrected β-values would be on the order of 0.52–0.56. It is quite feasible that the second and third correction terms are underestimated. This would imply that β is close to 0.5 and possibly equals $\frac{1}{2}$ exactly.

As was demonstrated in Subsection 5.3.4, crystal plasticity is characterized by large intrinsic spatio-temporal fluctuations with scale-invariant characteristics. Condensed-matter physicists consistently obtain the power-law size distribution with the probability density exponent close to 1.5, that is, the exponent β should be equal to $1/2$.

Another geophysical phenomenon, the non-volcanic seismic tremor (Schwartz and Rokosky 2007; Beroza and Ide 2009, and references therein), may be explained by the same physical mechanism. The tremor represents a long duration (minutes to hours) seismic signal, appearing similar to many small concatenated earthquake signals (Shelly *et al.* 2007). The tremor signals are sometimes quite pulsed in nature. For example, the temporal cumulative plot of seismic moment increase for tremors (Fig. 4d in Hiramatsu *et al.* 2008) looks similar to curves of crystal micro-deformation due to dislocation avalanches (Fig. 14 in Zaiser 2006): both diagrams show that the displacement increases in discrete steps, each step is followed by a plateau.

Such seismic tremors have been registered in diverse tectonic environments recently (Japan, Cascadia, New Zealand, Costa Rica, Taiwan, California). Tremor and other slow-slip events are typically found on the deep extension of faults, just below the region that produces the more familiar, "ordinary" earthquakes.

This recent observation of tremors has resulted in a flurry of research across many geophysical and geologic disciplines.

If tremors are a transitional feature between real earthquakes and seismic signal bursts, they are similar to the dislocation avalanches described earlier. Both phenomena occur in the conversion from the brittle to plastic mode of solid deformation. Seismic tremors, which are interpreted as small, continually occurring earthquakes, may also have the same scale-invariant, power-law features as earthquakes in brittle crust. Hiramatsu *et al.* (2008) and Hirose *et al.* (2010) measured the moment-frequency relation for tremors and found that it can be approximated by an exponential distribution rather than a power-law. However, because of low signal-to-noise ratio for tremors, only the upper tail of tremor size distribution can be observed, and the upper tail for earthquake size distribution (Eq. 5.5) is also exponential (Kagan 2002a; Bird and Kagan 2004, see Fig. 5.1). The higher end of the size distribution for dislocation avalanches also exhibits a non-power-law dependence (5.24) which is close to the exponential law. We may conjecture that as with dislocation avalanches, the size distribution of smaller tremor events would be the power-law with the universal value of the exponent constant ($\beta = 0.5$). Further study of tremors should answer this question.

If we are correct about the universality of the β-value constant ($\beta \equiv 0.5$), the observed variations in the b-parameter result from systematic and random effects not properly accounted for (see Preamble, items 1–3). Therefore, all attempts to connect the β-value variability with various physical parameters and conditions are eventually bound to fail. However, studying the b- or β-values in local and regional earthquake catalogs may still be useful, especially if such investigations are needed to evaluate seismic hazard and seismicity forecasts that would be prospectively tested with the same catalogs (Chapter 11). In addition, when the seismic activity or earthquake productivity level is calculated for large earthquakes, the regular β-estimates can be used.

If the conjecture that the power-law exponent is a universal constant and the corner moment is variable is correct, then it would provide a new theoretical approach to features of earthquake occurrence and account for the transition from brittle to plastic deformation. Thus, the difference between brittle fracture and plastic deformation amounts only to the size of the maximum earthquake or dislocation avalanche each system allows. More extensive investigation of corner moment behavior may afford new insights into regular earthquake occurrence and recently discovered slow deformation and seismic tremor events at the brittle-plastic crust boundary. As often happens in complex systems, new laws and features may be found to illuminate the transition from brittle fracture to plastic flow.

5.4 Seismic moment sum distribution

5.4.1 Simulation and analytical results

The global distribution of the seismic moment for small and moderate earthquakes is well approximated by a power-law (Pareto) distribution with index

$\beta \approx 2/3$ (5.3). This is a heavy-tailed distribution: it has the infinite mean and the standard deviation. Thus, if one uses the pure Pareto model, the Central Limit Theorem does not describe the distribution of the sum of seismic moments. The tapered (Eqs. 5.5–5.6) or truncated Pareto distribution appears to eliminate all summation problems. However, a detailed analysis shows that the Gaussian limit is reached for only a large number of observations. For a realistic number of events, the tapered or truncated Pareto still exhibits all the properties of a heavy-tailed distribution (Zaliapin *et al.* 2005a,b).

The cumulative seismic moment released in a region can be used as a proxy for the total regional deformation of the Earth's surface due to earthquakes. Formally, the strain rate for the volume of the deforming crust is proportional to the sum of the *tensor* moments of individual earthquakes (Kostrov 1974, see also Subsection 5.3.3). Thus, evaluating the observed seismic moment rates is an important problem connected to the regional earthquake rate assessment.

Below we briefly discuss the properties of the sum of variables distributed according to a Pareto law, for a more complete exposition of this complex subject, see Zaliapin *et al.* (2005). Let $X_i, i = 1, \ldots, n$ be independent identically distributed random variables with a common Pareto distribution (5.3), and let S_n denote their sum

$$S_n = \sum_{i=1}^{n} X_i. \tag{5.26}$$

If the exponent β of the power-law distributed variable is less than 2.0, according to the Generalized Central Limit Theorem (Zolotarev 1986; Samorodnitsky and Taqqu 1994; Uchaikin and Zolotarev 1999), the sum converges to a stable distribution with the probability density function

$$\phi\,(X, \beta, \gamma, \mu, \sigma), \tag{5.27}$$

where γ is a symmetry parameter (for positive variables $\gamma = 1$, that is, the sum is maximally-skewed), and μ, σ are shift and width parameters. For the Gaussian law (see Eq. 5.33) that is a limit case of stable distributions ($\beta = 2$), only the two last parameters are valid.

An arbitrary quantile z_q of the sum S_n can be approximated as

$$z_q \approx z_q^{(1)} \equiv n^{1/\beta}\, x_q\, C_\beta + b_n, \tag{5.28}$$

where $z_q^{(1)}$ is the most general approximation for the quantile z_q (Zaliapin *et al.* 2005a, pp. 1197–1198), and x_q solves the equation for the cumulative distribution F_β of the sum

$$F_\beta(x_q) = q. \tag{5.29}$$

For $\beta < 1$, $b_n = 0$ and

$$C_n = [\Gamma(1 - \beta)\cos(\pi\beta/2)]^{1/\beta}. \tag{5.30}$$

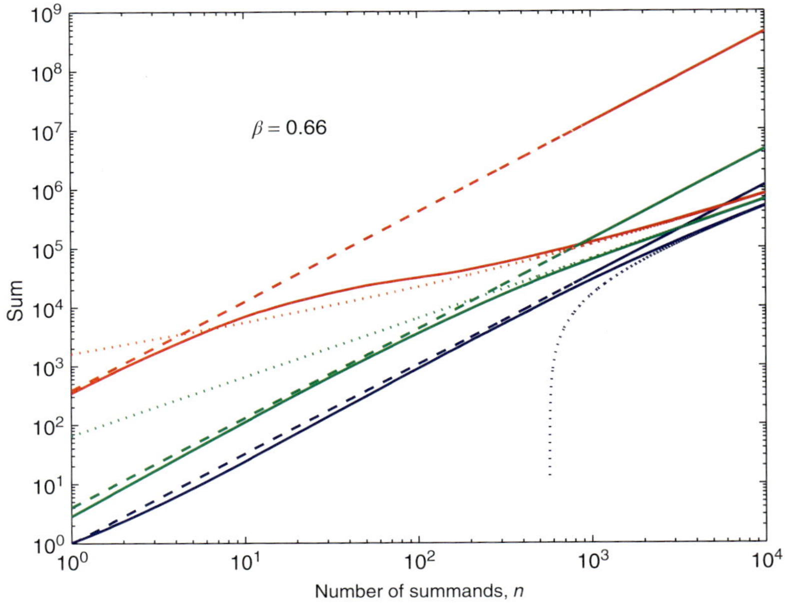

Fig. 5.12 Quantiles for the sum S_n of truncated Pareto variables (upper limit $y = 3.4 \times 10^4$, Eq. 5.6) and their approximations as functions of the number of summands, n. Two approximations are considered: via the stable distribution, Eq. 5.28 (dashed lines) and Gaussian, Eq. 5.33 (dotted lines). Solid lines represent quantiles of simulated Pareto sums. Three red upper curves are for the 0.98 quantile, three green middle curves are for the median, and three blue lower curves are for the 0.02 quantile. *Source:* Zaliapin *et al.* (2005), Fig. 8.

Fig. 5.12 displays an example of simulated sums (S_n) for the Pareto distribution truncated at $y = M_x/M_t = 3.4 \times 10^4$ compared to the stable distribution quantiles. In this example we take the threshold moment $M_t = 10^{17}$ Nm or $m_t = 5.33$, the threshold of the recent GCMT catalog (Kagan 2003; Section 2.3) and the maximum magnitude $m_x = 8.35$ (Section 5.2).

According to (5.28), quantiles of the stable distribution for $\beta = 0.66$ increase as $n^{1/0.66}$, thus, for example, the median of the sum of 40 variables μ_{40} compared to μ_{20} is equal to

$$\mu_{40} \approx 2.86 \times \mu_{20} \quad \text{or} \quad \mu_{40} > \mu_{20} + \mu_{20}. \tag{5.31}$$

This behavior of the stable distribution sums may seem counter-intuitive, as is the behavior of their other properties.

If the exponent β is less than 1.0, the sum of power-law distributed variables is comparable to the value of the largest observation moment M_n

$$E(S_n) = M_n/(1-\beta), \tag{5.32}$$

where E is a mathematical expectation sign (Feller 1971).

Therefore, Eq. 5.31 means that in a sample of 40 earthquakes, there is a higher chance of encountering a large event which would significantly increase their sum than in a sample of only 20 earthquakes. Pisarenko (1998), Pisarenko and Rodkin (2010) as well as Huillet and Raynaud (2001) also note that for the heavy-tailed distributions, sum quantiles increase non-linearly with n.

The upper quantiles of the Pareto sum generally approach the stable distribution limit faster than do the lower quantiles (Zaliapin *et al.* 2005a). However, in Fig. 5.12 the upper quantiles depart from the theoretical curve for the stable distribution starting with $n = 2$ because of the upper limit truncation. The behavior of the lower quantile is essentially unaffected by the truncation until n exceeds 10^3.

When the number of summands is large, the truncation point y dominates the behavior of the quantiles. The sum is then distributed asymptotically according to the Gaussian law:

$$\lim_{n\to\infty} F_{S_n}(x) = \Phi\left(\frac{x - n\,\mu_y}{\sigma_y\,\sqrt{n}};\ 0,\ 1\right), \tag{5.33}$$

where Φ is the normal cumulative distribution (5.34), and the parameters μ_y and σ_y are given below by (5.35), (5.36).

The cumulative distribution function (CDF) of the standard Gaussian (normal) law with expectation μ and standard deviation σ is given by

$$\Phi(x; \mu, \sigma^2) = \frac{1}{\sigma\,\sqrt{2\,\pi}} \int_{-\infty}^{x} \exp\left(-\frac{(y-\mu)^2}{2\,\sigma^2}\right) dy, \tag{5.34}$$

whereas

$$\mu_y = \frac{\beta}{1-\beta}\,(y^{1-\beta} - 1)/(1 - y^{-\beta}), \quad \beta \neq 1, \tag{5.35}$$

and

$$\sigma_y^2 = \frac{\beta}{2-\beta}\,(y^{2-\beta} - 1)/(1 - y^{-\beta}) - \mu_y^2, \quad \beta \neq 2, \tag{5.36}$$

are the conditional mean and variance of each summand (Kagan 2002a; Zaliapin *et al.* 2005a), given the restriction on the maximum ($X < y$).

To evaluate the approximation quality we define the approximation relative error as

$$\Delta_q^{(1)} = z_q^{(1)}/z_{q-1}. \tag{5.37}$$

where z_q is defined by Eq. 5.28. Table 5.3 collects the simulation results for selected values of β (Zaliapin *et al.* 2005a) Synthetic values of the sum are compared to stable quantiles as tabulated in McCulloch and Panton (1997). The stable distribution approximates the Pareto sum upper 0.98 quantile quite well: only for β approaching 1.0 and 2.0 from below does the number of summands necessary to achieve 10% relative accuracy exceeds two. The approximation deteriorates substantially for the lower quantile: even in the best cases several dozens of summands are needed to yield < 10% relative error.

5.4.2 Applications to seismicity analysis

From the beginning of the plate tectonics hypothesis, it was assumed that earthquakes are due to plate boundary deformation. Calculations for global tectonics and large seismic regions justified such an approach. However, applying this

Table 5.3
Number n of summands necessary to approximate the sum S_n by stable distributions $z_q^{(1)}$ with error $\Delta_q^{(1)} < 0.1$.

β	Quantile 0.02	Quantile 0.50	Quantile 0.98
0.50	20 ↑	2 ↑	2 ↓
0.60	25 ↓	3 ↓	2 ↓
0.66	270 ↓	18 ↓	2 ↓
2/3	320 ↓	25 ↓	2 ↓
0.80	> 10000 ↓	3000 ↓	2 ↓
0.90	≫ 10000 ↓	> 10000 ↓	8 ↓
0.94	≫ 10000 ↓	≫ 10000 ↓	2000 ↓
0.98	≫ 10000 ↓	≫ 10000 ↓	≫ 10000 ↓
1.00	55 ↑	2 ↓	2 ↓
1.02	65 ↑	2 ↓	2 ↓
1.10	75 ↑	2 ↓	2 ↓
1.20	85 ↑	2 ↓	2 ↓
1.30	100 ↑	2 ↓	2 ↓
1.50	140 ↑	2 ↓	3 ↓
1.66	230 ↑	2 ↓	8 ↓
1.80	300 ↑	3 ↓	30 ↓
1.90	1000 ↑	6 ↓	35 ↓
1.94	1800 ↑	7 ↓	500 ↓
1.98	6500 ↑	9 ↓	4000 ↓

Note: ↑ means that the value is approached from below; ↓ means that the value is approached from above. *Source:* Zaliapin *et al.* (2005), Table 1.

assumption to smaller regions has usually been inconclusive, given the high variability of seismic moment sums.

Holt *et al.* (2000) compared the observed seismic moment release with the tectonic release inferred from the joint inversion of the GPS and the Quaternary rates of strain for South-East Asia. They also compared strain release with the earthquake record from 1965–1998. Figure 5.13 shows the seismic coupling χ (the ratio between observed and predicted seismic moment) in four large regions and 42 subregions (Zaliapin *et al.* 2005b; Schoenberg *et al.* 2006). The coupling is calculated as

$$\chi = \frac{1}{T} \sum_{i=1}^{n} m_i / \dot{M}_T, \tag{5.38}$$

where T is the total catalog time and $\dot{M}_T$ is the rate of tectonic deformation.

It is clearly seen that the model describes very closely the available data over the wide range of parameters: the number n of events within a region changes from 1 to about one thousand, while the coupling χ changes from 10^{-2} to 10. In simulations with the TGR distribution we used the following model parameters suggested by the data set from Holt *et al.* (2000): $m_c = 8.6$, $m_t = 5.0$.

By gray contours Fig. 5.13 displays the approximate quantiles for the ratio between the sum of n random variables and the corresponding mean (which is finite in the truncated model). The conventional Gaussian picture (the sum

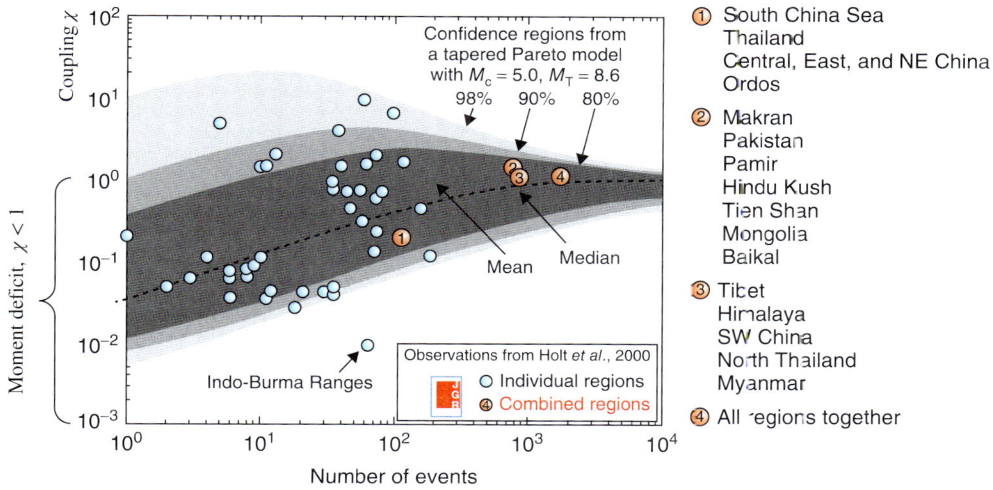

Fig. 5.13 Modeling seismic moment release in Asia during the twentieth century. Seismic coupling χ as a function of the total number of events in 42 small (blue balls) and 4 combined regions (red balls). Distribution of coupling according to the tapered Pareto model is shown by shades. The model explains the main features of the data over the wide range of parameters and indicates the existence of three regimes in the seismic moment release.

is proportional to the mean) is valid only with a very large number of events exceeding m_t: $n > 1000$.

Three regimes are clearly seen, depending on the number of earthquakes in a region. These regimes are well reproduced by the truncated Pareto model.

- *Regime 1.* Moment deficit, $\chi \ll 1$ ($n \leq 5$). This regime is seen within the regions with small numbers of events. It corresponds to a heavy-tailed highly skewed total moment distribution. The total seismic moment is best described by a Pareto (or tapered Pareto) random variable.
- *Regime 2.* Irregular moment release, $\chi \ll 1$ and $\chi \gg 1$ are possible ($5 \leq n < 600$). This regime is seen within the regions with intermediate numbers of events; it is a transition between Regimes 1 and 3.
- *Regime 3.* Regular moment release, $\chi \approx 1$ ($n > 600$). This regime is only seen in large regions with many events; here the Central Limit Theorem applies, and the total seismic moment can be approximated by a Gaussian distribution.

Note that by definition in our simulations, $E(\chi) = 1$. In practice, however, observed values of χ can differ substantially from unity, especially for small samples, due to the skewness of the seismic moment distribution.

Although the boundaries between regimes are somewhat vague, it is possible to determine the principal seismic moment regime given the values of control parameters n and the magnitude range $\Delta m = m_c - m_t$. The regime location diagram is shown in Fig. 5.14.

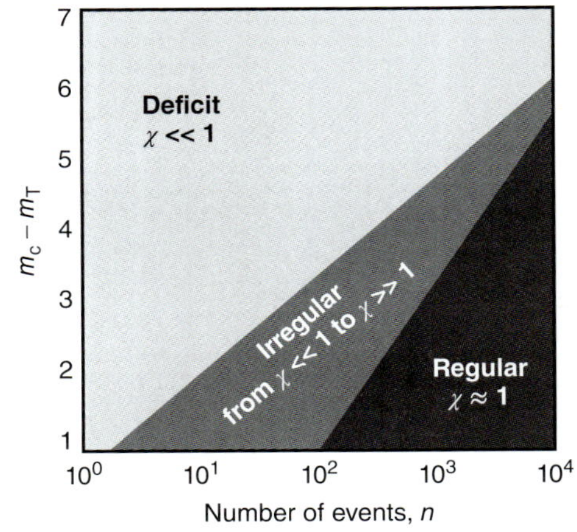

Fig. 5.14 Location of moment release regimes in the space of main control parameters: the number n of events and the magnitude range $\Delta m = m_c - m_t$.

As one can see from Fig. 5.14, the currently available quality and quantity of local regional data (with $\Delta m \approx 5$ and $n < 10^3$) imply that a substantial seismic deficit is not surprising. Regular moment release (that is, a nice match between long-term predictions and short-term observations) can only be expected in relatively large regions and over long time intervals. For instance, if one observes earthquakes with $m > 3$ in a region with corner magnitude 8.0, then regular release is expected after accumulating more than 10^5 earthquakes.

These conclusions may explain the often reported discrepancies between the observed seismic moment and its long-term predictions based on space-borne data (GPS, VLBI) and tectonic models. For example, Meade and Hager (2005) suggest that the seismic moment release in California during the twentieth century should have been twice as large as that observed in order to match the long-term strain rates. They use the discrepancy to advocate the possibility of a large ($m8$) impending earthquake in southern California. From Figs. 5.13 and 5.14 we infer that such discrepancy often occurs because of relatively small number of moderate and large earthquakes in a region.

5.5 Length of aftershock zone (earthquake spatial scaling)

Kagan (2002b) evaluated how the aftershock zone size for mainshocks $m \geq 7.0$ depends on the earthquake magnitude by approximating the aftershock epicenter maps through a two-dimensional Gaussian distribution. The mainshock focal zone length is the length of the 2-σ confidence area for the major ellipse axis (Kagan 2002b, Eq. 13).

In Fig. 5.15 we display the regression curves for GCMT/PDE earthquakes: all earthquakes offer three choices of focal mechanisms (updated Fig. 6a by Kagan 2002b). In regression curves we use $m = 8.25$ as a reference point. For the quadratic regression

$$L = \log_{10} \ell = a_0 + a_1 (m - 8.25) + a_2 (m - 8.25)^2, \tag{5.39}$$

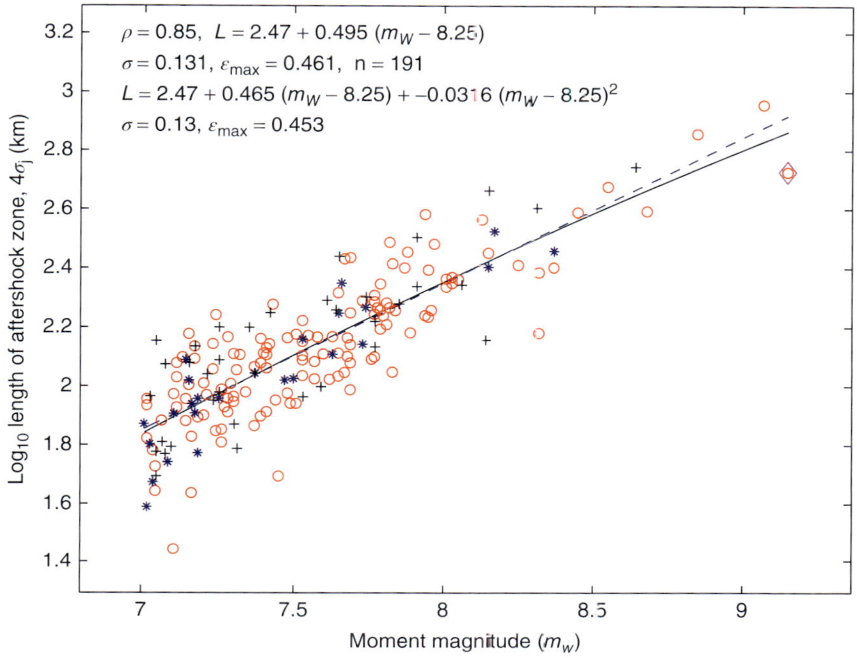

Fig. 5.15 Plot of the log aftershock zone length (L) against moment magnitude (Eq. 5.39). The GCMT/PDE catalogs time interval is 1977/1/1–2013/02. Magnitude values are shifted in formulas shown in the plot ($m_r = m - 8.25$). Rupture length is determined using a 1-day aftershock pattern. The diagram shows values of the correlation coefficient (ρ), coefficients for linear (dashed line) and quadratic (solid line) regression, standard (σ) and maximum (ϵ_{max}) errors, and the total number (n) of aftershock sequences. The Tohoku mega-earthquake is marked by a diamond sign. Two large strike-slip events $m8.6$ and $m8.2$ are Coast Sumatra (Wharton Basin) earthquakes occurring 2012/04/11. Red circles – thrust mainshocks; blue stars – normal mainshocks; black pluses – strike-slip mainshocks.

where ℓ is the length of the aftershock zone in km. For the linear regression we set $a_2 = 0$. The value of a_1 (scaling parameter) close to 0.5 means that the aftershock zone length is proportional to the cube root of the moment ($\ell \propto M^{1/3}$).

Comparing the regression results from Fig. 5.15 with those of Fig. 6a by Kagan (2002b) demonstrates that the scaling parameter estimates are robust. Since 2000, $m \geq 7.0$ earthquake numbers have increased by more than a factor of two. Furthermore, in Fig. 5.15 there are three major ($m \geq 8.8$) events, whereas the largest earthquake in the 2002 study was $m8.4$. However, the values of regression coefficients in both datasets are essentially the same. The plotted signs for two Coast Sumatra (Wharton Basin) earthquakes are close to the general regression line, showing that these events are not "rogue" earthquakes (McGuire and Beroza 2012), and their rupture size may be expected on the basis of Fig. 6a by Kagan (2002b).

Figire 5.16 displays the regression in a similar format for active continental tectonic zones (Kagan *et al.* 2010; Section 9.2). Table 5.4 summarizes the results of regression analysis for all global earthquakes, as well as for events in subduction zones (trenches) and active continental zones. Earthquakes are also subdivided by their focal mechanisms. Other tectonic zones lack a sufficient number of $m \geq 7.0$ mainshocks to carry out this statistical analysis.

The following conclusions can be made from Table 5.4: (a) aftershock zones exhibit similar scaling; (b) zone length (ℓ) on average is proportional to moment $M^{1/3}$; and (c) the value of a_0 parameter (zone length for the $m8.25$ earthquake) is close to $10^{2.5}$ (316) km for all cases. Normal earthquakes (rows 5 and 6) exhibit

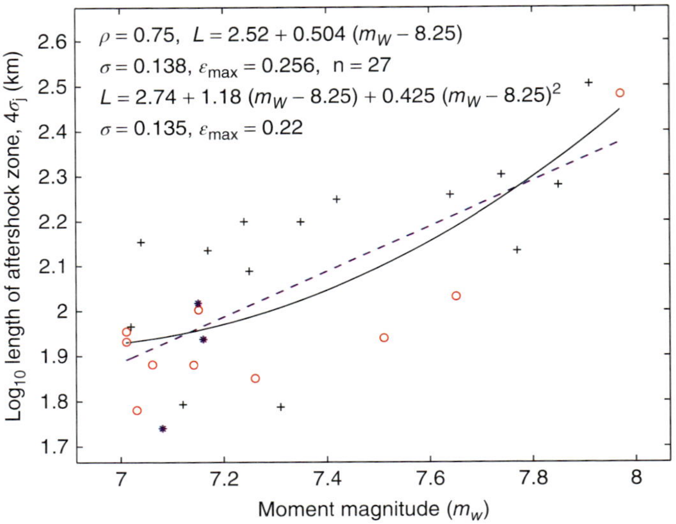

Fig. 5.16 Plot of log aftershock zone length (L) against moment magnitude for earthquakes in active continental zones (Section 9.2). The GCMT / PDE catalogs time interval is 1977/1/1–2010/09/21. For notation see, Fig. 5.15. *Source:* Kagan (2011), Fig. 14.

slightly different scaling; zone length (ℓ) for the linear regression is proportional to moment $M^{1/2.8}$. Scaling for strike-slip earthquakes (rows 7 and 8) also differs a little from average: zone length (ℓ) is proportional to moment $M^{1/3.5}$. However, the earthquake numbers in these subsets are small, thus it is possible that these variations are due to random fluctuations.

Only three subsets show a substantial nonlinearity: (a) trenches with strike-slip focal mechanisms (rows 15 and 16); (b) active continents with all focal mechanisms (rows 17 and 18, see also Fig. 5.16); and (c) continents with strike-slip focal mechanisms (rows 21 and 22). However, the earthquake numbers are small in all these plots, and although (b) and (c) aftershocks display zone lengths which increase strongly for the largest earthquakes, (a) earthquakes exhibit the opposite behavior.

In all diagrams the standard errors (σ) are almost the same for the linear and quadratic regression. The maximum errors (ϵ_{max}) follow the same pattern. This pattern suggests that the linear regression is sufficient to approximate the data. Although the quadratic regression fit yields no statistically significant improvement in almost any diagram, the sign of the quadratic correction term is negative for most cases. The negative value of the a_2 regression coefficient means that an increase in the aftershock zone length is weaker for the largest earthquakes. This feature contradicts those often quoted in other studies of length scaling (see Kagan 2002b and Wells and Coppersmith 1994, as well as subsequent publications citing the latter paper). Thus, the slope of the regression curve is either stable or decreases at the high magnitude end. No saturation effect for large earthquakes occurs in the data. Results in Table 5.4 imply that the major ellipse axis a (length) of an earthquake focal zone can be approximated by

$$a = 316 \times 10^{(m-8.25)/2} \text{ km}. \tag{5.40}$$

Table 5.4
Aftershock zone log length versus mainshock moment magnitude m.

#	Tectonic zone	Focal mech.	a_0	a_1	a_2	σ	ϵ_{max}	n
1	All	All	2.48	0.492	–	0.134	0.468	160
2	All	All	2.48	0.493	0.0013	0.134	0.468	160
3	All	Thrust	2.48	0.501	–	0.132	0.457	115
4	All	Thrust	2.48	0.499	0.0022	0.132	0.458	115
5	All	Normal	2.47	0.532	–	0.076	0.132	15
6	All	Normal	2.46	0.427	−0.0884	0.075	0.137	15
7	All	Str.-Slip	2.47	0.437	–	0.153	0.276	30
8	All	Str.-Slip	2.49	0.490	0.0038	0.153	0.278	30
9	Trench	All	2.47	0.499	–	0.131	0.454	129
10	Trench	All	2.47	0.482	−0.0177	0.131	0.449	129
11	Trench	Thrust	2.48	0.500	–	0.135	0.460	104
12	Trench	Thrust	2.48	0.488	−0.0142	0.135	0.455	104
13	Trench	Normal	2.47	0.543	–	0.067	0.120	12
14	Trench	Normal	2.46	0.440	−0.0915	0.065	0.114	12
15	Trench	Str.-Slip	2.38	0.409	–	0.146	0.302	13
16	Trench	Str.-Slip	2.29	0.034	−0.2790	0.142	0.268	13
17	Active Cont.	All	2.52	0.504	–	0.138	0.256	27
18	Active Cont.	All	2.74	1.180	0.4250	0.135	0.220	27
19	Active Cont.	Thrust	2.44	0.480	–	0.112	0.176	10
20	Active Cont.	Thrust	3.00	2.210	1.0800	0.064	0.130	10
21	Active Cont.	Str.-Slip	2.50	0.419	–	0.145	0.314	14
22	Active Cont.	Str.-Slip	2.65	0.889	0.2990	0.143	0.295	14

Notes: a_0, a_1, a_2 are regression coefficients in Eq. 5.39; σ is the standard uncertainty; ϵ_{max} is the maximum error, i.e., the largest difference between the data and the fit shown in Figs. 5.15, 5.16; n is the number of aftershock sequences. The GCMT/PDE catalogs time interval is 1977/1/1–2010/09/21.
Source: Kagan (2010b), Table 2.

We conclude that the earthquake rupture length is proportional to the cube root of moment, which implies that width and slip should scale the same way. Otherwise, one of them increases less strongly with moment and the other more strongly. For either case that would pose the problem of "inverse saturation."

We assume that the majority of aftershocks are concentrated within an ellipse having a 2-a major axis. The probability that a point will lie inside a 2-a ellipse is shown in Eq. 5 by Kagan (2002b). If we know the length of an earthquake focal zone, we can calculate its area. If the ratio of the major ellipse axis to the minor axis is 4, then area S of the focal zone is

$$S = \pi a^2. \tag{5.41}$$

Using the derived scaling law and moment-frequency relation, we can derive the distribution of earthquake slip not only for a region, but also for a specific place on a fault (Kagan 2005a). This distribution depends on the linear size of the earthquake rupture. For example, if the rupture is relatively short, a particular spot on a fault would be ruptured less frequently but would have a larger slip. Kagan and Jackson (2013) discuss how this distribution can be applied to determine the maximum earthquake size for a specific fault point.

5.6 Maximum or corner magnitude: 2004 Sumatra and 2011 Tohoku mega-earthquakes

5.6.1 Maximum moment for subduction zones

The 11 March 2011 Tohoku, Japan, magnitude 9.2 earthquake and the ensuing tsunami near the east coast of Honshu caused nearly 20,000 deaths and more than 300 billion dollars in damage, ranking as one of the worst natural disasters ever recorded (Hayes *et al.* 2011; Simons *et al.* 2011; Geller 2011; Stein *et al.* 2011). The great difference between the expected and observed earthquake magnitudes contributed to this enormous amount of damage. The estimated maximum magnitude for the Tohoku area (around 7.7) was proposed in the official hazard map (Headquarters for Earthquake Research Promotion 2005; Simons *et al.* 2011).

Several quantitative estimates of the maximum possible earthquakes in the subduction zones had been published before the Tohoku event (Kagan 1997a; Kagan and Jackson 2000; Bird and Kagan 2004; McCaffrey 2007, 2008; Kagan *et al.* 2010). In these publications, the upper magnitude parameter was determined to be within a wide range from 8.5 to 9.6.

Two quantitative methods have been deployed to estimate the upper magnitude limit: a statistical determination of the magnitude–moment/frequency parameters from earthquake data alone, and a moment conservation principle in which the corner/maximum moment is estimated from comparison of the seismic moment rate with geologic or geodetic deformation rate (Kagan 1997a; Bird and Kagan 2004). However, for individual zones the statistical method is usually ineffective in estimating the maximum magnitude; only the lower limit can be evaluated (Kagan and Jackson 2013). Figures. 5.1 and 9.3 show that the upper m_c limit may be $+\infty$.

5.6.2 Seismic moment conservation principle

We try to estimate the upper bound of the seismic moment-frequency relation, using the moment conservation principle as another, more effective method for determining the maximum/corner magnitude. Quantitative plate tectonics and space geodetic methods currently provide a numerical estimate of the tectonic deformation rate for all major tectonic plate boundaries and continental regions of significant distributed deformation (Bird and Kagan 2004; Kagan *et al.* 2010). We compare these estimates with a similar one for the seismic moment release.

Equations 5.5 –5.7 demonstrate how the seismic rate can be calculated. The rate of earthquakes of any magnitude and the total seismic moment rate can be computed from three observable parameters: the rate of earthquakes at the lower moment threshold; the β-value or asymptotic slope of the moment distribution; and the maximum or corner moment. Conversely, the maximum or corner moment may be determined if the threshold rate, β-value, and total moment rate are known.

The tectonic moment rate $\dot{M}_t$ depends on the following three variables which are not well known:

1. The seismogenic zone width (W – 30–104 km).
2. The seismic efficiency (coupling) coefficient (χ – 50–100%).
3. The value of the shear modulus (μ – 30–49 GPa).

Therefore,

$$\dot{M}_T = \chi \; \mu \; W \; \mathcal{L} \; \dot{u}, \tag{5.42}$$

where $\dot{u}$ is the slip rate, and $\mathcal{L}$ is the length of a fault (compare Eq. 13 by Kagan 2002c and Eq. 5.8).

Figure 5.17 shows Flinn-Engdahl (FE) zones (Gutenberg and Richter 1954, Fig. 1; Flinn *et al.* 1974; Young *et al.* 1996) in the Pacific area. These 18 FE regions correspond to major subduction zones and they have been selected by us because the FE regionalization had been defined before the GCMT catalog started, thus eliminating earthquake selection bias.

Figure 5.18 displays the β-values determined for these 18 Flinn-Engdahl regions. Comparison with Kagan's (1997a) results demonstrates that (a) the β-values do not depend significantly on the catalog duration, though their standard errors decrease as the duration and earthquake numbers increase;

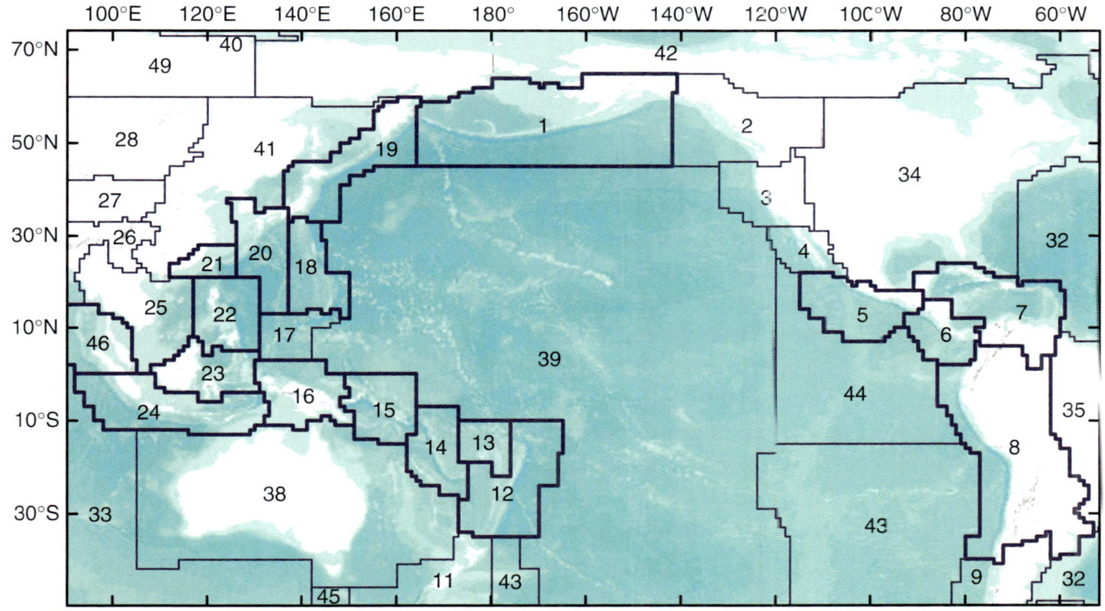

Fig. 5.17 Flinn-Engdahl seismic zones in the circum-Pacific area. The 18 major subduction zones from Table 5.5 are delineated by thick lines.

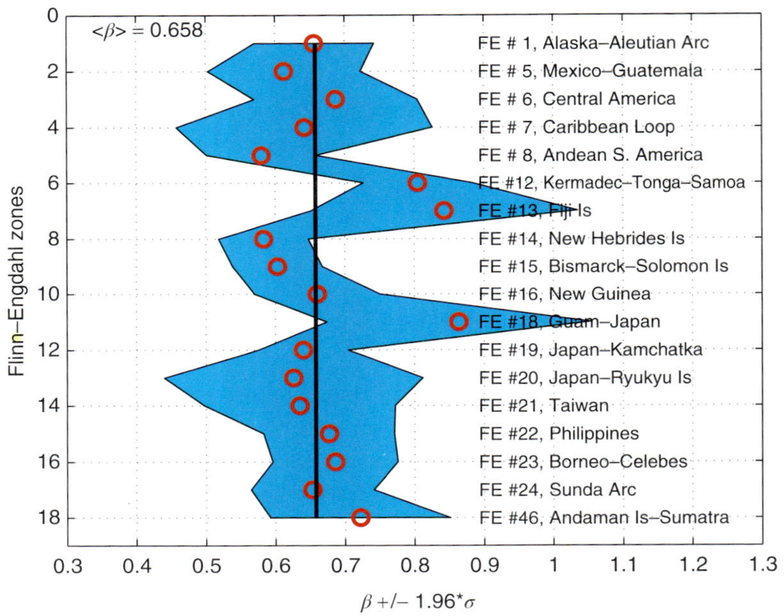

Fig. 5.18 Parameter β distribution in the Flinn-Engdahl (FE) subduction zones. The GCMT catalog 1977–2011 is used. The ordinate numbers are sequential numbers of subduction zones considered, the FE numbers and names for these zones are shown in the right-hand part of the diagram. The average region's β (red circles) and ± 1.96 standard deviations (cyan polygon) are shown; the solid black line corresponds to the average $< \beta >= 0.658$ for all subduction zones. *Source:* Kagan and Jackson (2013), Fig. 4.

(b) the β-values are approximately the same for all the zones, and the hypothesis of the values equality cannot be statistically rejected with 95% confidence (Kagan 1997a, pp. 2843–2844).

The estimate of m_c depends on the parameter values used to calculate the tectonic moment rate: for $\beta = 2/3$, the change of any parameters (such as W or χ) by a factor of two implies an increase or decrease of the m_c by about 0.6 (Kagan 2002c, Eq. 17). We see this influence by comparing Table 5.5 with results for the subduction zones in the similar Table 1 by Kagan (1997a), where the parameters used for calculating tectonic rate were $W = 30$ km, $\mu = 30$ GPa, $\chi = 1.0$. The difference in the m_c estimates for the two tables is caused mainly by changes in the above parameters. The average m_c-value for 18 subduction zones in Flinn-Engdahl regions is 9.35. For the TGR distribution, Bird and Kagan (2004, Table 5) determined $m_c = 9.58^{+\infty}_{-0.23}$ in all global subduction zones, and the approximate 95% upper limit is $m_c = 10.1$. Why are these estimates for the corner magnitude different? A possible reason for the difference is that here all earthquakes are included for each of the Flinn-Engdahl zones. Bird and Kagan (2004) used a refined tectonic definition of subduction zones compared to the Flinn-Engdahl regions; mechanisms were included in their categorization, and the tectonic deformation data have been updated. Kagan *et al.* (2010) estimated m_c for the "trench" zones, in which less stringent criteria were applied for earthquake selection (see Section 9.3). Assuming a TGR distribution, they obtained a maximum likelihood estimate $m_c = 8.8^{+\infty}_{-0.35}$, that is, a much smaller value than the $m_c = 9.6$ in Bird and Kagan (2004).

The last column in Table 5.5 shows the ratio of seismic to tectonic moment rates ($\psi = \dot{M}_s/\dot{M}_T$). Significant ψ variability confirms our results for seismic

Table 5.5 FE subduction seismic zones, GCMT 1977–2011, $m_t = 5.8$.

No	FE No	Flinn-Engdahl seismic region name	N	$\beta \pm \sigma_\beta$	$\dot{M}_T$ $\times 10^{27}$	$m_c \pm \sigma_M$	m_o	$\dot{M}_s$ $\times 10^{27}$	ψ
1	1	Alaska-Aleutian Arc	289	0.656±0.044	5.096	9.350±0.281	8.0	1.674	0.328
2	5	Mexico-Guatemala	169	0.613±0.056	2.378	9.172±0.286	8.0	1.314	0.552
3	6	Central America	163	0.687±0.060	2.491	9.237±0.286	7.8	0.573	0.230
4	7	Caribbean Loop	61	0.642±0.094	1.048	9.329±0.308	7.4	0.128	0.122
5	8	Andean S. America	305	0.580±0.040	8.493	9.711±0.280	8.8	9.314	1.097
6	12	Kermadec-Tonga-Samoa	459	0.805±0.040	5.945	9.106±0.278	8.1	2.231	0.375
7	13	Fiji Is	82	0.843±0.098	2.293	9.714±0.299	6.8	0.061	0.026
8	14	New Hebrides Is	445	0.583±0.033	4.813	8.964±0.278	7.9	1.765	0.357
9	15	Bismarck-Solomon Is	459	0.603±0.033	4.926	8.958±0.278	8.1	2.402	0.438
10	16	New Guinea	270	0.660±0.046	8.493	9.807±0.281	8.3	1.440	0.169
11	18	Guam-Japan	88	0.864±0.097	2.888	9.840±0.298	7.8	0.328	0.114
12	19	Japan-Kamchatka	507	0.640±0.033	8.493	9.310±0.277	9.2	20.769	2.445
13	20	S.E. Japan-Ryukyu Is	58	0.626±0.095	1.812	9.802±0.310	7.2	0.096	0.053
14	21	Taiwan	110	0.635±0.070	1.529	9.162±0.293	7.7	0.336	0.220
15	22	Philippines	248	0.677±0.048	3.539	9.183±0.282	7.7	0.719	0.203
16	23	Borneo-Celebes	273	0.686±0.046	4.162	9.235±0.281	7.9	1.060	0.255
17	24	Sunda Arc	285	0.653±0.045	6.512	9.555±0.281	8.6	4.535	0.696
18	46	Andaman Is-Sumatra	147	0.722±0.066	2.661	9.371±0.288	9.1	14.825	5.571
		1977–2011/12/31 ZONES	4418	0.658±0.011	77.572	9.347±0.273	9.2	63.568	0.819
		1977–1995/6/30 ZONES	2127	0.63±0.02	27.40	3.60±0.27	8.4	20.67	0.754

Notes: FE – Flinn-Engdahl seismic region; N – earthquake number, β – parameter of the power-law distribution of earthquake sizes, $\dot{M}_T$ – annual tectonic moment rate, m_c – corner magnitude; m_o – maximum moment magnitude observed in 1977–2011, $\dot{M}_s$ – annual seismic moment rate, $\psi = \dot{M}_s/\dot{M}_T$ – ratio of seismic to tectonic moment rate; W – seismogenic width, μ – elastic shear modulus, χ – seismic coupling coefficient. Seismic moment and moment rate are measured in *dyne-cm* and *dyne-cm/yr*, respectively. Tectonic rate for 1977–2011 period is calculated by using Bird and Kagan's (2004) parameters: $W = 104$ km, $\mu = 49$ GPa, $\chi = 0.5$. In the last row of the table we show the subduction zones total calculation results for Kagan (1997a, Table 1), where the following parameters have been used: $W = 30$ km, $\mu = 30$ GPa, $\chi = 1.0$. *Source:* Kagan and Jackson (2013), Table 1.

moment rate release in Asia (Fig. 5.13). Several thousand events in each subduction zone would be needed to see the ψ ratio stabilize.

The hypothesis that the m_c-values are the same within their uncertainties for all subduction zones considered cannot be rejected with statistical significance (cf. Kagan 1997a, p. 2846). Thus, the conjecture that $m_c \approx 9.0 - 9.7$ in all such zones is supported by comparing the theoretical estimates with the measured maximum observed magnitude values in several subduction zones. For 110 years of the instrumental seismic record, five zones have experienced earthquakes with a magnitude about 9 or larger. Figure 5.19 also shows that for a longer catalog, the average maximum observed magnitude (m_o) approaches the average estimate of m_c. This suggests that if the available earthquake record duration were

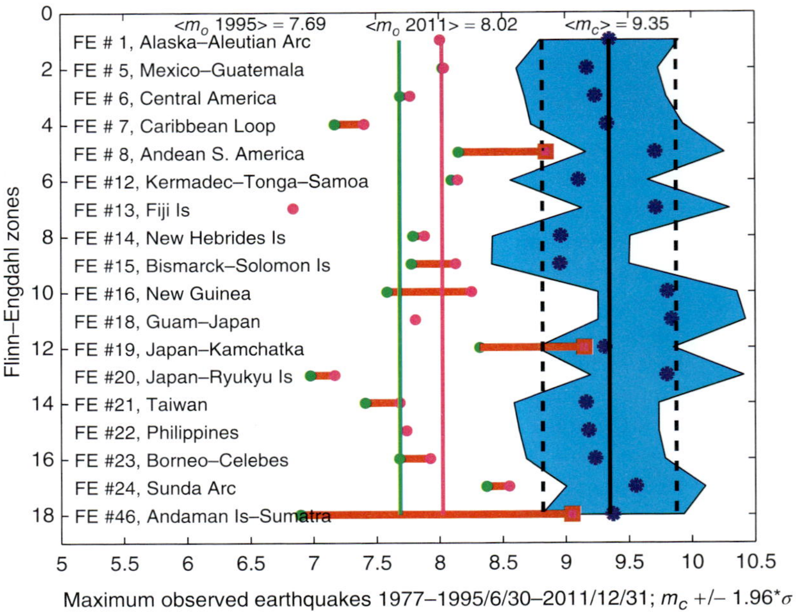

Fig. 5.19 The corner moment magnitude m_c distribution in the Flinn-Engdahl subduction zones. GCMT catalog 1977–1995/6/30 and 1977–2011 is used. The region's m_c (blue stars) and ± 1.96 standard deviations (cyan polygon) are shown, the black solid line corresponds to the average $< m_c >= 9.35$ for all subduction zones, black dashed lines show $< m_c > \pm 0.27 \times 1.96$, i.e., 95% confidence limits for the average corner magnitude estimate. In m_c calculations we use the parameters of the tectonic motion as proposed by Bird and Kagan (2004): the width of seismogenic zone $W = 104$ km, the shear modulus $\mu = 49$ GPa, $\chi = 0.5$. Green circles show earthquakes with the maximum magnitude m_o which occurred in the regions during 1977–1995/6/30. The green line corresponds to the average $< m_o >= 7.69$ for all subduction zones for the same time. Magenta x's and the line show the maximum magnitude m_o data for the time interval 1977–2011, the average is $< m_o >= 8.02$. Red squares indicate the magnitudes of giant earthquakes during 1995/6/30–2011.

comparable to the recurrence time of the largest earthquakes (a few hundred years), the difference between the observed maximum magnitude and m_c would largely disappear.

Using the parameter values for the moment-frequency distribution determined by Bird and Kagan (2004, Table 5) for all the subduction zones ($b = 0.96$, $M_t = 3.5 \times 10^{17}$ Nm, $\alpha_t = 76.74$ eq/y, $m_c = 9.58$), we calculate the number of $m \geq 9$ events expected to occur worldwide over a century

$$N(m > 9) = 100.0 \times \alpha_t \times 10^{-0.96\ (9.0-5.696)} = 5.2. \tag{5.43}$$

In fact five large earthquakes with magnitude 9 or greater did occur in the last 100 years. The distribution parameters (Bird and Kagan 2004) were estimated before two recent giant earthquakes struck.

Figure 5.19 shows the distribution of the corner magnitude obtained for the GCMT earthquake catalog. Estimates of m_c in all the subduction zones are statistically indistinguishable. This means that all such zones should be assumed to have the same maximum or corner magnitude. We also display in the plot the uncertainties of m_c estimates (Kagan 2002a). In contrast to the uncertainties for the purely statistical corner magnitude estimates (see Fig. 5.1), the upper m_c limit for the moment conservation technique cannot go to infinity, because it would imply the infinite moment rate.

6 Temporal earthquake distribution

This chapter is essentially based on findings of two publications (Kagan and Houston 2005; Kagan 2011).

6.1 Omori's law

Omori (1894) showed that the aftershock rate decays approximately as

$$n(t) = \frac{K}{t + c}, \tag{6.1}$$

where K and c are coefficients, t is the time since the mainshock origin, and $n(t)$ is the aftershock frequency measured over a certain interval of time. Presently a more complicated equation is used to approximate the aftershock rate

$$n(t) = \frac{K}{(t + c)^p}. \tag{6.2}$$

This expression with the additional exponent parameter p is called the "modified Omori formula" or the Omori-Utsu law (Utsu *et al.* 1995). Omori's law has been incorporated in many phenomenological and physical models of earthquake occurrence.

Utsu *et al.* (1995) show that the aftershock rate decay according to Eqs. 6.1 and 6.2 still continues in the focal zone of the 1891 Nobi, Japan, earthquake described by Omori (1894). Statistical analysis of earthquake catalogs indicates that a power-law dependence characterizes the occurrence of both foreshocks and aftershocks. As has been mentioned in previous chapters, a mainshock may be considered as an aftershock which happens to be stronger than the previous event (Kagan and Knopoff 1981; Agnew 2005; Gerstenberger *et al.* 2005).

The parameter "c" in (6.1) is almost always found to be positive and typically ranges from 0.5 to 20 hours in empirical studies (Utsu 1961; Reasenberg and Jones 1989, 1994; Utsu *et al.* 1995). It was introduced to explain the seeming saturation of aftershock rate close to the origin time of a mainshock. The saturation is usually interpreted as a delay between mainshock rupture end and the start of aftershock activity (Rundle *et al.* 2003; Kanamori and Brodsky 2004). Kagan and

---

Earthquakes: Models, Statistics, Testable Forecasts, First Edition. Yan Y. Kagan.

Houston (2005, Figs. 1 and 2) display Omori's law curves in the linear scale and in the more common log–log format.

No reliable empirical regularities in the behavior of c have been found. Positive c in (6.1) means that the singularity in (6.1) occurs *before* the mainshock, which is not physically true. Negative c means that the singularity occurs after the mainshock. The latter case is a more physically natural assumption. In this case, $n(t)$ is not defined for the period $t \leq -c$. This could correspond, for example, to the period of mainshock rupture, during which individual aftershocks cannot be defined, identified, or counted. However, Eq. 6.1 assumes that earthquakes are instantaneous, therefore, for times comparable to the rupture time of mainshocks Omori's law breaks down, since earthquake counting is not possible for such small time intervals. Moreover, Omori's law in its regular form (6.1) predicts that for time $t \to 0$, the aftershock rate stabilizes around K/c. Again, aftershock counting is not feasible at the time of mainshock rupture and its coda, hence some time limit (Ogata 1983) needs to be introduced in (6.1) and (6.2). Kagan and Houston (2005, their Section 4.2) discuss in more detail the reasons for the appearance of the non-zero c-value in many publications.

6.2 Seismic moment release in earthquakes and aftershocks

To understand temporal earthquake distribution, in this section we compare source-time functions (seismic moment release rates) for large events in California and global shallow large earthquakes with the seismic moment release rate of aftershock sequences. Because we are interested in the transition between the mainshock rupture process and the beginning of the aftershock sequence, we need to use data from regional and local earthquake catalogs, rather than global catalogs. The former catalogs are based on the interpretation of high frequency seismograms, and they record aftershocks which are closer in time to the mainshock rupture end than global catalogs (Kagan 2004, see Figs. 6.1, 6.2 below). Global catalogs are also analyzed to confirm the results obtained for the southern California earthquakes.

6.2.1 Temporal distribution of aftershocks

Kagan (2004) and Kagan and Houston (2005) argue that the observed saturation in the aftershock numbers described by the "time offset" parameter c in Omori's law (Eqs. 6.1 and 6.2) is likely to be an artifact due to the under-reporting of small aftershocks. This under-counting comes from the difficulty of detecting large numbers of small aftershocks in the mainshock coda, as well as other factors (Kagan 2004; see also Subsection 6.2.4). Figure 6.1 displays the aftershock time-magnitude distribution for the 2004 great Sumatra earthquake. The similar general aftershock pattern is seen in many other aftershock sequences (Kagan 2004): larger aftershocks begin early in the sequence, and the starting time for the occurrence of weaker events is progressively delayed.

Two displays in Fig. 6.1 exhibit an important property of earthquake catalogs: in the PDE catalog, aftershocks start at about 10^{-2} days after the mainshock,

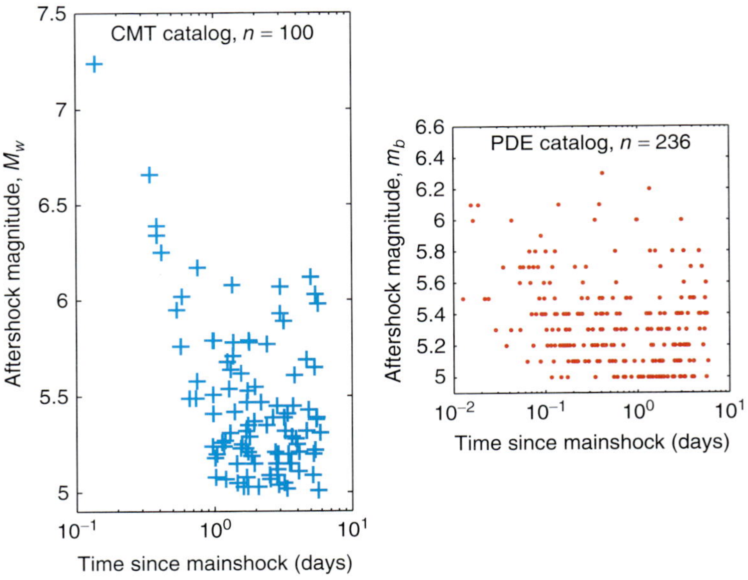

Fig. 6.1
Time-magnitude distribution of the Sumatra aftershocks; n is the aftershock number. The GCMT and the PDE earthquake catalogs (Section 2.3) are used. Events in the six days following the 2004/12/26 $m = 9.1$ mainshock and between latitude 0°N and 15.0°N and longitude 90.0°E and 100.0°E were selected. *Source:* Kagan (2006), Fig. 10.

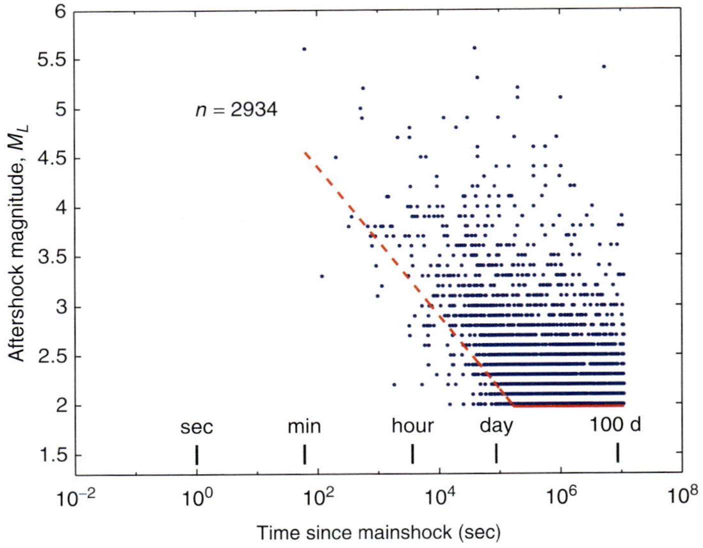

Fig. 6.2
Time-magnitude distribution of Northridge, California, aftershocks. Events in the 128 days following the 1994/1/17 $m = 6.7$ mainshock and occurring between latitude 34.0°N and 34.5°N and longitude 118.35°W and 118.80°W were selected. The CIT earthquake catalog is used. The red dashed line shows an approximate estimate of the completeness threshold (Eq. 6.4), which can be used to correct aftershock frequency and moment release rate for missing aftershocks. *Source:* Kagan and Houston (2005), Fig. 4.

whereas in the GCMT catalog they start at about 10^{-1} days. The total number of aftershocks as well as their magnitude range also significantly differ in these two diagrams. The main reason for the difference is the frequency range of the seismograms used in compiling both catalogs: in the PDE catalog the aftershocks are determined using waves with 1 s period, whereas the GCMT catalog uses low frequency (period 50 s and greater) waves. The magnitude estimates in the PDE catalog saturate at about $m_b = 6.0 - 6.5$ (Section 5.1): therefore, we see no large magnitude aftershocks in its display. On the other hand, long-period coda waves of the mainshock and large aftershocks in the GCMT catalog extend over a longer time. They make it difficult to discern smaller events in the seismograms.

In Fig. 6.2 we show the aftershock distribution for the 1994 Northridge, California, earthquake. Similarly to Fig. 6.1, aftershocks in any magnitude band after they start at the left-hand side of the diagram seem to be distributed almost uniformly over log time above the threshold. This would correspond to aftershock rate decay according to Omori's law (6.1).

Thus, Figs. 6.1 and 6.2 as well as the arguments in Kagan (2004) and Kagan and Houston (2005) demonstrate that the c-value depends on the methods of seismogram interpretation. Therefore, depending on the frequency characteristics of a seismographic network, the number of stations, and the seismogram processing technique, the same earthquake sequence could be variously identified. In one catalog it could be identified as one complex earthquake with some subevents, but in another as a foreshock–mainshock–aftershock sequence with many "individual" events (Kagan 2003, 2004). Thus "an individual earthquake" results from interpretations and selections made by the catalog compilers. It is not in itself a physical entity, as is tacitly assumed in most statistical analyses of seismicity (see Section 2.2).

6.2.2 Southern California earthquakes and their aftershocks

Figure 6.3 shows the moment release rates during the 1994 Northridge, California, earthquake and during its aftershock sequence. We subdivide time after the mainshock origin into intervals increasing by a factor of 2, and sum the scalar seismic moments of its recorded aftershocks (Kagan 2004). For most of the aftershocks, the seismic moment was not determined. We assume that their local magnitude is equivalent to the moment magnitude m (Hutton and Jones 1993) and calculate the moment M as the inverse of Eq. 2.2.

Figure 6.3 suggests that the aftershock moment rate $\dot{M}(t)$ can be approximated by a power-law time dependence similar to Omori's law

$$\dot{M}(t) = \frac{k\ \tau_{pk}\ \dot{M}_{pk}}{t - c}\,, \tag{6.3}$$

where t is time after mainshock origin, c is a coefficient similar to that in (6.1), but possibly different in value, $\dot{M}_{pk}$ is the peak moment release rate of a mainshock and τ_{pk} is the time the peak occurs. The coefficient k characterizes the ratio of peak mainshock moment rate ($\dot{M}_{pk}$) and aftershock moment rate extrapolated to τ_{pk} (with $c = 0$). We do not yet know how close the end of mainshock moment release comes to the beginning of the aftershock process; it is possible that there is no actual temporal gap between these two phenomena.

During the occurrence of a mainshock, the rupture process is often punctuated by significant changes in moment rate amplitude, momentarily stopping or

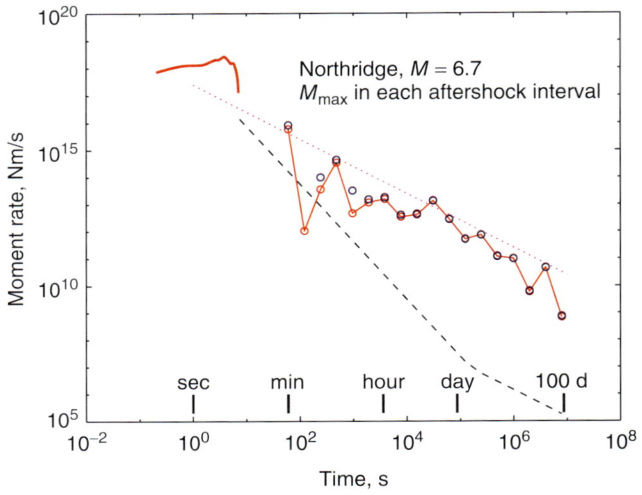

Fig. 6.3 Source-time function for 1994 *m*6.7 Northridge, California, earthquake (thick red line) compared to moment release of its immediate aftershocks (red circles), averaged over logarithmic time intervals. Blue circles show aftershock moment release corrected for under-reported small aftershocks (Eq. 6.6). For correction we use the aftershock moment threshold (black dashed line) divided by the duration of the corresponding time interval to yield quantities comparable to the moment rate (Eq. 6.4). The thin red dotted line shows a power-law approximation for aftershock moment release rates, analogous to Omori's law (Eq. 6.1). *Source:* Kagan and Houston (2005), Fig. 5.

restarting rupture and by other rupture complexities (see Subsection 5.3.3). As a result, large earthquakes in a detailed analysis are often subdivided into several sub-events. Aftershock moment release, on the other hand, is calculated here by summing the moments of several separate events. It seems possible that in the transition time interval after the mainshock rupture end and the beginning of the recorded aftershock sequence, the moment release could exhibit intermediate features – quasi-continuous rupture episodes which are supplanted by more discrete events. In part, our recognition of distinct events is affected by the limited frequency content of seismograms, the presence of various kind of noise, etc. Without ideal recording, the difference between mainshock and aftershock moment release rates may not be clear, abrupt, or well-defined.

An advantage of moment summation of aftershocks as opposed to the more usual counting earthquake numbers, is that early in an aftershock sequence many small events may be missing from the catalog as in the left-hand parts of Figs. 6.1 and 6.2. This under-count of small earthquakes gives an impression of aftershock rate saturation or rate decay when approaching the mainshock rupture end (i.e., going backwards in time towards the mainshock). In contrast, most of the seismic moment in a sequence is carried by the strongest aftershocks, hence the bias in moment summation is less significant. However, the summation of seismic moments carries a significant price – random fluctuations of the sum are very large (Zaliapin *et al.* 2005; see Section 5.4), hence we need more summands to yield more reliable results.

Helmstetter *et al.* (2005) found the following approximate relation between the aftershock magnitude completeness threshold $m_a(t, m)$ at time t (in days) after a mainshock of magnitude m

$$m_a(t, m) = m - 4.5 - 0.75 \log_{10}(t). \tag{6.4}$$

For several recent (1980–2004) southern California mainshocks (see Fig. 6.2 as an example), the magnitude completeness threshold has been as high as 4.5 shortly after the mainshock, dropping only to about 2 later in the sequence. The equation is plotted in Fig. 6.2.

Assuming that the aftershock size distribution follows the G-R relation (Kagan 2004), we can calculate the moment rate of the undetected, or missing weak aftershocks and thus compensate for an incomplete catalog record. The part of the total seismic moment M_f in an aftershock time interval, which is missing due to the incompleteness of the small earthquake record, can be obtained by modifying Eq. 21 in Kagan (2002c).

$$F(M_f) = \left[\frac{M_a}{M_{xp}}\right]^{1-\beta}, \tag{6.5}$$

where β is the parameter of earthquake size distribution (Eqs. 5.4, 5.5), M_a is the lower moment threshold for the aftershock sequence, and M_{xp} is the maximum moment. The threshold M_a depends on time according to Eq. 6.4 . If $M_a = M_{xp}$, all moment is missing, whereas for $M_a \to 0$, the moment sum is complete. We use the above two equations to correct aftershock moment release curves for under-reporting small events.

For aftershock sequences we assume $\beta = 2/3$ (Kagan 2002a,c; Chapter 5) and take the maximum moment (M_{xp}) to be the moment of the mainshock: if an earthquake stronger than the mainshock occurs during aftershock sequence, then the former mainshock would be re-classified as a foreshock. As alternative possibilities, we equate M_{xp} to the moment of the largest aftershock in a time interval or to the largest aftershock in the sequence.

After calculating the moment threshold (6.4), we estimate the multiplicative correction coefficient

$$[1 - F(M_f)]^{-1}. \tag{6.6}$$

Calculating the correction term (6.6) for different mainshocks and various choice of M_{xp} (see above), we found out that, as one would expect from expression (6.4), the correction is largest for the smallest time intervals. Even for these intervals the correction is less than 50%. In Fig. 6.3 as well as in all the calculations below we take as the maximum moment M_{xp} in (6.6) as the value of the largest aftershock in each time interval. Only for time interval closest to the mainshock rupture end is the difference between non-corrected and corrected values observable. In the second time interval in Fig. 6.3, the largest aftershock was smaller than the assumed threshold value (6.4), thus no correction coefficient is calculated.

Figure 6.4 shows scaled moment release rates for three California earthquakes and their aftershocks as well as the averages of both sets. In averaging datasets here and below, we divide the sum either by the number of curves, or by the number of non-zero entries in a dataset. The reason for the latter version is that source-time functions for some earthquakes were not defined at the same time intervals. Here the difference between two methods of averaging is quite minor.

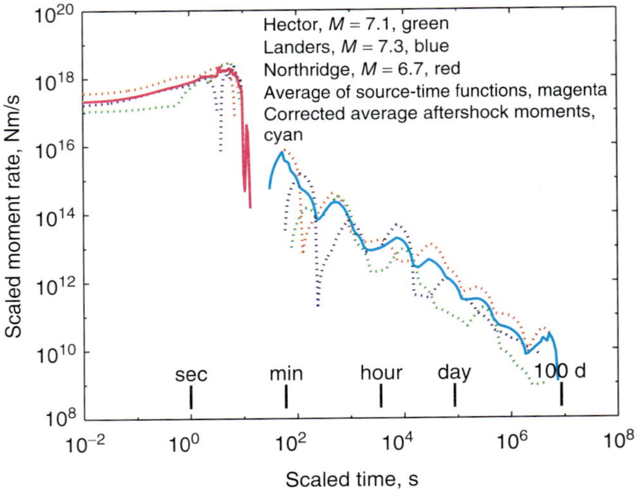

Fig. 6.4 Scaled source-time functions for three California earthquakes: 1992 *m*7.3 Landers, 1994 *m*6.7 Northridge, and 1999 *m*7.1 Hector Mine, and the moment release of their immediate aftershocks, corrected for missing small aftershocks (6.6) and averaged over logarithmic time intervals. *Source:* Kagan and Houston (2005), Fig. 6.

To average source time functions (STF) together, it is necessary to normalize for the effect of their varying seismic moments of mainshocks. Figure 3 in Kagan and Houston (2005) illustrates that the apparent duration of earthquake rupture increases with earthquake size. Houston *et al.* (1998) and Houston (2001) scaled STFs to a common moment of 10^{19} Nm as follows. Scaled moment rate $\dot{M}_{sc}$ and scaled time t_{sc} are given by

$$\dot{M}_{sc}(t_{sc}) = \dot{M}(t_{sc}) \times (10^{19}/M_m)^{2/3} \quad \text{Nm/s}, \tag{6.7}$$

and

$$t_{sc} = t \times (10^{19}/M_m)^{1/3}, \tag{6.8}$$

where M_m is the moment of a mainshock, $\dot{M}(t)$ and t are unscaled seismic moment rate (i.e., the original STF) and unscaled time respectively. These transformations are equivalent to normalizing the source-time function for an $m = 6.67$ earthquake. The variables τ_{pk}, $\dot{M}_{pk}$, t, and c in the formula for moment rate decay with time (6.3) can be scaled similarly.

As in Fig. 6.3 the extrapolation of aftershock moment release rates according to Omori's law is approximately 1.5 orders of magnitude below the maximum of the scaled source-time function at about 5 scaled seconds. This would mean $k \approx 1/30$ in (6.3).

The seismic moment scaling with the $2/3$ exponent as in (6.7) and averaging the obtained quantities correspond to the summation of earthquake rupture areas

$$S \propto M^{2/3}. \tag{6.9}$$

When we sum aftershock numbers in the standard application of Omori's law, small events dominate the sum. In a sum of seismic moments of events with a

G-R distribution with $\beta = 2/3$, the largest earthquake on average carries 1/3 of the total moment (Zaliapin *et al.* 2005a; Section 5.4). For $\beta = 2/3$, the average of earthquake rupture areas balances the influence of large and small earthquakes (Rundle 1989).

6.2.3 Global shallow earthquakes

To make a more general comparison between moment release rates during mainshocks and those during aftershock sequences, we compare average scaled source time functions for several sets of global large shallow earthquakes with scaled aftershock sequences of six California mainshocks. Houston (2001) studied 255 source time functions determined by inversions of teleseismic body waves by Tanioka and Ruff (1997) and colleagues. Figure 6.5 shows the average of the scaled time functions. In the scaling and averaging procedure, these time functions were truncated (i.e., assumed to be zero) after the duration picked as the end of rupture by Tanioka and Ruff and others (Houston 2001). After its maximum, the average decreases exponentially with time (dashed yellow and solid blue lines, Fig. 6.5). For the non-truncated time functions, aftershock moment release at the end of mainshock rupture seems to decay as $1/t$ as suggested by Omori's law. Kagan and Houston (2005) discuss in more detail possible biases in averaging STFs.

Figure 6.6 compares the average source-time function with scaled and corrected aftershock moment rates for six California earthquakes, similar to Fig. 6.4. These sequences are for the 1952 Kern County, the 1992 Joshua Tree, the 1992 Landers, the 1992 Big Bear, the 1994 Northridge, and the 1999 Hector Mine earthquakes. Although the Big Bear earthquake was an aftershock of the Landers

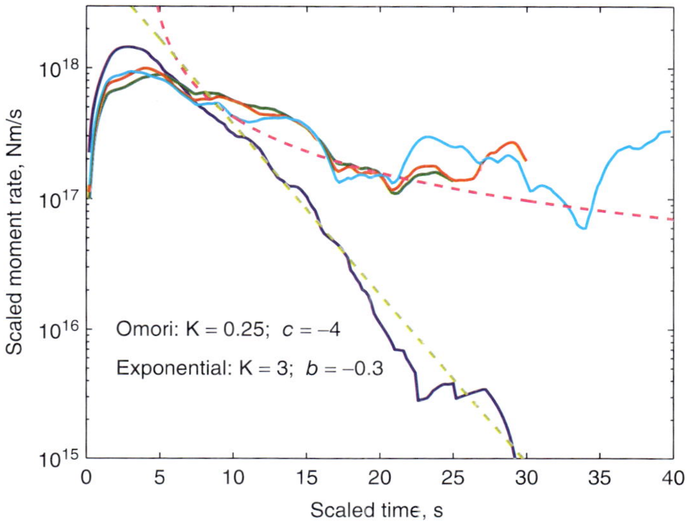

Fig. 6.5 Average scaled seismic moment source-time functions for 143 shallow (depth 5–40 km) global earthquakes, ranging in size from m 6.2 to 8.3, and their approximation by an exponential function (yellow) and a power-law (Omori's) distribution (magenta). For the latter approximation we use $c = -4.0$ s which is close to the rupture time of $m = 6.67$ earthquake. The blue curve shows the average of scaled time functions truncated at the inferred end of rupture. Green, red, and cyan curves show the averages of subsets of source-time functions comprising those non-truncated time functions for which the STF inversion result was available for at least 25, 30, and 40 scaled seconds (i.e., including a sufficient interval after the apparent end of mainshock rupture). For the non-truncated time functions, moment release at the end of mainshock rupture seems to decay as $1/t$ similar to Omori's law. *Source:* Kagan and Houston (2005), Fig. 7.

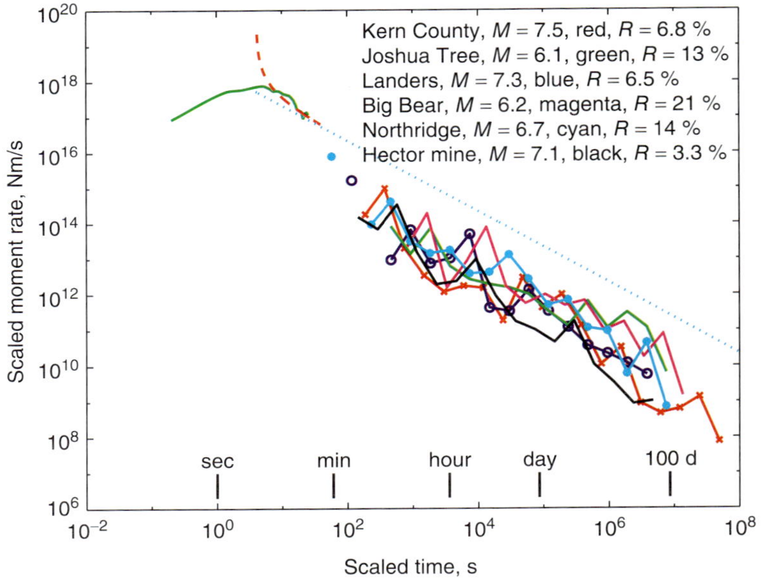

Fig. 6.6 Average scaled source-time function for shallow global earthquakes compared to scaled aftershock moment release rates for six California aftershock sequences. The average includes only mainshocks with inversion results available for at least 25 s of scaled time. Both types of moment rates were scaled to a magnitude 6.67 event. Two Omori's law approximations to the source time functions are shown with $c = 0$ s (dotted blue line) and $c = -4.0$ s (dashed magenta line), both lines coincide starting at about 1 min. The California aftershock sequences were corrected for missing small aftershocks (following Eq. 6.6). The coefficient R in the figure is the percent of total seismic moment released by immediate aftershocks compared to the mainshock scalar moment. The moment activity level in the aftershock sequences extrapolates to about 30 times less than the peak rate in the average scaled global time function. *Source:* Kagan and Houston (2005), Fig. 8.

event, it has an extensive aftershock sequence of its own. We calculated the ratio of total moment release in the aftershock sequences to the seismic moment of the mainshock. The percentages are 6.3%, 12%, 5.8%, 19%, 12%, and 2.8%, respectively. After correction for missing small aftershocks, they are 6.8%, 13%, 6.5%, 21%, 14%, and 3.3%, respectively. As explained earlier, the correction is in general small; only for aftershocks in those time intervals closest to the end of mainshock rupture, does it reach several tens of percent.

In Fig. 6.7 the scaled moment release rates of the aftershock sequences are averaged. The average behavior is similar to that of Fig. 6.4: an extrapolation of the average aftershock moment rate according to Omori's law (assuming $c = 0$) is about 1.5 orders of magnitude below the maximum of the source-time function, i.e., $k \approx 1/30$ in (6.3).

Figures 6.4 and 6.6 compare the seismic moment release of mainshocks and aftershocks. Such a comparison can be made only retrospectively, that is, only after an aftershock sequence has ended do we know that the first (main) event is not followed by an even stronger shock. For example, the $m6.1$ 1992 Joshua Tree, California, earthquake which occurred on April 23, 1992, was followed 66 days later by the $m7.3$ June 28 Landers event. Hence, were we to predict its aftershock decay, the forecast would be significantly wrong (Kagan 2004). Therefore, our results relate to *typical* aftershock sequences, that is, such that no earthquake comparable or stronger than the mainshock occurs in the sequence. Similarly the ratio of mainshock moment to the total moment of aftershock sequence, discussed above, would substantially change if unusual earthquake clusters are considered.

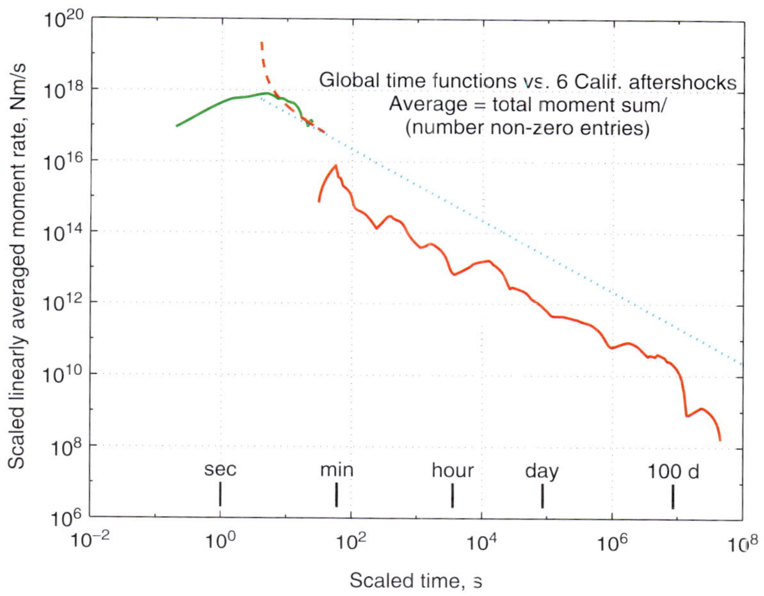

Fig. 6.7 Average scaled source-time function for global shallow earthquakes with non-truncated time functions (green line in Fig. 6.5) compared to the corrected scaled average for six California aftershock sequences (red line). Approximations of average time function by power-law (Omori) distributions (dashed and dotted lines) are also shown. As before, two Omori's law approximations are given with $c = -4.0$ s, and with $c = 0$ s, the lines coincide starting at about 1 min. The average aftershock moment rates fall about 8 orders of magnitude over about 200 days. Extrapolating back in time according to Omori's law yields a level of aftershock moment activity about 1.5 orders of magnitude less than the maximum mainshock moment rate. *Source:* Kagan and Houston (2005), Fig. 9.

6.2.4 Comparison of source-time functions and aftershock moment release

In previous subsections we compared average source-time functions for large shallow earthquakes with the ensuing moment release of immediate aftershocks. The global and California earthquakes were plotted against the average aftershock curves of California events. In both of these cases the pattern is similar: aftershock moment release follows Omori's law with the p-value (i.e., exponent in Eq. 6.2) close to 1.0. If the average curve is extrapolated toward the earthquake origin time, its continuation is about 30 times below the maximum of the average source-time function at the scaled time of about 5 s, which corresponds to the maximum release of seismic moment for a $m = 6.67$ earthquake. Taking into account the size of the focal zone for such an earthquake (about 20 km) and the average rupture velocity (2–3 km/s), this time seems to be reasonable for a bilateral rupture. The value of k in (6.3) found here, $k \approx 1/30$, is consistent with the rule of thumb known as Båth's law (Console *et al.* 2003a) which holds that the magnitude of the largest aftershock in a sequence is roughly 1–1.2 units smaller than that of the mainshock.

However, as we explained at the end of previous subsection (6.2.3), this k determination is based on the retrospective identification of a mainshock and aftershocks. In real time such identification is not possible, thus an earthquake may be followed by a stronger event, with the seismic moment by the order of magnitude higher than all the previous events. Thus, if we use Eq. 6.3 to estimate future activity, the average k-value may be close to unity. However, as shown in Fig. 4.1 some strong shallow earthquakes produce no aftershocks. Additional studies are needed to resolve this problem.

Around scaled time interval 20–60 s there is no moment release activity either in the source-time functions or in aftershock curves. This gap is most likely caused by the mainshock coda, which hinders aftershock detection. If this conjecture is true, one can extrapolate aftershock curves right to the end of earthquake rupture. Because of the coda wave interference we cannot extend the aftershock moment rate right to the end of mainshock rupture, but it seems likely that the transition of mainshock rupture into the aftershock process is smooth.

What might explain the difference between the rate of seismic moment release during mainshock rupture and that extrapolated from the aftershock moment release via Omori's law? The earthquake rupture process is most likely controlled by dynamic stresses: a rupture front is concentrated in a pulse (Heaton 1990) with a strong stress wave initiating rupture. In contrast, the aftershock process is essentially static: dynamic waves generated by an aftershock have almost always left the mainshock focal region before the occurrence of a subsequent registered aftershock. According to various evaluations (Antonioli *et al.* 2002; Kilb *et al.* 2002; Gomberg *et al.* 2003), the amplitude of the dynamic stress wave is at least an order of magnitude stronger than the amplitude of the incremental static stress. If the number and total moment release of both aftershocks and the rupture events comprising the mainshocks are proportional to the stress increase, we would expect the source-time function to be higher than the appropriately scaled aftershock moment release rate. This may explain the difference in moment release rates for mainshocks and aftershock sequences.

Moreover, it is likely that the temporal and spatial properties of earthquake rupture differ significantly from those of an aftershock distribution. Spatially, aftershock patterns are no different from the general earthquake distribution; they seem to be fractally distributed with a correlation dimension close to 2 (Robertson *et al.* 1995; Guo and Ogata 1997, see Chapter 7). Although it seems likely that the aftershock cloud is slowly expanding with time after a strong earthquake, there is no obvious strong order in the space-time distribution of aftershocks. Earthquake rupture, on the contrary, has a clear pattern associated with rupture driven by the propagating seismic waves (e.g., Heaton 1990). Although the propagation of rupture has many complex features, such as temporary stops, change of slip direction, jumping from one fault segment to another, in general, the spatio-temporal evolution of rupture exhibits significantly more orderly behavior than that of an aftershock sequence. Unfortunately, presently there is an insufficient amount of data in earthquake source inversions for statistical analysis of the rupture propagation complexity (see Subsection 5.3.3).

What is the time interval between the end of the mainshock rupture and the beginning of the aftershock sequence? Our results shown in Figs. 6.4, 6.6, and 6.7 indicate that the interval is small, no longer than 20–60 s of scaled time, perhaps it is effectively close to zero. The end of mainshock rupture is defined by a relatively low level of moment release. If the release is still high, this is considered as a continuation of the earthquake rupture process. Hence, the low-level interval is assumed in the definition of the rupture duration during the retrospective interpretation of seismic records. As we suggested, an objective way to study the late part of the mainshock moment release and the beginning of the aftershock sequence, would be to process all the seismic records to a pre-arranged scaled time interval.

How then can one self-consistently identify an individual earthquake event? One criterion is to define the end of an individual event by a rare (low probability) time interval without strong aftershocks. Another is to look at dynamic stress waves at a certain amplitude (or other characteristic property) in the source zone of an earthquake. An earthquake is considered to end when all such waves have ceased. Both of these definitions depend on some quantitative criterion, and that most likely, the number and properties of such identified individual events would depend strongly on the value adopted.

6.3 Random shear stress and Omori's law

Our aim in this and following Sections (6.4–6.5) is to investigate the connection between random shear stress and Omori's law of aftershock occurrence (see more in Kagan 2011). Several attempts to explain Omori's law have been published (see, for example, Kagan 1982; Kagan and Knopoff 1987a; Dieterich 1994). Since shear stress in the Earth's interiors cannot be easily measured or calculated, these studies usually consider stress as a scalar stochastic variable, ignoring for a while its tensorial nature. In these sections we also regard the stress as a scalar.

Kagan (1982) considered shear stress time history as a Brownian motion, i.e., randomly fluctuating stress acting on the stressed environment of an earthquake focal zone. Then the probability density for the time intervals between earthquake events has the Lévy distribution with a power-law tail having the exponent $-3/2$ (see Eq. 6.10). Kagan and Knopoff (1987a) added a tectonic drift component to the Brownian motion. For such a case, the distribution of inter-event times is the Inverse Gaussian Distribution (IGD) which, depending on the value of the initial shear stress drop and velocity of tectonic motion, can exhibit occurrence patterns ranging from the Lévy distribution to a quasi-periodic occurrence.

Matthews *et al.* (2002) also proposed the IGD (they called it the BPT – Brownian Passage-Time) as a law for inter-earthquake intervals; however, they considered only a limiting, quasi-periodic long-term distribution. The authors suggest (ibid, abstract) that

> *... [the Inverse Gaussian] distribution has the following noteworthy properties: (1) the probability of immediate [large earthquake] rerupture is zero; (2) the hazard rate increases steadily from zero at $t = 0$ to a finite maximum near the mean recurrence time ...*

and real earthquake occurrence follows the same pattern. For both practical (evaluation of the seismic hazard) and theoretical reasons it is important to consider arguments for and against the quasi-periodicity of earthquake occurrence (see also Kagan *et al.* 2012 and Section 2.4).

Ellsworth (1995) and Ellsworth *et al.* (1999) present several sequences of characteristic, quasi-periodic earthquakes, mostly in the western U.S., which, in their opinion, confirm a smaller coefficient of variation than the Poisson process for earthquake recurrence intervals. Nadeau *et al.* (1995) and Nadeau and Johnson (1998, and its references) show a similar quasi-periodic pattern for micro-earthquakes in the Parkfield area.

There is ample evidence that after almost any earthquake, subsequent events follow Omori's law. This has been observed for smaller earthquakes (aftershocks) as well as large events comparable in size to the original shock or even exceeding it (Kagan and Jackson 1999; Parsons 2002; see Subsection 6.5.2). Omori's law pattern of power-law rate decaying seismicity is observed at the decadal and century-long scales (Utsu *et al.* 1995; Ebel 2009; Stein and Liu 2009).

Omori's law aftershock behavior is observed even for acoustic emission events (micro-earthquakes) with the decay parameter value similar to that determined for large and moderate earthquakes (Ojala *et al.* 2004; Nechad *et al.* 2005). Thus, Omori's law is a general feature of earthquake occurrence observed for any brittle deformation of solids.

The presumed pattern of earthquake quasi-periodicity depends on the argument that shear stress drops to zero or close to zero in the focal zone of an earthquake as suggested, for example, by Shimazaki and Nakata (1980). Thus, it would be necessary to wait for some time before a critical shear stress level is reached. Kagan and Jackson (1999) showed that large earthquakes follow after a short time interval (often just a few days) within a focal zone of similar large events (see Table 6.1). Such an inter-earthquake interval is clearly insufficient for shear stress to be replenished by tectonic motion.

Thus, for relatively short time intervals, earthquakes are clustered in time with the coefficient of variation of interevent times significantly greater than that for the Poisson process (Kagan and Jackson 1991a). Why is it widely believed that large earthquakes are quasi-periodic in time? Evidence mostly comes from specifically selected sequences (Bakun and Lindh 1985; Ellsworth 1995) or from paleo-seismic investigations (Ellsworth *et al.* 1999). However, paleo-seismic investigations have poor time resolution: two or more events occurring closely in time are likely to be identified as one event (Kagan *et al.* 2012). Hence, their coefficient of variation estimates should be biased towards smaller values: a more periodic pattern. Some paleo-seismic investigations (Marco *et al.* 1996; Rockwell *et al.* 2000; Ken-Tor *et al.* 2001; Dawson *et al.* 2003) also suggest that earthquakes follow a long-term clustering pattern at least in regions of slow tectonic deformation.

Additionally, paleo-seismicity studies the Earth's displacement only at the surface. But temporal and spatial distributions of earthquakes rupturing the surface at one site substantially differ from general earthquake statistics (Kagan 2005a). The earthquake size distribution is also significantly different for site- and area-based statistics. Rigorously reliable statistical properties that are relevant for theoretical studies as well as for seismic occurrence estimates can only be obtained by analyzing instrumental earthquake catalogs.

Table 6.1
Pairs of shallow earthquakes $m \geq 7.5$ in the GCMT catalog

Notes: R – centroid distance, Φ – 3-D rotation angle between focal mechanisms (see Subsection 8.4.2), Δt – time interval between events, η – degree of zone overlap (6.23), the ratio of earthquake focal zone sizes to twice their distance, see Eqs. (2,3) in Kagan and Jackson (1999). The total earthquake number with magnitude $m \geq 7.50$ for the period 1976/1/1–2010/10/25 is 121. The maximum epicentroid distance is 250 km.

	First Event				Second Event				Difference			
No	Date	Coord. Lat.	Coord. Long.	m	Date	Coord. Lat.	Coord. Long.	m	R km	Φ °	Δt day	η
1	1977/06/22	−22.9	−174.9	8.1	2009/03/19	−23.1	−174.2	7.7	75	55	11593.26	1.4
2	1978/03/23	44.1	149.3	7.6	1978/03/24	44.2	149.0	7.6	25	7	1.69	2.3
3	1980/07/08	−12.9	166.2	7.5	1980/07/17	−12.4	165.9	7.8	62	18	8.35	1.1
4	1980/07/08	−12.9	166.2	7.5	1997/04/21	−13.2	166.2	7.8	33	42	6130.53	2.0
5	1980/07/08	−12.9	166.2	7.5	2009/10/07	−12.6	166.3	7.7	37	13	10682.95	1.6
6	1980/07/17	−12.4	165.9	7.8	2009/10/07	−12.6	166.3	7.7	41	14	10674.10	1.8
7	1980/07/17	−12.4	165.9	7.8	2009/10/07	−11.9	166.0	7.9	65	12	10674.11	1.3
8	1983/03/18	−4.9	153.3	7.8	2000/11/16	−4.5	152.8	8.1	83	72	6452.83	1.3
9	1983/03/18	−4.9	153.3	7.8	2000/11/16	−5.0	153.2	7.9	47	91	6452.94	1.8
10	1985/09/19	17.9	−102.0	8.0	1985/09/21	17.6	−101.4	7.6	71	14	1.51	1.3
11	1987/03/05	−24.4	−70.9	7.6	1995/07/30	−24.2	−70.7	8.1	33	7	3068.83	2.8
12	1990/04/18	1.3	123.3	7.7	1991/06/20	1.0	123.2	7.6	37	29	427.65	1.6
13	1995/08/16	−5.5	153.6	7.8	2000/11/16	−5.0	153.2	7.9	76	74	1918.89	1.1
14	1997/04/21	−13.2	166.2	7.8	2009/10/07	−12.6	166.3	7.7	70	30	4552.42	1.0
15	2000/06/04	−4.7	101.9	7.9	2007/09/12	−3.8	101.0	8.6	150	85	2655.78	1.3
16	2000/11/16	−4.6	152.8	8.1	2000/11/16	−5.0	153.2	7.9	67	89	0.12	1.7
17	2000/11/16	−4.6	152.8	8.1	2000/11/17	−5.3	152.3	7.8	93	88	1.67	1.2
18	2001/06/23	−17.3	−72.7	8.5	2001/07/07	−17.5	−72.4	7.7	34	8	13.54	4.7
19	2005/03/28	1.7	97.1	8.7	2010/04/06	2.0	96.7	7.8	58	7	1835.25	4.0
20	2006/11/15	46.7	154.3	8.4	2007/01/13	46.2	154.8	8.2	70	82	58.71	2.7
21	2007/09/12	−3.8	101.0	8.6	2007/09/12	−2.5	100.1	7.9	176	11	0.53	1.2
22	2007/09/12	−3.8	101.0	8.6	2010/10/25	−3.7	99.3	7.9	189	8	1139.15	1.1

Sequences of micro-earthquakes in the Parkfield area (Nadeau *et al.* 1995; Nadeau and Johnson 1998) exhibit certain properties of characteristic, quasi-periodic earthquakes: regularity of occurrence and nearly identical waveforms. However, these micro-events are not characteristic earthquakes in a strict sense: they do not release almost all the tectonic deformation on a fault segment while real characteristic earthquakes are assumed to do so (Kagan *et al.* 2012). As with large characteristic earthquakes, attempts using micro-earthquake quasi-periodicity for their prediction have not yet been successful. While retrospective tests indicated that the forecasts based on recurrence times were better than random, the forward test results were unsuccessful or inconclusive. Moreover, these micro-events apparently occur on isolated asperities surrounded by creeping zones. Therefore, such an asperity exists as a secluded entity and may produce characteristic quasi-periodic rupture events, similar to fracture in many laboratory experiments. But tectonic earthquakes occur in an environment that is never isolated. This may explain their ubiquitous clustering in time and space.

The quasi-periodic model for characteristic earthquakes (see, for example, McCann *et al.* 1979; Bakun and Lindh 1985; Nishenko 1991) has been put to a prospective test using new data. Testing the validity of this hypothesis ended in failure (Rong *et al.* 2003 and its references; Bakun *et al.* 2005; Jackson and Kagan 2006). Kagan and Jackson's (1999) results and Table 6.1 may explain why testing of the seismic gap model predictions (see also Section 2.4) demonstrates that large earthquakes occur more often in red (recent occurrence of large events – low earthquake potential) rather than in green zones (McCann *et al.* 1979).

How can we investigate earthquake occurrence patterns in various tectonic zones? Our statistical studies of seismicity (see for example, Kagan 1991b; Kagan and Jackson 1991b; Kagan *et al.* 2010; see Section 9) are biased to some degree because global earthquake rate is dominated by places with a high tectonic strain. Thus, the likelihood function maximum for our stochastic model (see more in Section 9.4.4) is mainly determined by these, mostly subduction zones, earthquakes. However, we should study earthquake patterns in continental areas (active and non-active) where the earthquake rate is low, but the vulnerable human population is large (see Stein and Liu 2009; Parsons 2009). A naïve extrapolation of the aftershock sequence rates by assuming that the present rate would continue, may exaggerate the seismic hazard. Instead, we need a convincing tool to produce a truer estimate.

6.4 Aftershock temporal distribution, theoretical analysis

6.4.1 Lévy distribution

Kagan (1982, see also Subsection 3.3.2) proposed a heuristic model for an earthquake fracture as follows. At the moment that any given earthquake rupture stops, the shear stress near the edge of the rupture is lower than the critical breaking stress for extension of the fracture. The subsequent stress history near the earthquake rupture tip depends on other fractures in the neighborhood and additional random factors. In this case, the time history of the stress might resemble a Brownian random walk. The stress-time function is thus given as a solution to the diffusion equation. When the shear stress reaches a critical threshold level, a new earthquake begins. The PDF of the time intervals of the "first passage" (Feller 1971) is the Lévy distribution

$$f(t) = \frac{\sigma}{\sqrt{2\,\pi D\, t^3}} \exp\left(-\frac{\sigma^2}{2Dt}\right), \qquad (6.10)$$

where D is the diffusion coefficient, t is the time, and σ is the threshold or barrier stress: the difference between initial breaking stress and the stress when previous rupture ceases. This distribution as a function of stress is the Rayleigh law (see Eq. 9.10); as a function of time it is the Lévy distribution (cf. Zolotarev 1986). In this model, the shear stress is taken to be a scalar, which corresponds to the addition of aligned stress tensors; the interaction of misaligned tensors is briefly discussed in Subsection 5.3.3 (Figs. 5.8–5.10). The Lévy distribution

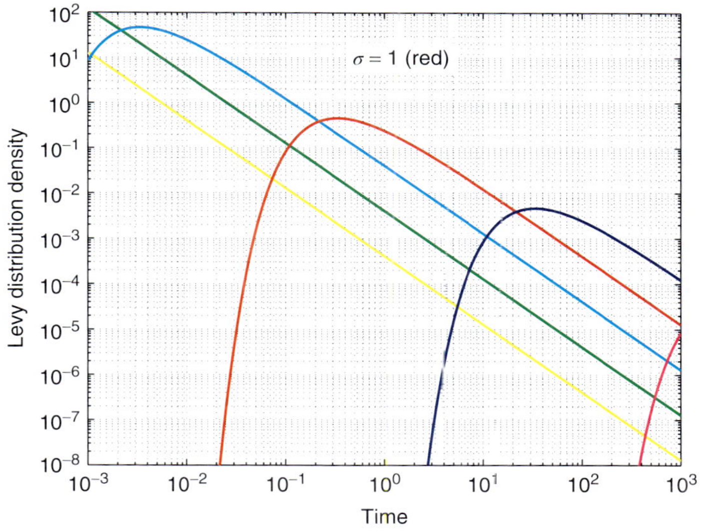

Fig. 6.8 Plot of PDFs for the Lévy distribution (6.10), $D = 1.0$. From left- to right-hand $\sigma =$ 0.001 (yellow), 0.01 (green), 0.1 (cyan), 1.0 (red), 10.0 (blue), 100.0 (magenta). *Source:* Kagan (2011), Fig. 1.

(6.10) was used by Kagan (1982, Eqs. 8 and 9) to model the time-dependent part of the space-time sequence of micro-earthquakes. Here we address only the time sequence of events.

Figure 6.8 displays several Lévy curves for various values of stress drop, σ. The curve's tail is a power-law similar to that exhibited by Omori's law. However, for small values of time the curves decay to zero: depending on the stress drop, Brownian motion takes time to reach a critical stress level. This latter feature, though observed in aftershock sequences, is most likely caused by effects from coda waves of both the mainshock and stronger aftershocks that prevent the identification of smaller aftershocks (Kagan 2004; Section 6.2). Measurements of total seismic moment rate decay in aftershock sequences (Kagan and Houston 2005; Section 6.2) suggest that aftershocks start right after the end of a mainshock rupture (within one minute for a *m*6.7 earthquake). Therefore, the left-hand decay of the curves in Fig. 6.8 is not realistic.

There is another problem with the comparison of theoretical distributions with aftershock observations. The Lévy curves may explain the inter-event time statistics only for the first generation (parent-child) of clustered earthquakes. Most observed aftershocks or clustered events are generally separated from their "parents" by multiple generations. For example, the present aftershocks of the 1811–12 New Madrid earthquakes or the 1952 Kern County, California, earthquake are mostly distant relatives of their progenitors (mainshock and earlier aftershocks). According to Omori's law (6.1–6.2), we count all aftershocks, disregarding their parentage, whereas the Lévy law describes the time distribution for the *next* event: when the shear stress level reaches a critical value to trigger a new rupture. When we count all events, we decrease the power-law exponent from 1.5 (as in Eq. 6.10) to 1.0 (Kagan and Knopoff 1987a).

We assume that the next earthquake may occur in any place in the focal zone of a previous event. Attempts to localize the initiation of strong earthquakes have not been successful; for example, the 2004 Parkfield earthquake occurred well outside the previously specified area (Jackson and Kagan 2006). We also assume that the stress drop after an earthquake has a power-law distribution density:

$$f(\sigma) \propto \sigma^{-1-\psi}, \text{ for } 0 \leq \psi < 1.0. \tag{6.11}$$

Kagan (1994b), Marsan (2005), Helmstetter *et al.* (2005), Lavallée (2008), Powers and Jordan (2010) support this statement.

Then for the modified Lévy distribution in earthquake focal zone we obtain

$$\phi(t) \propto \int_0^\infty f(t)\, \sigma^{-1-\psi} d\sigma = \frac{1}{2\, t\, \sqrt{\pi}\, (2\, t\, D)^{\psi/2}}\, \Gamma\left(\frac{1-\psi}{2}\right), \tag{6.12}$$

where Γ is a gamma function. The new Lévy PDF, as it depends on time t, is a power-law with the exponent $-(1+\psi/2)$. The density should be truncated or modified at the left side, because aftershocks cannot be observed at short-time intervals close to a mainshock (Kagan 2004; Section 6.2). Normalizing the distribution would depend on this truncation. Kagan (1991b, Eq. 8) introduced a minimum time (t_M), where M is a seismic moment of a preceding event (see also Eq. 9.5). Before that time, the aftershock rate is measured with a substantial under-count bias (Section 6.2).

6.4.2 Inverse Gaussian distribution (IGD)

As Matthews *et al.* (2002) indicate, the Inverse Gaussian Distribution (IGD) has been known since 1915. However, this distribution only acquired its name in 1945. Kagan and Knopoff (1987a) proposed using this distribution (without calling it IGD) to describe the inter-earthquake time distribution. Their major impetus was to explain aftershock statistics (Omori's law) by stress dynamics.

Here we discuss the appropriate pairwise interval law for a model in which a steadily increasing source of stress, which we take to be due to plate tectonics, is added to or subtracted from a random or diffusion component, that is, Lévy distribution (6.10). If the rate of tectonic loading is a constant V, the distribution density $f(t)$ is modified to become the Inverse Gaussian distribution (Feller 1971)

$$f(t) = \frac{\sigma}{\sqrt{2\,\pi D\, t^3}} \exp\left[-\frac{(\sigma - V\, t)^2}{2D\, t}\right]. \tag{6.13}$$

For tectonic loading velocity $V = 0$, this equation transforms to (6.10). Figure 6.9 displays several IGD curves for various values of stress drop. For small values of σ and t, the curves are similar to the values of the Lévy distribution (6.10): if the stress drop is insignificant, or the tectonic loading influence is small in the initial stages of stress behavior.

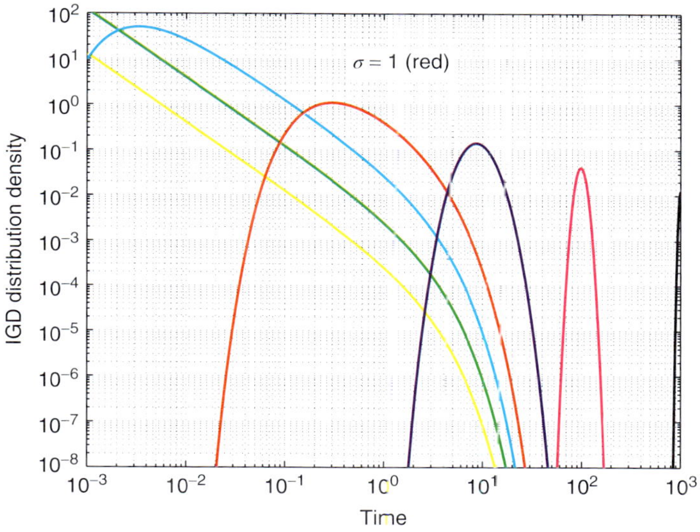

Fig. 6.9 Plot of PDFs for the IGD distribution (6.13), $V = \sqrt{D}$ $D = V^2$. From left- to right-hand $\sigma = 0.001$ (yellow), 0.01 (green), 0.1 (cyan), 1.0 (red), 10.0 (blue), 100.0 (magenta), 1000.0 (black). *Source:* Kagan (2011), Fig. 2.

Assuming again that the stress in a fault neighborhood is distributed according to a power-law (Eq. 6.11), we obtain a new modified distribution density of inter-earthquake time, based on the IGD (6.13)

$$\phi\ (t) \propto \int_0^\infty f(t)\ \sigma^{-1-\psi} d\sigma = \frac{\exp\ [-V^2\ t/(2D)]}{2\ t\ \sqrt{\pi}(2\ t\ D)^{\psi/2}} \left[\Gamma\left(\frac{1-\psi}{2}\right) \right.$$
$$\left. \times\ {}_1F_1\left(\frac{1-\psi}{2},\ \frac{1}{2},\ \frac{V^2\ t}{2D}\right) + V\sqrt{\frac{2t}{D}}\ \Gamma\left(1-\frac{\psi}{2}\right) {}_1F_1\left(1-\frac{\psi}{2},\ \frac{3}{2},\ \frac{V^2\ t}{2D}\right)\right], \tag{6.14}$$

where ${}_1F_1$ is a Kummer confluent hypergeometric function (Abramowitz and Stegun 1972, p. 504; Wolfram 1999). Another expression for the density is

$$\phi\ (t) \propto \frac{\exp\ [-V^2\ t/(2D)]}{2\ t\ \sqrt{\pi}(2\ t\ D)^{\psi/2}} \Gamma\left(\frac{1-\psi}{2}\right) \Gamma\left(1-\frac{\psi}{2}\right) U\left(\frac{1-\psi}{2},\ \frac{1}{2},\ \frac{V^2 t}{2D}\right), \tag{6.15}$$

where U is a confluent hypergeometric function (ibid.). As in Eq. 6.12 , both of the above distributions should be truncated or modified as $t \to 0$ and can be normalized after it.

For certain values of the exponent ψ the expressions (6.14 and 6.15) can be simplified. For example, for $\psi = 0$

$$\phi\ (t)\ \propto\ \frac{1}{2t}\left[1 + \text{erf}\left(V\sqrt{\frac{t}{2D}}\right)\right]. \tag{6.16}$$

For $\psi = 0.5$ we obtain two equations. For a positive V, using Eq. 13.6.3 by Abramowitz and Stegun (1972), we transform (6.14) into

$$\phi(t) \propto \frac{1}{2\,t}\sqrt{\frac{\pi\,V}{4D}}\;\exp\left(-\frac{V^2\,t}{4D}\right)\left[I_{-1/4}\left(\frac{V^2\,t}{4D}\right)+I_{1/4}\left(\frac{V^2\,t}{4D}\right)\right], \quad (6.17)$$

where $I_{-1/4}$ and $I_{1/4}$ are modified **I**-Bessel functions (Abramowitz and Stegun 1972, p. 374; Wolfram 1999, p. 767).

For a negative V using Eq. 13.6.21 by Abramowitz and Stegun (1972), we transform (6.15) into

$$\phi(t) \propto \frac{1}{2t}\sqrt{-\frac{V}{\pi D}}\;\exp\left(-\frac{V^2\,t}{4D}\right)K_{1/4}\left(\frac{V^2\,t}{4D}\right), \quad (6.18)$$

where $K_{1/4}$ is a modified **K**-Bessel function (Abramowitz and Stegun 1972; Wolfram 1999).

In Fig. 6.10 we display the new IGD curves for various values of the ψ parameter. Although the general behavior of the curves remains power-law, the curves change their slope at the time value of about 1.0. The curves with $V = \pm\sqrt{D}$ show the distribution difference for the tectonic loading sign; in the positive case tectonic movement is opposite to the fault displacement during an earthquake, whereas the negative sign corresponds to motion consistent with the earthquake mechanism. In the latter case, random fluctuations can bring the fault to rupture only during the early period of development.

To show the difference of curve slopes more clearly, the PDF values of Fig. 6.10 are multiplied by $t^{1+\psi/2}$ in Fig. 6.11. For small time intervals the

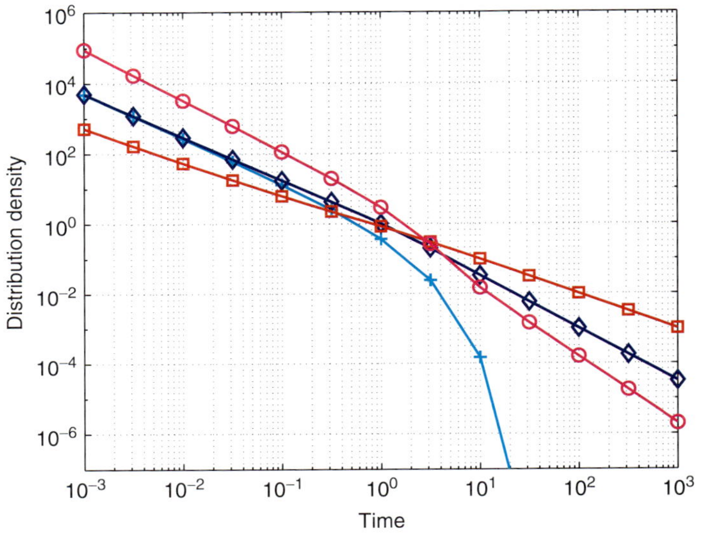

Fig. 6.10 Plot of PDFs for the modified IGD distribution (Eqs. 6.14 –6.18). Notes: Red curve with squares (6.16): $V = \sqrt{D}$, $\psi = 0.0$; blue curve with diamonds (6.17): $V = \sqrt{D}$, $\psi = 0.5$; cyan curve with crosses (6.18): $V = -\sqrt{D}$, $\psi = 0.5$; magenta curve with circles (6.14): $V = \sqrt{D}$, $\psi = 0.9$. *Source:* Kagan (2011), Fig. 3.

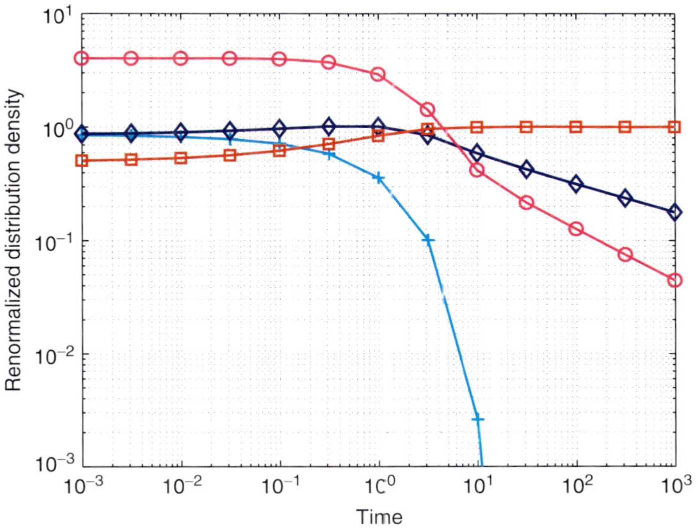

Fig. 6.11 Plot of PDFs for the modified IGD distribution (Eqs. 6.14 –6.18), multiplied by $t^{1+\psi/2}$, $D = 1.0$. Notes: Red curve with squares (6.16): $V = \sqrt{D}$, $\psi = 0.0$; blue curve with diamonds (6.17): $V = \sqrt{D}$, $\psi = 0.5$; cyan curve with crosses (6.18): $V = -\sqrt{D}$, $\psi = 0.5$; magenta curve with circles (6.14): $V = \sqrt{D}$, $\psi = 0.9$. *Source:* Kagan (2011), Fig. 4.

curves are horizontal, suggesting that the modified IGD is similar to the Lévy distribution in Eq. 6.12.

For small time values, the power-law exponents in Fig. 6.10 are essentially the same as for the compounded Lévy law (6.12). This can be seen from the modified Bessel function approximations for small values of the argument (Abramowitz and Stegun 1972, Eqs. 9.6.7 and 9.6.9),

$$I_{-1/4}(t) \propto \left(\frac{2}{t}\right)^{1/4} \text{ and } K_{1/4}(t) \propto \left(\frac{2}{t}\right)^{1/4}. \tag{6.19}$$

In the general case, the same result can be obtained with Eq. 13 5.5 by Abramowitz and Stegun (1972). Then for $t \to 0$ Eq. 6.14 transforms into

$$\phi(t) \propto t^{-1-\psi/2}. \tag{6.20}$$

It is obvious from Figs. 6.10 and 6.11 that, except for the $\psi = 0$ curve, the slope of curves for large values of the time increases when compared with t close to zero. Using Eq. 13.1.27 of Abramowitz and Stegun (1972) we add an exponential term in (6.14) within the hypergeometric function ${}_1F_1$. Then for $t \to \infty$ we obtain

$$\phi(t) \propto t^{-1-\psi}, \tag{6.21}$$

(ibid., Eq. 13.5.1). Therefore, in Eq. 6.17 the exponent would be −1.5 (ibid., Eq. 9.7.1), whereas in Eq. 6.18 the power-law term (ibid, Eq. 9.7.2) is multiplied by an exponential decay term

$$\phi(t) \propto t^{-1.5} \exp\left(-\frac{V^2 t}{4D}\right), \tag{6.22}$$

see Figs. 6.10 and 6.11.

6.5 Temporal distribution of aftershocks: Observations

Many observations of Omori's law behavior have been published (see Utsu *et al.* 1995 and its references). There are some problems with these measurements. The standard interpretation of Omori's law is that all aftershocks are caused by a single mainshock. However, the mainshock is often followed by strong aftershocks and those are clearly accompanied by their own sequence of events, and so on. The second problem is that some earthquakes have very few or no aftershocks (see Fig. 4.1); such sequences cannot be included in the naïve study of individual earthquake sequences, but can be analyzed as stochastic processes.

Thus, three techniques can be applied to study the temporal distribution of real earthquakes: (1) traditional, phenomenological techniques based on observing individual aftershock sequences; (2) using statistical moments of earthquake occurrence, considered as a point process; (3) applying stochastic process modeling to infer the parameter values of earthquake temporal interaction.

6.5.1 Aftershock sequences

Beginning with Omori (1894), the temporal distribution of aftershock numbers has been studied for more than one hundred years (Utsu *et al.* 1995). The aftershock rate decay is approximated as t^{-p} with the parameter p value close to 1.0 (ibid.; Kagan 2004, 2011).

But the simple, superficial study of aftershock rate decay often encounters serious problems. First, only relatively rich aftershock sequences can be investigated by direct measurements; if there are too few aftershocks, their properties can be studied only by combining (stacking) many sequences. Second, to isolate individual sequences, one should exclude any cases when one sequence is influenced by another, an arbitrary procedure which may introduce a selection bias. Third, an aftershock sequence often contains one or several large events which are clearly accompanied by a secondary aftershock sequence. Taking the influence of secondary earthquakes into account is not simple (see Section 9.4.4 for more detail). Fourth, some sequences start with a strong foreshock which is sometimes only slightly weaker than a mainshock. Again, handling this occurrence presents a serious problem.

Therefore, a strong bias may result from directly measuring Omori's law exponents. Two other statistical methods, considered below, enable analysis of the whole earthquake occurrence as a point process, to minimize the problem of data and interpretation technique selection bias.

6.5.2 Temporal distribution for earthquake pairs

Kagan and Jackson (1991a) investigated space-time pairwise earthquake correlation patterns in several earthquake catalogs. They showed that earthquake pairs follow a power-law distribution for small time and distance intervals. Kagan and Jackson (1999) showed that, contrary to the seismic gap model (McCann *et al.* 1979; Nishenko 1991; Matthews *et al.* 2002), the clustering pattern continues for strong ($m \geq 7.5$) earthquakes. Large shallow earthquakes can re-occur after a small time interval and follow the Omori-type temporal distribution. Michael (2011) shows that earthquakes as large as $m8.5$ are clustered in time and space: thus, such a big event does not protect its focal area from the next giant shock.

Table 6.1 shows the location and focal mechanism difference for $m \geq 7.5$ global shallow earthquakes in the GCMT catalog (Ekström 2007; Ekström *et al.* 2012) from 1976–2010. The table format is similar to Table 1 in Kagan and Jackson (1999). However, here we keep only those pairs in the table for which their focal zone overlap (the η-parameter) is greater than 1.0:

$$\eta = \frac{L_1 + L_2}{2R}, \tag{6.23}$$

where R is the distance between the centroids and L_1 and L_2 are the respective rupture lengths for the first and second earthquakes in the pair (Eq. 3 in Kagan and Jackson 1999). The length L can be estimated by using the results of Section 5.5. Therefore, if $\eta \geq 1.0$ the earthquake focal zones would intersect. For several doublets $\eta \geq 2$, implying that the smaller event should be mainly within the focal zone of the larger earthquake. Inspecting the time difference and the 3-D rotation angle between focal mechanisms (see Chapter 8) suggests that these high η pairs may occur after very short time intervals and have very similar double-couple mechanisms.

All earthquakes in the table occur in subduction (trench) zones as defined in Kagan *et al.* (2010). However, even with relatively high deformation velocity at these plate boundaries, the inter-earthquake time is in most cases substantially lower than the time necessary for tectonic motion to restore the critical stress conditions by the occurrence time of the second earthquake (see the last column in Table 1 by Kagan and Jackson 1999).

Figure 6.12 shows how the normalized number of $m \geq 6.5$ shallow earthquake pairs depends on the tectonic deformation rate as defined by Bird *et al.* (2010). Three curves are shown: all earthquakes from the GCMT catalog, earthquakes from subduction or trench zones (Kagan *et al.* 2010), and events from active continental zones. As expected, earthquakes in the trench zones occur in higher rate zones compared to all earthquakes, while most (about 80%) of the earthquake pairs in active continents are concentrated in zones with the deformation rate less than 20×10^{-20} events/(m^2 s).

In the following displays we would like to show the distribution of temporal intervals between earthquakes when measured in a catalog of a limited duration, T. Figure 6.13 shows the integration domain for calculating earthquake pair rates. Figure 6.13 is a square with a side length equal to a catalog duration, T; since the plot is symmetric, only the lower-right portion of the square is shown. The first event shown as a filled circle, is supposed to be at the square diagonal, the second one at the end of a hatched area. We assume that the time difference between earthquakes cannot be less than t_0 (similar to t_M in Eq. 8).

For the Poisson process, the interval pair density is uniform (see Kagan and Jackson 1991a, Eq. 1). Thus, the rate is proportional to the hatched area. For the

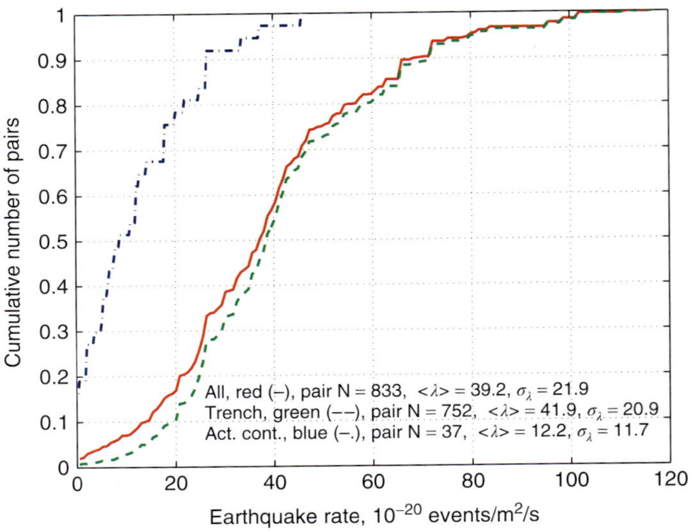

Fig. 6.12 Normalized cumulative distribution of earthquake rate for $m \geq 6.5$ shallow earthquake pairs in all zones, trench (subduction) zones, and active continental zones. The pair numbers N, average rate $< \lambda >$ and its standard deviation σ_λ are also shown. Earthquake rate is taken from a table by Bird *et al.* (2010) for a magnitude threshold $m_t = 5.66$. *Source:* Kagan (2011), Fig. 5.

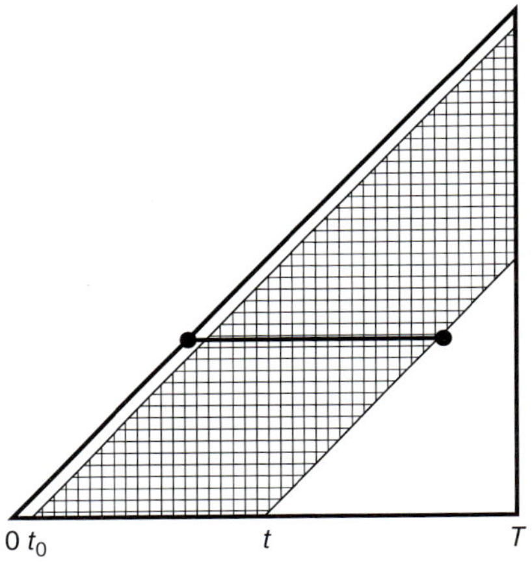

Fig. 6.13 Integration domain for calculating inter-earthquake rates in a catalog of duration T. The minimum time interval, t_0, corresponds to the coda duration time of the first event. *Source:* Kagan (2011), Fig. 6.

normalized survival rate

$$n_p = \left(\frac{T - t}{T - t_0}\right)^2, \tag{6.24}$$

where t_0 is the minimum time interval, T is the catalog duration and t is the inter-earthquake time interval.

For the power-law time distribution with distribution density $\phi(t) \propto t^{-1-\theta}$, we obtain the normalized survival rate by integrating over the domain shown in Fig. 6.13:

$$n_\theta = \left(\frac{t^{-\theta} - T^{-\theta}}{t_0^{-\theta} - T^{-\theta}}\right). \tag{6.25}$$

Unfortunately, the number of $m \geq 7.5$ earthquakes shown in Table 6.1 is too small to find the variation of the earthquake pair time differences in various tectonic zones. Such an analysis can be performed only for $m \geq 6.5$ earthquakes. For such events, their spatial separation becomes comparable with the location errors ($\sim$20 km) in the GCMT catalog (Kagan and Jackson 1999; Kagan *et al.* 2010). Using results from Fig. 5.15 we calculate the rupture size for $m6.5$ earthquake $\ell \approx 42$ km and from Eq. 6.23 we obtain $\eta \geq 0.84$. The results are less reliable when rupture areas for earthquake pairs can be superimposed due to the location errors.

However, even for $m \geq 6.5$ events the pair numbers are usually too small, especially for longer time intervals, to carry out rigorous statistical procedures. A simulation or a bootstrap procedure can be used as an alternative method to evaluate uncertainties. However, our model of earthquake occurrence is a branching process governed by fractal distributions. A simulation has not yet been developed in such a case (see more discussion in Kagan 2007a; Chapter 7, as well as in description of Fig. 9.6). Thus, we use a largely qualitative analysis of the pairs' temporal distributions.

In Figs. 6.14–6.17 the temporal distribution of the inter-earthquake times for all $m \geq 6.5$ event pairs in the GCMT catalog is shown as it depends on earthquake rate determined by Bird *et al.* (2010). Several normalized approximations for pair intervals are also given: the temporal Poisson distribution of earthquakes (Eq. 6.24) in the time span 1976–2010 and several power-law interval (Eq. 6.25) dependencies.

The observational distribution curves consist of two parts: for small time intervals, they follow a power-law and for larger intervals the distribution is parallel to the Poisson rate. Similar results for smaller earthquakes are obtained by Touati *et al.* (2009). The transition from one behavior to another occurs sooner for zones

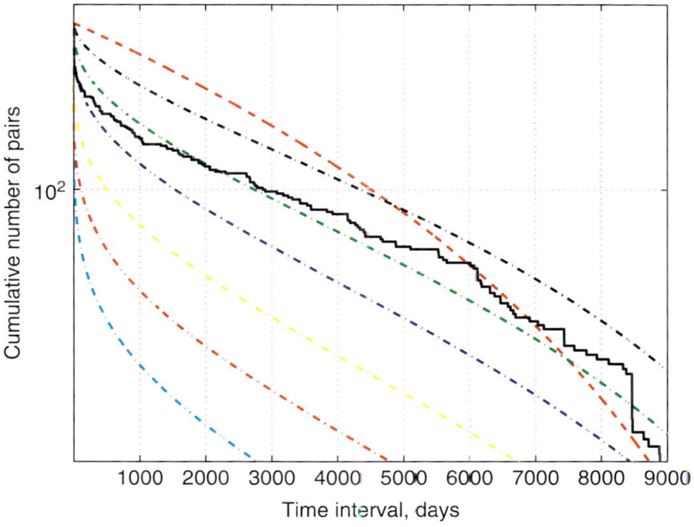

Fig. 6.14 Earthquake pair numbers in all zones with rates $0 - 26 \times 10^{-20}$ events/(m^2 s). Earthquake rate is taken from a table by Bird *et al.* (2010) for a magnitude threshold $m_t = 5.66$. Pair number is 240. The maximum epicentroid distance is 50.0 km. The solid curve is the earthquake pair numbers; the dashed red curve is the Poisson approximation (6.24), dash-dotted curves are for power-law approximations (6.25), the θ-value is 0.5, 0.65, 0.75, 0.85, 0.925, and 0.99 from top to bottom. The catalog's time interval is 1976/1/1–2010/11/14. Average recurrence time ($\bar{t}$) for earthquakes is $\bar{t} = 3272 \pm 3596$ days. The coefficient of variation C_v is 1.10. *Source:* Kagan (2011), Fig. 7.

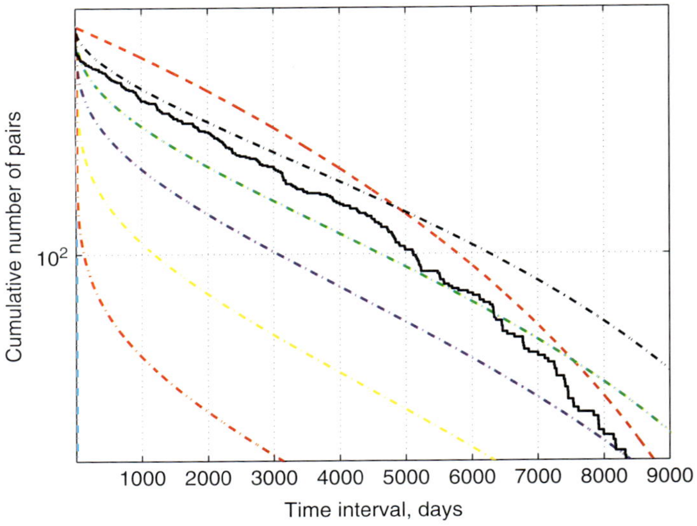

Fig. 6.15 Earthquake pair numbers in all zones with rates $26 - 45 \times 10^{-20}$ events/(m^2 s). Pair number is 333. Average recurrence time ($\bar{t}$) for earthquakes is $\bar{t} = 3507 \pm 3242$ days. The coefficient of variation C_v is 0.925. For notation, see Fig. 6.14. *Source:* Kagan (2011), Fig. 8.

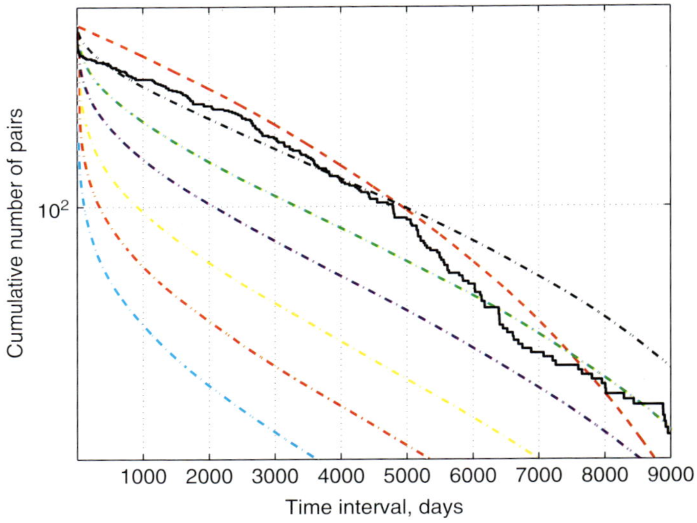

Fig. 6.16 Earthquake pair numbers in all zones with rates $\geq 45 \times 10^{-20}$ events/(m^2 s). Pair number is 264. Average recurrence time ($\bar{t}$) for earthquakes is $\bar{t} = 3947 \pm 3319$ days. The coefficient of variation C_v is 0.841. For notation, see Fig. 6.14. *Source:* Kagan (2011), Fig. 9.

with a higher tectonic deformation rate: by the curves inspection we find that in Fig. 6.14 the transition is observed for the time of about 6000 days; in Fig. 6.15 it is about 4000–5000 days, and in Fig. 6.16 it is less than 3000 days.

Figure 6.17 shows the time interval distribution for active continental areas (Kagan *et al.* 2010). In this case a visual inspection suggests that the best approximation is the power-law; no transition to the Poisson rate is observable. This absence can be explained by a low deformation rate in these zones (see Fig. 6.12). As was observed for many aftershock sequences in continental and slowly deforming areas, they continue according to Omori's law for decades and even

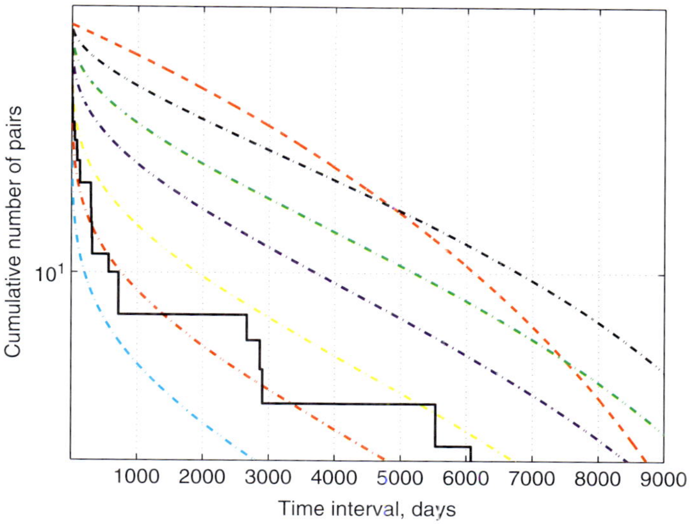

Fig. 6.17 Earthquake pair numbers in active continental zones. Pair number is 37. Average recurrence time ($\bar{t}$) for earthquakes is $\bar{t} = 1304 \pm 2578$ days. The coefficient of variation C_v is 1.977. For notation, see Fig. 6.14. *Source:* Kagan (2011), Fig. 10.

centuries (Utsu *et al.* 1995; Ebel 2009; Stein and Liu 2009). Here the span of 34 years covered by the GCMT catalog is likely insufficient to demonstrate the transition from an aftershock sequence to a background, Poisson rate.

We compute an average recurrence time (t) for earthquakes in Figs. 6.14–6.17. The smallest value for t is observed for distributions with a significant component of the power-law: Figs. 6.14 and 6.17. We also calculate the coefficient of variation (C_v) of earthquake inter-occurrence time as a ratio of the standard deviation (σ_t) to the average time $\bar{t}$ (Kagan and Jackson 1991a, Fig. 1). A completely random Poisson occurrence has the coefficient of variation equal to one, whereas quasi-periodicity yields a coefficient of less than one. For clustered earthquakes the coefficient is larger than one. Although the C_v estimates are biased downwards when determined at a relatively short catalog time span, their mutual relations are indicative of occurrence patterns. For Figs. 6.14–6.17, the C_v-values are 1.10, 0.925, 0.841, and 1.977, respectively. These evaluations again suggest that earthquakes in areas with a smaller tectonic rate become more clustered, and their Poisson component is diminished.

6.6 Example: The New Madrid earthquake sequence of 1811–12

To illustrate the arguments and results of the previous sections, we calculate seismicity parameters of the New Madrid earthquake sequence (1811–12) and its consequences. There is substantial literature on this sequence (Hough 2004; Stein and Liu 2009; Calais *et al.* 2010, and their references).

Three or four large earthquakes with magnitudes on the order of 7.3–7.8 occurred over a few months of 1811–12 in the New Madrid area; aftershocks of these events are still registered. As an illustration, we would assume that only one $m8$ event occurred at that time. If in reality earthquakes were smaller than such an event, their total focal zone and combined aftershock sequence at the present time would be equivalent to about one $m8$ mainshock (Section 5.5). The size of the focal zone can be evaluated by using regression equations in Figs. 5.15 and 5.16. The first plot contains many earthquakes but most of these events are in subduction zones. The second diagram uses earthquakes in active continental zones; the focal size of these earthquakes is likely to resemble the New Madrid area which can be classified as plate-interior (Kagan *et al.* 2010). Too few large earthquakes are available within the plate-interior to obtain their features. The difference between regression parameters in Figs. 5.15 and 5.16 is small; therefore the size of earthquake focal zones either does not change in various tectonic zones or changes only slightly.

For an $m8$ earthquake, calculations yield 227 km and 259 km as the length of the focal zone, defined as the 4σ major axis of an ellipse comprising a majority of aftershocks (Kagan 2002b). The linear regression coefficients are used in both figures: the former value corresponds to Fig. 5.15 and the latter to Fig. 5.16. The two estimates are similar and roughly correspond to the size of the present aftershock zone, as shown, for example, in Calais *et al.* (2010). For the Cottonwood fault that is about 110–120 km length (ibid., Fig. 1), the magnitude estimate from Fig. 5.16 is $m \sim 7.2–7.4$.

To calculate the surface area of an aftershock zone, we assume that the minor axis of the ellipse is 1/4 of the major axis, taken as 240 km. Then we obtain the $m8$ earthquake focal area as 11300 km^2 (Eqs. 5.40 and 5.41). Taking the stationary strain rate as $\dot{\epsilon} = 10^{-9}\text{yr}^{-1}$ (Calais *et al.* 2006), we compute the tectonic moment rate (Eq. 5.8): 3.4×10^{15} Nm/year. Assuming that 50% of the tectonic rate is released seismically (Bird and Kagan 2004), we obtain the background rate $\alpha_0 = 1.27 \times 10^{-3}$ $m \geq 5$ earthquakes per year (we take in Eq. 5.9 $M_c = 10^{21}$ Nm, $W = 20$ km, $\beta = 2/3$; then $\Gamma(4/3) = 0.893$).

We use Eq. 5.5 to calculate the recurrence time for an $m \geq 8$ earthquake in the focal zone of the New Madrid events: more than two million years. In this computation any $m8$ earthquake with an epicenter or centroid in the focal zone counts: in Eq. (5.9) we do not request that the entire rupture of such an earthquake be contained in the zone. The recurrence time is an average value; even for events as large as $m \geq 8$ the earthquake occurrence is clustered (see Section 6.5.2). Thus, a new large earthquake can follow after a relatively short time period, as exemplified by the 1811–12 New Madrid sequence.

A similar calculation of the $m8$ earthquake rate could be carried out by using the results of Bird *et al.* (2010, p. 188). They calculate an estimate of the mean intraplate seismicity rate of 4.27×10^{-22} $m^{-2}\ s^{-1}$ for $m_0 = 5.66$ earthquakes. Bird *et al.* use a slightly different formula for the conversion of the scalar moment to magnitude. They use 6.033 instead of our 6.0 coefficient in (2.2). Thus, their estimate would yield the rate for the New Madrid area $\alpha_0 = 7.5 \times 10^{-4}$ $m \geq 5$ earthquakes per year, the value of the same order of magnitude as computed above.

We would like to calculate the duration of an aftershock sequence up until the aftershock rate decays to the background level. The results in Fig. 9.6 can be applied for this purpose. However, we need to make a correction for the mainshock and aftershock magnitudes ($m8$ instead of $m8.15$ and $m \geq 5.0$ instead of $m_b \geq 4.9$ in the plot, respectively). In the diagram the aftershock rate per one interval (the interval duration increases consequently by a factor of two) is 7 events. This translates into 4.55 $m \geq 5$ events per interval for our choice of magnitudes (Kagan *at al.* 2010). After comparing the background and aftershock rates (we take 4.55 $m \geq 5$ aftershocks per the first day, decaying according to Omori's law, with $1/t$ rate with time), a computation shows that the aftershock sequence would reach the background rate in about 3,600 years. This duration estimate agrees roughly with Stein and Liu's (2009) value. In these calculations, we presume that no independent large earthquake clusters would occur during the aftershock sequence. A possible occurrence of spontaneous events makes any evaluation of aftershock sequence duration largely approximate.

Stein and Liu (2009) obtained aftershock duration values for several sequences, using Eq. 14 from Dieterich (1994). This equation employs parameters whose values for actual earthquake focal zones are not known. Generally, the parameters have been adjusted based on the statistics of earthquake occurrence. This may explain an apparently reasonable fit of Dieterich's formula to aftershock sequences.

In contrast, we obtain the aftershock sequence duration by extending the well-known TGR and Omori's laws and using measured geometrical features of tectonic deformation. Furthermore, according to Stein and Liu (2009, Fig. 1c) the New Madrid aftershock rate for the last 50 years was about 0.5 $m \geq 4$ events per year. Computations based on Omori's law similar to those shown above, yield the rate of about 0.26 $m \geq 4$ events per year 175 years after the mainshock occurrence. This number is close to the one evaluated above.

6.7 Conclusion

Since 1894 it has been known that temporal distribution of aftershock sequences is described by Omori's law. However, in the formulation of this dependence (Eqs. 6.1 and 6.2) a c-parameter was introduced. This parameter was often interpreted as having a physical meaning. In Section 6.2 we argue that the seeming saturation of aftershock numbers close to the occurrence time of a mainshock is due to deficiencies in registration and interpretation of seismograms. We also show that the aftershock *moment rate* follows a time decay similar to Omori's law. Therefore, moment release during mainshock and aftershocks represents the same process, confirming our continuum model of earthquake rupture process (Section 3.3).

In Sections 6.3–6.5 we also derived the theoretical time distributions for earthquake occurrence; the aftershock distribution is shown to be controlled by power-laws. How can the parameters of these distributions be determined? If one excludes the interiors of plates, the tectonic deformation rate V is reasonably well known for plate boundaries and for active continents (Bird *et al.* 2010; Kagan *et al.* 2010). The diffusion rate D is presently unknown. If we could obtain the earthquake temporal distribution as shown in Figs. 6.8 and 6.9, the D evaluation would be easy. These distributions are derived for a particular area within an earthquake fault zone. However, if the stress in the focal zone of an earthquake is distributed according to the power-law with an exponent ψ (see Eq. 6.11), the problem becomes more difficult.

Figures 6.10 and 6.11 suggest that the distribution temporal behavior changes when $V = \sqrt{D}$. This change relates to the first generation of offspring. Thus, we should not be able to see it in regular Omori plots which combine many generations of aftershocks. The inversion of earthquake occurrence parameters based on stochastic branching processes yields first generation effects. In the present models (both CBM and ETAS), as discussed in Section 9.4.4, temporal dependence is parameterized by just one exponent. These models should in principle demonstrate changes in the temporal pattern, if a more complicated temporal function is applied. However, the results of statistical analysis are very uncertain even for one-parameter time decay. Given the contemporary quantity and quality of earthquake catalogs, more complicated models are unlikely to be effective in resolving this issue.

New laboratory experiments (Zaiser 2006) may help solve the problem of diffusion rate evaluation, but it is not clear whether such measurements are possible. The acoustic emission event rate exhibits fore- and aftershock sequences associated with the dynamic failure of the test specimen (see, for example, Ojala *et al.* 2004). These and similar tests can be used to infer the dependence of the Omori law parameters on spatial scale, deformation rate, and stress diffusion rate.

The results from the statistical analysis of earthquake occurrence in our previous publications (Bird and Kagan 2004; Kagan *et al.* 2010; Chapter 9), as well as the results reported above, suggest that the earthquake process in all tectonic provinces can be described by the same model. We advocate the Poisson cluster process with clusters controlled by a critical branching process and an Omori-type power-law time dependence. As shown in Sections 6.4–6.5, the transition time from Omori's law decay to a quasi-stationary rate clearly increases with a decrease in the strain rate. As the result of these studies, we propose that the essentially same statistical model can be applied to regions of both fast and slow tectonic deformation. Combined with the earthquake size distribution approximation by the TGR law, such a model allows a quantitative forecast of a spatially variable, time-independent (long-term) earthquake rate. It optimally smoothes the seismicity record (Kagan and Jackson 2011, 2012) or translates the plate-tectonic and geodetic rate into a seismic rate estimate (Bird and Liu 2007; Bird *et al.* 2010). A short-term forecast can be performed by using the temporal properties of earthquake clusters, an extrapolation which uses a variant of Omori's law to estimate future earthquake rate (Kagan and Jackson 2011, 2012). In Section 6.6 we presented an example of such calculations.

7 Earthquake location distribution

7.1 Multipoint spatial statistical moments

Discussion in the chapter follows mainly Kagan and Vere-Jones (1996), Kagan (2007a), as well as several of our earlier papers. In the early 1980s, Kagan and Knopoff (1980) and Kagan (1981a,b) investigated the spatial moment structure of earthquake hypo- and epicenters for global (PDE and others) and several local earthquake catalogs.

The two-, three-, and four-point statistical moment functions were obtained and analyzed in these studies. The quality and quantity of the earthquake data were relatively poor at that time, and computer capacity was sufficient to study only a small subset of data at a time, especially when computing the higher moments. However, it became apparent that the spatial distribution of earthquakes has a scale-invariant pattern.

Spatial moment functions are basic to the investigation of hypocentral patterns. The principal quantities of the study are the proportions of k-tuples ($k = 2, 3, 4$) points from the catalog (Fig. 7.1) with the property that the maximum distance between any two points in the k-tuple does not exceed r, as a function of r, and the joint density function of the distances between the points forming such a k-tuple.

We write

$$q_k(r) = N_k(r)/N_k, \tag{7.1}$$

where $N_k(r)$ is the number of k-tuples with the stated property, and N_k is the total number of k-tuples from the catalog. The quantities $q_k(r)$ are computed first for the epicenters, as points in $\mathbf{R}^2$, and then for the hypocenters, as points in $\mathbf{R}^3$. This function can be interpreted as the average number of k-tuples within a distance r of an "average" point of the catalog.

To overcome the biases in such estimates, which arise from boundary effects, the ratios $q_k(r)$ may be compared to the corresponding values for simulated Poisson catalogs. The simulated catalogs are of the same size and extent as the original catalog, but their epicentral coordinates are uniformly distributed over the region, and the depth distribution is matched to that of the actual catalog.

Earthquakes: Models, Statistics, Testable Forecasts, First Edition. Yan Y. Kagan.

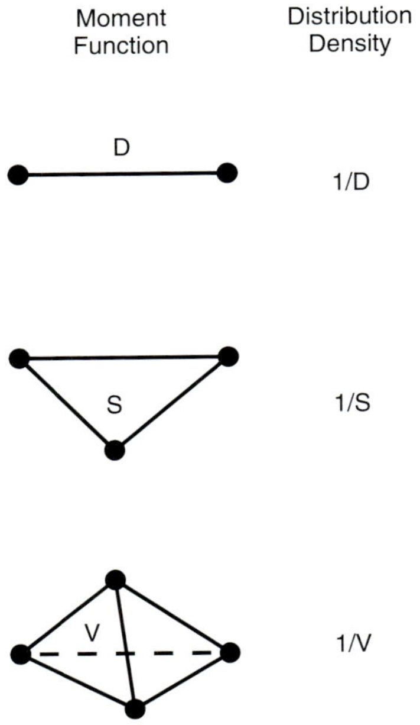

Fig. 7.1 Schematic representation of 2-, 3-, and 4-point spatial moment functions and their suggested approximate densities. *Source:* Kagan and Vere-Jones (1996), Fig. 2.

This results in the ratios

$$Q_k(r) = q_k(r)/\tilde{q}_k(r), \tag{7.2}$$

where the tilde refers to the simulated catalog. Values of $Q_k(r)$ have been tabulated and graphed in various ways. Kagan (1981a,b) shows graphs for 3-point $Q_3(r)$ and 4-point $Q_4(r)$ moment functions.

The graphs of the ratios $Q_k(r)$ against distance r typically display three ranges: the initial range, the middle range over which the $1/r$ behavior is observed, and the final range, in which the ratio approaches 1.0, as r approaches the diameter of the observed region. We interpret the first range as dominated by measurement errors. The second range illustrates self-similar behavior, and the third range is dominated by boundary effects. Kagan (2007a, see also Section 7.2) provides extensive analysis of various errors and biases in the 2-point moment evaluation.

Thus, our key results can be summarized as follows. The growth rates of the moment functions are consistent with a dimensional deficit of approximately 1.0. With an order of magnitude over different radial and angular combinations:

(i) the distribution of pairs of points selected at random from the catalog is consistent with the density inversely proportional to the distance $1/D$ (Kagan and Knopoff 1980; Kagan 1991a, 2007a) – this distribution exponent is called the correlation dimension δ, it is extensively analyzed in the following sections of this chapter;

(ii) the distribution of triplets of points selected at random from a catalog is consistent with the density inversely proportional to the area of a triangle, $1/S$ (Kagan 1981a);

(iii) the distribution of quadruplets of points selected at random from a catalog is (for the hypocenters only) consistent with the density inversely proportional to the volume of a tetrahedron, $1/V$ (Kagan 1981b).

7.2 Sources of error and bias in estimating the correlation dimension

There is an extensive bibliography on statistical estimation of the fractal spatial dimension (for example, Smith 1988; Nerenberg and Essex 1990; Ogata and Katsura 1991; Pisarenko and Pisarenko 1995; Vere-Jones *et al.* 1997; Harte 1998, 2001; De Luca *et al.* 1999; De Luca *et al.* 2002; Vere-Jones 1999; Molchan and Kronrod 2005, 2009; Kagan 2007a; and their references). However, these publications insufficiently consider the systematic effects which significantly influence estimation of the fractal or scaling dimensions for earthquakes. Many of these publications consider methods for estimating correlation dimension with little quantitative discussion of various biases, which, as we see later, may notably alter the dimension value even if evaluated by a seemingly efficient statistical method. Some of the above publications estimated several effects by using synthetic catalogs. As we discuss later in this section, such simulations are insufficient for fully understanding various biases.

Smith (1988) and Nerenberg and Essex (1990) considered the boundary effect by calculating the correlation dimension for a hypercube or a hypersphere. These region geometries are not often encountered in earthquake statistics. De Luca *et al.* (1999, 2002) investigated boundary bias and location errors influence by simulation. Harte (1998, p. 602, see also Harte 2001, his Section 11.3) discusses several effects that may cause bias in the correlation dimension estimates: "boundary effect, lacunarity, rounding effect, and noise or location error," but again these effects are only analyzed by simulation. We consider lacunarity or intermittency of spatial earthquake distribution as a natural consequence of its statistical self-similar pattern, for other biases analytic formulas are derived below to describe their influence.

These systematic effects are largely caused by geometrical factors. Many stochastic geometrical problems for point distribution in various figures are considered in http://mathworld.wolfram.com/topics/GeometricConstants.html (see also Wolfram 1999).

For some of these geometrical distributions we need to make assumptions which are not fully realistic. As a rule, such assumptions are needed to obtain an analytic result. More general distributions can be obtained by simulations. However, theoretical expressions are still useful: first, they often are sufficient to understand the extent of an error or a systematic effect influence on measurement results; second, the closed-form solutions are needed to check simulation programs which may contain hidden flaws. The theoretical expressions below are usually represented in a form accessible to computation by standard software packages: FORTRAN, MATLAB, and MATHEMATICA.

7.2.1 The number of earthquakes in a sample

Clearly the δ-value cannot be determined for distances less than the smallest distance (r_{min}) between points. Strictly speaking, the fractal dimension of a point set is zero (Vere-Jones 1999), but in catalog measurements $\delta = 0$ for for distances $r \leq r_{min}$ and should increase for greater distances. Nerenberg and Essex (1990) call this effect "depopulation" and estimate the critical minimum distance as

$$r_{min} = 2R \times (1/N)^{1/D}, \tag{7.3}$$

where N is the number of points in the cube of size $2R$, and D is the dimension of embedding Euclidean space.

Modern earthquake catalogs contain thousands and tens of thousands of events. Thus, unless we are interested in the distribution of only the strongest earthquakes or places of low seismicity level, samples are well "populated." As we will see, the correlation dimension estimate is more likely to be strongly biased by other effects.

7.2.2 Earthquake location error

The location error randomizes earthquake positions and, as a result, the correlation dimension estimate δ approaches space dimension D for distances less than or comparable to the average location uncertainty. In early catalogs, earthquakes and especially aftershocks, were often assigned a common epicenter and depth. This designation reduces the δ-value (an example is shown in Fig. 7.10).

We assume that earthquake location errors are normally distributed and that errors for both earthquake points are independent of each other. The obtained earthquake locations for close events are most likely influenced by similar biases, hence they are not statistically independent. However, in this work we measure the pairwise distances between earthquakes, thus only relative errors influence the result.

The distribution of distance r between the two epicenters or hypocenters whose actual separation is ρ obeys a "non-central χ-distribution" (Fisher 1928; Kendall and Moran 1963, their Chapter 5.19). Below we analyze this distribution for 3-D and 2-D earthquake spatial patterns (hypocenters and epicenters).

7.2.2.1 Earthquake location error: the 3-D case

Horizontal error is usually much smaller than vertical error. However, the general case of unequal errors cannot be analytically solved. If we assume for simplicity in a 3-D case that horizontal and vertical location errors (σ) are equal, the probability density of the 3-D non-central χ-distribution for hypocenters is

$$\phi(r|\rho) = \frac{r}{\rho\sqrt{2\pi}}\left[\exp -\frac{(r-\rho)^2}{2} - \exp -\frac{(r+\rho)^2}{2}\right]. \tag{7.4}$$

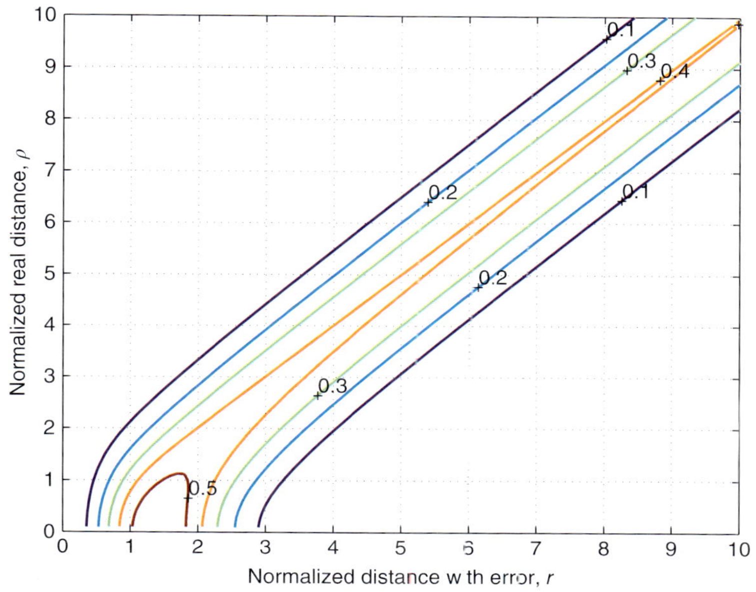

Fig. 7.2 Density of non-central χ-distribution in 3-D. *Source:* Kagan (2007a), Fig. 2.

Both distances r and ρ are scaled by σ. If $\rho \to 0$, the distance distribution becomes the Maxwell law, with probability density function

$$\phi(r) = r^2 \sqrt{\frac{2}{\pi}} \exp\left(-\frac{r^2}{2}\right). \tag{7.5}$$

The Maxwell law corresponds to the distribution of vector length in three dimensions, if the components of a vector are statistically independent and have a Gaussian distribution with the zero mean and the standard error σ (because of scaling, we take $\sigma = 1$). For large ρ distribution (7.4) becomes the Gaussian one (Fisher 1928, p. 664). The density is shown in Fig. 7.2.

If we assume that hypocenters are distributed over a fractal set with the correlation dimension δ_h, then the number of events in a sphere of the radius R, centered on one of the points, that is, the number of pairs in the 3-D space, $N_3(R)$, is

$$\begin{aligned} N_3(R) &\propto \int_0^R dr \int_0^\infty \phi(r|\rho)\, \rho^{\delta-1}\, d\rho \\ &= \frac{2^{\delta/2}}{\sqrt{2\pi}}\, \Gamma\left(\frac{\delta}{2}\right) \int_0^R r^2 \exp\left(-\frac{r^2}{2}\right) {}_1F_1\left(\frac{\delta}{2}, \frac{3}{2}, \frac{r^2}{2}\right) dr, \end{aligned} \tag{7.6}$$

where Γ is a gamma function and ${}_1F_1$ is the Kummer confluent hypergeometric function (Abramowitz and Stegun 1972; Wolfram 1999).

For $\delta = 3$ the function ${}_1F_1\left(\frac{3}{2}, \frac{3}{2}, \frac{r^2}{2}\right)$ is $\exp\left(\frac{r^2}{2}\right)$, and for $\delta = 2$

$$ {}_1F_1\left(1, \frac{3}{2}, \frac{r^2}{2}\right) = \frac{1}{r}\sqrt{\frac{\pi}{2}}\exp\left(\frac{r^2}{2}\right)\mathrm{erf}\left(\frac{r}{\sqrt{2}}\right), \tag{7.7} $$

where erf(R) is an error function. Then

$$ N_3(R) \propto \int_0^R r\,\mathrm{erf}(r)\,dr = R\sqrt{\frac{2}{\pi}} \times \exp[-R^2/2] + \mathrm{erf}\left(\frac{R}{\sqrt{2}}\right)[R^2 - 1]. \tag{7.8} $$

The ratio $RN_3(R)/V_R$ can be estimated as

$$ RN_3(R)/V_R = \frac{2}{R^2}\,N_3(R) = \sqrt{\frac{2}{\pi}} \times \frac{\exp[-R^2/2]}{R} + \mathrm{erf}\left(\frac{R}{\sqrt{2}}\right)[1 - R^{-2}], \tag{7.9} $$

where V_R is a sphere volume. For other values of the dimension δ, the integrals (7.6) and Eq. (7.14) below can be estimated numerically (Wolfram 1999).

Using (7.8) we estimate the correlation dimension for $\delta = 2$ in the presence of location errors as

$$ \hat{\delta} = \frac{\partial\,[\log N_3(R)]}{\partial\,(\log R)} = \frac{2\,R^2\,\mathrm{erf}\left(\frac{R}{\sqrt{2}}\right)}{N_3(R)}. \tag{7.10} $$

For large distances $\hat{\delta} \to 2$. For small values of the argument

$$ \mathrm{erf}\left(\frac{R}{\sqrt{2}}\right) \approx R\sqrt{\frac{2}{\pi}} \times \exp[-R^2/2] \times \left[1 + \frac{R^2}{3}\right], \tag{7.11} $$

(Abramowitz and Stegun 1972, Eq. 7.1.6). Hence putting (7.11) in (7.10), we obtain that for small distances $\hat{\delta} \to 3$.

In Fig. 7.3 we show the dependence of $\hat{\delta}$ on the distance, both for simulation and computation according to Eq. (7.10). In a synthetic catalog, we simulated 200,000 points on a 500×1000 km plane fault and then perturbed the location by adding the Gaussian errors. When calculating the $\hat{\delta}$-value, we compared the correlation function at distances separated by a factor $2^{1/4} = 1.189$.

7.2.2.2 Earthquake location error: the 2-D case

For the 2-D case of epicenters with location errors equal to σ, the non-central χ-distribution density is

$$ \phi(r|\rho) = r\exp\left(-\frac{r^2 + \rho^2}{2}\right)\,I_0\,(r\rho), \tag{7.12} $$

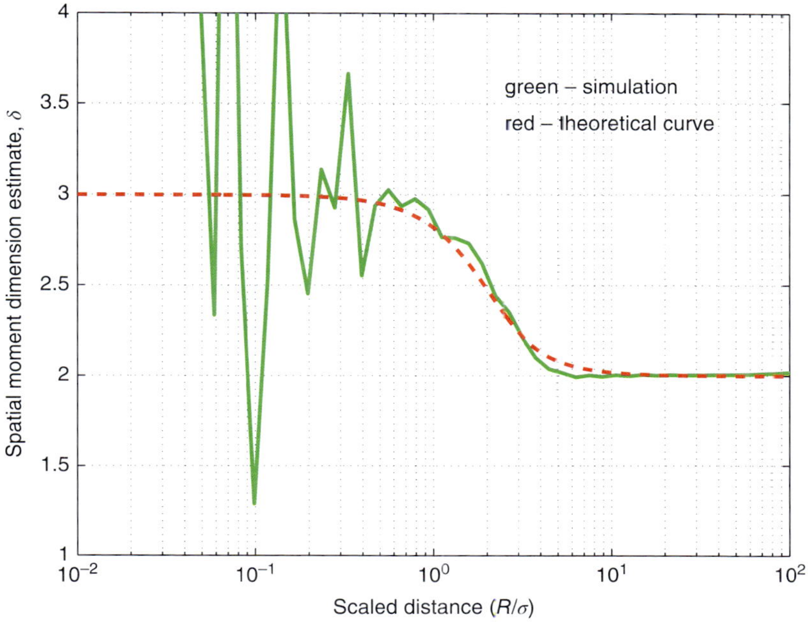

Fig. 7.3 Dependence of the correlation dimension estimate $\hat{\delta}$ on distance scaled with the location error σ in 3-D. Solid line is simulation; dashed line shows $\hat{\delta}$ change according to Eq. (7.10). *Source:* Kagan (2007a), Fig. 3.

where I_0 is the modified I-Bessel function of zero-th order (Wolfram 1999, p. 767; Abramowitz and Stegun 1972, p. 374). If $\rho \to 0$, the distance distribution becomes the Rayleigh law

$$\phi(r) = r \exp\left(-\frac{r^2}{2}\right), \tag{7.13}$$

which is a distribution of vector length in two dimensions (cf. Eq. 7.5) with $\sigma = 1$.

As in the 3-D case, we assume that epicenters are distributed over a set with the fractal dimension δ. Then the number of the events measured over the Earth's surface in a circle of the radius R, again centered on one of the points, that is, the number of pairs in the 2-D space, $N_2(R)$, is

$$N_2(R) \propto \int_0^R dr \int_0^\infty \phi(r|\rho)\ \rho^{\delta-1} d\rho$$
$$= 2^{\delta/2-1}\ \Gamma\left(\frac{d}{2}\right) \int_0^R r \exp\left(-\frac{r^2}{2}\right) {}_1F_1\left(\frac{\delta}{2}, 1, \frac{r^2}{2}\right) dr, \tag{7.14}$$

see Eq. (7.6).

For $\delta = 1$

$${}_1F_1\left(\frac{1}{2}, 1, \frac{r^2}{2}\right) = \exp\left(\frac{r^2}{4}\right)\ I_0\ \left(\frac{r^4}{4}\right), \tag{7.15}$$

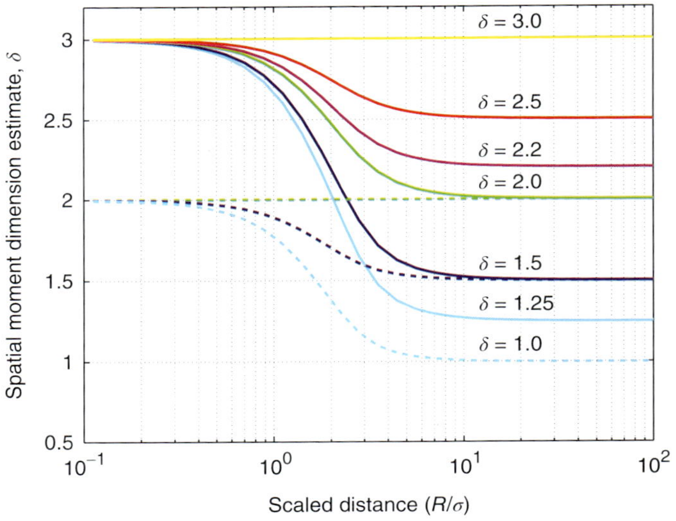

Fig. 7.4 Dependence of the correlation dimension estimate $\hat{\delta}$ on distance scaled with the location error σ in 2-D and 3-D. Solid lines are for the 3-D distribution (hypocenters); dashed lines are for the 2-D distribution (epicenters). The initial fractal point dimension indicated as $\delta = 3$, etc. *Source:* Kagan (2007a), Fig. 4.

and for $\delta = 2$

$$ {}_1F_1\left(1, 1, \frac{r^2}{2}\right) = \exp\left(\frac{r^2}{2}\right). \tag{7.16}$$

In Fig. 7.4 we show the dependence of $\hat{\delta}$ on the scaled distance for hypocenters or epicenters displaced by the Gaussian errors with the standard deviation σ. We compute $\hat{\delta}$ according to Eqs. (7.6) and (7.14). There is little difference between these curves as the underlying dimension of the point pattern changes. In the 2-D case, however, the curves are slightly displaced toward smaller values of the scaled distance.

7.2.3 Projection effect for epicentral scaling dimension

Assuming, for example, that hypocenters cover a fault-plane ($\delta = 2$), the epicentral correlation dimension may fluctuate from $\delta_e = 1$ to $\delta_e = 2$, depending on the dip angle of a fault. Thus, for distances comparable to the thickness of the seismogenic zone, the correlation dimension value should depend on the style of the earthquake deformation pattern. In the general case, projecting a 3-D distribution onto a 2-D plane is called "grading" after Matheron (1971).

For illustrative purposes we have calculated the 2-D spatial moment, assuming both that the original seismicity is isotropic and the 3-D distance pair number function $N_3(R)$ is proportional to R^δ. Then a horizontal layer of seismicity of thickness W is projected on a horizontal plane:

$$G(R) = RN_2(R)/(\pi R^2) \propto (2/R) \int_0^R r dr \int_0^W (W-h)(r^2+h^2)^{(\delta-3)/2}\, dh. \tag{7.17}$$

This integral is a complex expression involving hypergeometric functions.

For integer δ simpler expressions can be found. For $\delta = 1$

$$G(R) = W \left[2 \arctan \left(A^{-1} \right) - \frac{1 - A^2}{A} \log (B) + A \log (A) \right], \tag{7.18}$$

where $A = R/W$ and $B = \sqrt{1 + A^2}$. For $\delta = 2$ we obtain

$$G(R) = W^2 \left[\frac{B - 1}{A} + A \log \frac{B + 1}{A} - \frac{2 \left(B^3 - A^3 - 1 \right)}{3 A} \right]. \tag{7.19}$$

And for $\delta = 3$

$$G(R) = W^2 \frac{A}{2}. \tag{7.20}$$

In this case the ratio (or an estimate of the correlation dimension) does not depend on distance. This is to be expected since projecting uniformly 3-D distributed hypocenters on a surface produces a uniformly 2-D distributed pattern.

Using expressions (7.18–7.20) we can calculate an estimate of the correlation dimension for the grading problem:

$$\hat{\delta} = 1 + \frac{\partial (\log G)}{\partial (\log R)}. \tag{7.21}$$

In particular for $d = 1$

$$\hat{\delta} = \frac{2 A \left[\arctan \left(\frac{1}{A} \right) - A \log \left(\frac{B}{A} \right) \right]}{2A \arctan \left(\frac{1}{A} \right) - A^2 \log \left(\frac{B}{A} \right) + \log (B)}, \tag{7.22}$$

for $d = 2$

$$\hat{\delta} = A^2 \frac{2 \log \left(\frac{1+B}{A} \right) + \frac{6A-4B-3}{3} + \frac{1+B-2A^2B+A^2}{3B(1+B)}}{A^2 \log \left(\frac{1+B}{A} \right) + \frac{2A^3-1+B(1-2A^2)}{3}}, \tag{7.23}$$

and for $d = 3$

$$\hat{\delta} \equiv 2. \tag{7.24}$$

In Fig. 7.5 we show the dependence of $\hat{\delta}$ on the distance scaled with the width of a seismogenic layer (W). Three distributions of the points are assumed in the layer: with $\delta = 1$; $\delta = 2$; and $\delta = 3$ (i.e., uniform spatial Poisson distribution). The correlation dimension is determined for the projection of the points on a horizontal plane (equivalent to the epicenters). We display both theoretical values of the dimension (Eqs. 7.22, 7.23, 7.24, respectively) and the simulated values again evaluated at distances separated by a factor $2^{1/4}$.

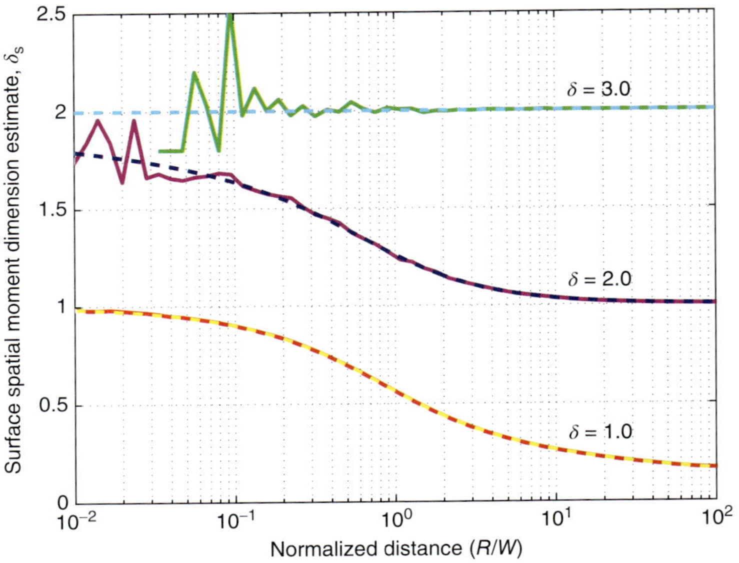

Fig. 7.5 Dependence of the correlation dimension estimate $\hat{\delta}$ on scaled distance for epicenter distribution (grading effect). Three distributions in a layer of thickness W are simulated in 3-D: $\delta = 3$ or uniform spatial Poisson distribution (upper two curves); with $\delta = 2$ (middle two curves); and with $\delta = 1$ (lower two curves). Dashed lines are theoretical curves (Eqs. 7.22–7.24); solid lines are simulation results. *Source:* Kagan (2007a), Fig. 5.

The curves show the expected behavior: for small scaled distances $\hat{\delta} \to \delta - 1$ as the result of projection (Mandelbrot 1983), but for large distances $\hat{\delta} \to \delta - 2$. For the Poisson point distribution in a layer, the point pattern projected on a surface has a uniform $\hat{\delta} = 2$.

7.2.4 Boundary effects

For practical measurements when the distance r exceeds the size (diameter) of the set explored, the correlation function stops increasing. Nerenberg and Essex (1990) call this effect "saturation" and estimate the critical minimum distance as

$$r_s = R/(D + 1). \tag{7.25}$$

As in (7.3) $2R$ is the side of a cube and D is the embedding dimension. If $r > r_s$ in the correlation function $C(r)$, the function is saturated: its value does not represent the scaling effect of a point pattern.

For local and regional earthquake catalogs, the spatial boundaries delineate the area of sufficient coverage. For distances comparable to the area size, the δ-value estimate is biased depending on how the fault system pattern relates to the area polygon. If, for example, a narrow rectangular box were oriented along the direction of the major faults or perpendicular to them, this bias would significantly differ.

Below we consider a few simple cases where $\hat{\delta}$ can be analytically derived. The distribution of the distances in more complicated polygons and other figures can be obtained by a simulation.

7.2.4.1 Boundary effects: a 2-D case

The simplest 2-D figure is a disk. For points in a disk of diameter d, the distribution density for normalized distance $y = r/d$ between two random points inside is (Hammersley 1950)

$$\phi(y) = \frac{16\,y}{\pi\,d}\,[\arccos(y) - y\sqrt{1-y^2}\,]. \tag{7.26}$$

The surface area for the polygon in Fig. 2.5 is $S \approx 233,300\ \text{km}^2$. Approximating it by a circle, we obtain its radius as 272.5 km. When calculating temporal correlation functions, 10% of the time interval is usually considered the upper limit for reliable estimation. In southern California, this would suggest that for distances over 25–50 km the $\hat{\delta}$ values are more questionable.

Garwood (1947) and Ghosh (1951) propose formulas to calculate the distribution density for distances r between random point pairs in a rectangular box $a \times b$ for $a > b$:

$$\phi_1(r) = \frac{2\,r}{a^2\,b^2}[a\,b\,\pi - 2\,r\,(a+b) + r^2] \quad \text{for} \quad r \le b, \tag{7.27}$$

$$\phi_2(r) = \frac{4\,r}{a^2\,b^2}\left[a\,\sqrt{r^2-b^2} - \frac{b^2}{2} - a\,r + a\,b\,\arcsin\left(\frac{b}{r}\right)\right]$$

$$\text{for} \quad b < r \le a, \tag{7.28}$$

and

$$\phi_3(r) = \frac{4\,r}{a^2\,b^2}\,[a\,\sqrt{r^2-b^2} + b\,\sqrt{r^2-a^2} - \frac{1}{2}(a^2+b^2+r^2)$$

$$- a\,b\,\arccos\left(\frac{a}{r}\right) + a\,b\,\arcsin\left(\frac{b}{r}\right), \tag{7.29}$$

for $a < r \le \sqrt{a^2+b^2}$.

7.2.4.2 Boundary effects: 3-D case

Here, we discuss a distribution of distances in a horizontal layer of width W in a 3-D space. This would correspond to the average number of hypocenter pairs within a distance R of an arbitrary point in a layer of uniformly distributed seismicity. Three cases need to be considered: a sphere of radius R not touching any layer boundary, a sphere intersecting only one boundary, and another intersecting both boundaries.

For various distance ranges, we obtain three expressions. For $R < W/2$

$$N_3(R) = \frac{4\,\pi\,(W - 2R)\,R^3}{3W} + \frac{2\,\pi}{3W}\int_0^R (2\,R - h)\,(R + h)^2 dh$$
$$= \frac{\pi\,R^3}{6W}\,(8\,W - 3\,R). \tag{7.30}$$

For $W/2 < R < W$ as in (7.30)

$$N_3(R) = \frac{2\pi}{3}\int_0^{W-R} (2\,R - z)\,(R + z)^2 dz + \frac{2\pi}{3}\int_{W/2}^{R} [4\,R^3 - (R - z)^2\,(2\,R + z)$$
$$- (W + 2\,R - z)\,(R - W + z)^2 dz = \frac{\pi\,R^3}{6W}\,(8\,W - 3\,R). \tag{7.31}$$

For $W < R$

$$N_3(R) = \frac{\pi}{3W}\int_0^W [4\,R^3 - (R - h)^2\,(2\,R - h)$$
$$- (W + 2\,R - h)\,(R - W + h)^2]dh = \pi\,W^2\left(R^2 - \frac{W^2}{6}\right). \tag{7.32}$$

For $R < W$ the estimate of correlation dimension is

$$\hat{\delta} = \frac{12\,(2\,W - R)}{8\,W - 3\,R} = \frac{12\,(2 - \rho)}{8 - 3\,\rho}, \tag{7.33}$$

where $\rho = R/W$. Therefore, for $R > W$ it is

$$\hat{\delta} = \frac{2\,R^2}{R^2 - W^2/6} = \frac{2\,\rho^2}{\rho^2 - 1/6}. \tag{7.34}$$

As expected, for $\rho \to 0$ the hypocentral correlation dimension estimate $\hat{\delta} \to 3$ and for $\rho \to \infty$ the dimension $\hat{\delta} \to 2$.

7.2.5 Inhomogeneity of earthquake depth distribution

Inhomogeneity of earthquake depth distribution influences the hypocentral fractal dimension. If this distribution were in fact uniform over depth, the dimension estimate would approach the real δ-value for distances smaller than the thickness of the seismogenic zone (see Eqs. 7.33–7.34). Otherwise, the apparent dimension value is effectively a convolution of the scale-invariant distribution with a non-uniform depth distribution.

We correct the hypocentral moment for the non-uniformity of the earthquake depth distributions:

$$G(R) = RN_2(R)/(\pi R^2) \propto (2/R) \int_0^R r dr \int_0^R K(z)(r^2 + z^2)^{(\delta-3)/2}\, dz. \tag{7.35}$$

Here, $K(z)$ is the depth covariance function

$$K(z) = (\Delta h)^{-2} \int_{h_1}^{h_2} N(h,\ h+\Delta h)\, N(h+z,\ h+z+\Delta h)\, dh, \tag{7.36}$$

where $N(h,\ h+\Delta h)$ is the number of hypocenters in Δh depth interval, h_1 and h_2 are depth limits: for example, for global shallow seismicity, $h_1 = 0$ and $h_2 = 70$ km. The integrals in Eqs. (7.35) and (7.36) can be evaluated numerically for a known distribution $N(h)$. The similarity between (7.17) and (7.35) is not accidental, since $C\,(W - h)$ is a correlation function (C is a normalizing coefficient) for the number of hypocenters in a layer W, if hypocenters are distributed uniformly over depth.

In Fig. 7.6 we display the distribution of the hypocenter numbers for the Hauksson and Shearer (2005) catalog as well as the correlation function (7.36). We used the magnitude threshold $m_t = 3$, since the depth accuracy of

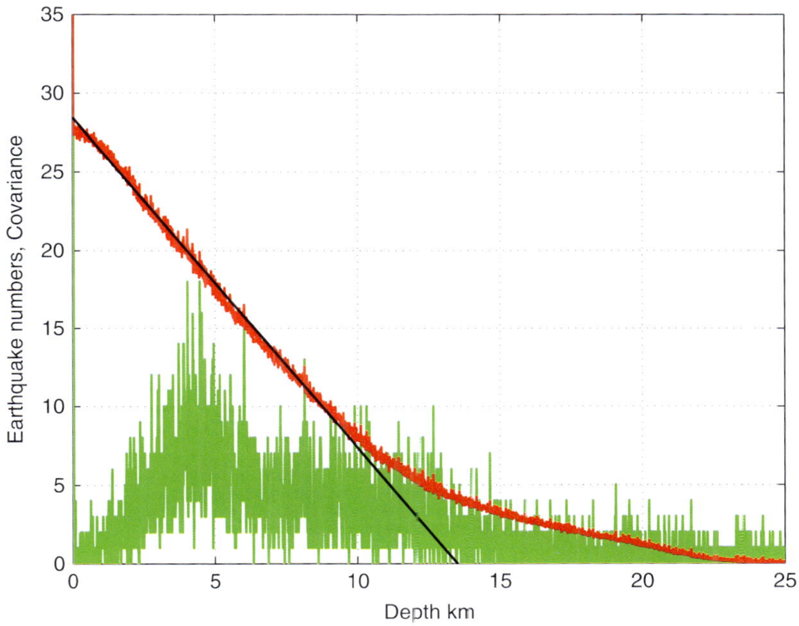

Fig. 7.6 Depth dependence histogram for the Hauksson and Shearer (2005) catalog ($m_t = 3$). Same events as in the polygon in Fig. 2.5. Depth correlation function (red line) and its approximation by linear function (black line), corresponding to a layer of thickness 13.5 km with uniform distribution of seismicity with depth. *Source:* Kagan (2007a), Fig. 6.

these earthquakes should be higher. Most of the correlation function can be approximated by a linear function $C\ (W - h)$ with $W = 13.5$ km.

For $d = 3$, that is, the Poisson 3-D distribution of hypocenters with depth density $N(h,\ h + \Delta h)$, the pair number distribution is

$$N_3(R) = \frac{4\pi}{S} \int_0^R r dr \int_0^r K(z)\, dz, \tag{7.37}$$

where S is the surface area spanned by a catalog. Putting $K(z) = C\ (W - z)$ in (7.37), corresponding to a covariance function for a layer with uniform seismicity, we recover (7.30). For distances longer than $W/2$, the covariance function can be padded by zeros. Thus, we do not need to calculate more complicated formulas (7.31) and (7.32).

Kagan (2007a, Fig. 7) demonstrates that generally the depth correlation function (7.37) correction works well. It may seem that we can correct for depth inhomogeneity without Eqs. 7.35–7.37. However, the depth distribution in other earthquake catalogs may be more difficult to approximate by a uniformly distributed depth point pattern. For example, in global catalogs of shallow earthquakes, many events are assigned the depth of 10 and 33 km (Kagan 2003), making the covariance function highly "spiky."

7.2.6 Temporal influence

For small time intervals, earthquakes are more clustered and their correlation dimensions are smaller (Kagan 1991a, 2007a; Helmstetter *et al.* 2005). This is perhaps the most important issue influencing the δ-value estimate. The previous items in this chapter influence the estimate for a limited distance range only and hence can be recognized and compensated, or alternatively these distance ranges can be ignored in the correlation dimension calculation. However, time dependence extends over all distance intervals and cannot be that easily corrected.

To demonstrate the influence of time limits on the correlation dimension, Fig. 7.7 shows the distribution of distances between accurately located hypocenters in southern California. We have estimated the distribution of distances between hypocenters $N_3(R)$, using the Hauksson and Shearer (2005) catalog (Section 2.2). We have selected only $m \geq 2$ earthquakes relocated with a location error of ϵ_h (horizontal) and ϵ_z (vertical) smaller than 0.1 km. In the catalog, there are 82,442 $m \geq 2$ earthquakes in the time period 1984–2002, of which 33,676 (41%) are relocated with $\epsilon_h < 0.1$ km and $\epsilon_z < 0.1$ km (see also Helmstetter *et al.* 2005, Fig. 5).

The distance number function $N_3(R)$ between the hypocenters is close to a power-law $N_3(R) \propto R^{\,\delta}$ in the range $0.1 \leq R \leq 5$ km. The correlation fractal dimension (measured by least-square linear regression of log (R) and log $[N_3(R)]$ for $0.1 \leq R \leq 5$ km) is $\hat{\delta} \approx 1.5$ (black lines in Figs. 7.7 and 7.8). The faster decay for $R < 0.1$ km is due to location errors (Section 7.2.2), and the roll-off for distances $R > 5$ km is due to the finite thickness of the seismogenic crust

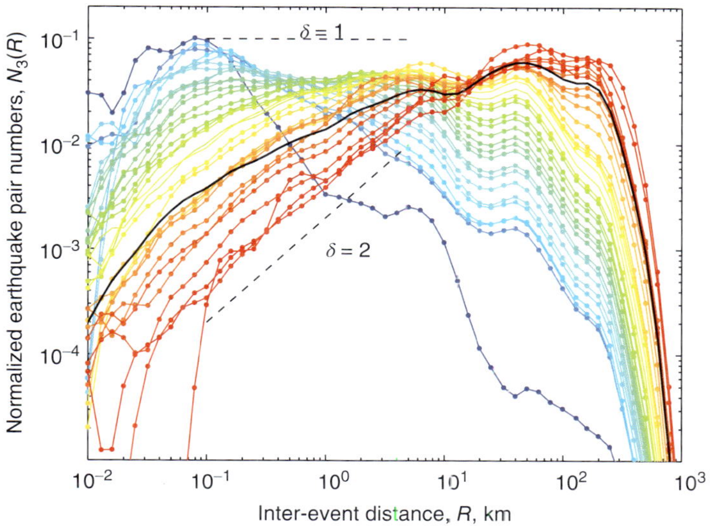

Fig. 7.7 Distribution of distances between hypocenters $N_3(R, t)$ for the Hauksson and Shearer (2005) catalog, using only earthquake pairs with inter-event times in the range $[t, 1.25t]$. Time interval t increases from 1.4 minutes (blue curve) to 2500 days (red curve). We divide the earthquake pair number by R so that the horizontal line would correspond to $\delta = 1$. The black line is the function $N_3(R)$ measured for all earthquake pairs; it has a fractal dimension $\hat{\delta} \approx 1.5$ for $0.1 \leq R \leq 5$ km. *Source:* Kagan (2007a), Fig. 8.

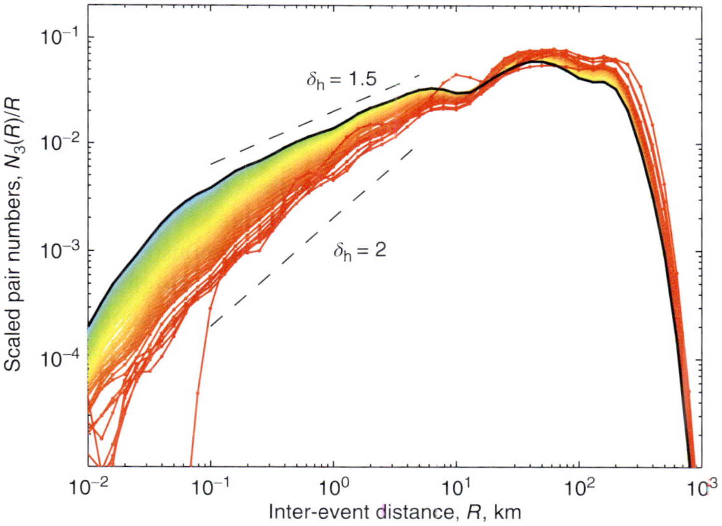

Fig. 7.8 This diagram is similar to Fig. 7.7 but for pairs with inter-event times larger than t. *Source:* Kagan (2007a), Fig. 9.

(Section 7.2.4). For larger distances ($R > 50$ km), the $\hat{\delta}$ decrease is caused by catalog boundaries (Kagan 2007a; Subsection 7.2.4).

In Figs. 7.7 and 7.8 we scale the number of earthquake pairs N_3 by dividing the number by distance R or by its square R^2. This is done to reduce the size of the plots and simplify them.

To estimate the time dependence of the spatial distribution of inter-event distances, we have measured the distribution $N_3(R, t)$, using only earthquake pairs with an inter-event time τ in the range $[t, t + dt]$ (Fig 7.7). We have also computed the distribution

$$N_3(R, \tau \geq t) = \int_t^T N_3(R, t')dt', \tag{7.38}$$

cumulated over all times larger than t, that is, only using events with inter-event times larger than t up to $T = 2500$ days, see Fig. 7.8.

As the minimum inter-event time increases, the fraction of small distances decreases. For inter-event times larger than 1000 days, the fractal dimension of the cumulative distribution $N_3(R, \tau \geq t)$ increases with t from the value measured for the whole catalog ($\hat{\delta} \approx 1.5$) to a maximum value close to 2. For $N_3(R, t)$, the correlation dimension $\hat{\delta}$ increases between $\hat{\delta} \approx 0$ at times $t = 5$ minutes up to $\hat{\delta} \to 2$ for $t = 2500$ days (Fig. 7.9). This maximum inter-event time of 2500 days is long enough so that earthquake interactions are relatively small compared to the tectonic loading. Only an insignificant fraction of earthquake pairs are triggering-triggered events. This value $\hat{\delta} = 2$, measured for $t = 2500$ days, can thus be interpreted as approaching the long-term fractal dimension of the active fault network.

However, Fig. 7.9 clarifies that $\hat{\delta} \to 2$ is not an asymptotic limit for the correlation dimension. The $\hat{\delta}$-value continues to increase. The time interval for the Hauksson and Shearer (2005) catalog (1984–2002) is too short for the $\hat{\delta}$ to reach the final value. As we mentioned earlier, in a local catalog the seismicity pattern for larger time intervals is strongly influenced by a few strong earthquakes and their aftershocks.

The results for global catalogs yield a better measure of the asymptotic $\hat{\delta}$-value. In such cases the range of the correlation dimension change dependent on time interval is also large. For shallow earthquakes, $\hat{\delta}$ varies from 1.2 to 2.1 (Table 2 in Kagan 1991a) for time intervals of 1.0–8575 days. Furthermore, whereas the results shown in Figs. 7.7–7.9 may be explained by a strong spatial concentration of aftershocks for several large Californian earthquakes, Figs. 2 and 3 in Kagan (1991) demonstrate that this temporal effect is present in the PDE global catalog, both the original and declustered one.

The great variability of the $\hat{\delta}$-value with time intervals is important. Whereas previous items (in Sections 7.2.1–7.2.5) are discussed in many spatial analyses of

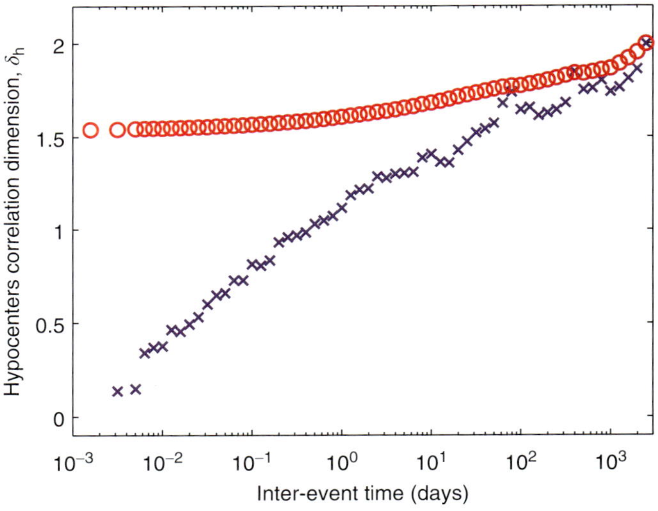

Fig. 7.9 Fractal dimension of Fig. 7.7 curves (crosses) and of Fig. 7.8 curves (circles) as a function of time interval t. Distance interval is $0.1 \leq R \leq 5$ km. *Source:* Kagan (2007a), Fig. 10.

earthquake patterns (see references in beginning of the chapter), little has been done to address the temporal aspect. Most likely, the full statistical analysis of this problem would require taking into account tectonic and seismic deformation rate in various types of plate boundaries (Bird and Kagan 2004).

7.2.7 Randomness

Finally, we mention the randomness of earthquake occurrence. Even when all other effects are taken into account, earthquakes occur randomly in space and time (Chapter 6 and this chapter).

Moreover, because of earthquake clustering, especially the short-term clustering which expresses itself in foreshock–mainshock–aftershock sequences, the effects of randomness are stronger than for a Poisson process (Vere-Jones 1999). Local catalogs often contain large aftershock sequences which make up a significant part of the total. Therefore, though such catalogs list many thousands of earthquakes, the effective size of a sample may amount to considerably fewer events. Consequently, random fluctuations can be quite strong.

7.3 Correlation dimension for earthquake catalogs

7.3.1 California catalogs

Figures 7.10 and 7.11 display the spatial moment curves for the CIT catalog (Hileman *et al.* 1973) in two periods: 1932–2001 and 1975–2001. The curves' behavior for small distances demonstrates the influence of location errors and catalog compiling procedures. In the first plot (Fig. 7.10) for distances smaller than 3 km, the fractal correlation dimension is less than 1.0. The most likely reason for this is that during the first years of the CIT catalog, the aftershocks were often assigned the same location as the mainshocks (see also Subsection 7.2.2). For such an aftershock sequence all pairwise distances would be zero. This would make the dimension of this aftershock sequence equal to zero. Because many sequences are present in this catalog, the combined dimension is small, while not being a zero.

In the second plot (Fig. 7.11), only recent earthquakes have been processed. The hypocentral correlation dimension for small distances (0–5 km) is close to 3.0. This value results from location errors which randomize the position of hypocenters and from projection effects (see Subsections 7.2.2 and 7.2.3).

Both curves' behavior at large distances (more than 100–200 km) is controlled by box boundary effects (see Subsection 7.2.4). In these plots we did not account for such effects as we do in Fig. 7.13. The scale-invariant part of the curves is in a distance range of 2–200 km, where the hypocentral correlation dimension is slightly over 2.0. As explained in Subsection 7.2.7, the curves' fluctuations are probably caused by large aftershock sequences of the 1992 Landers, the 1994 Northridge, and the 1999 Hector Mine earthquakes.

Figure 7.12 displays similar spatial distribution for the catalog of relocated earthquakes (Richards-Dinger and Shearer 2000). The higher location accuracy of these events is seen in the extension toward smaller distances of the

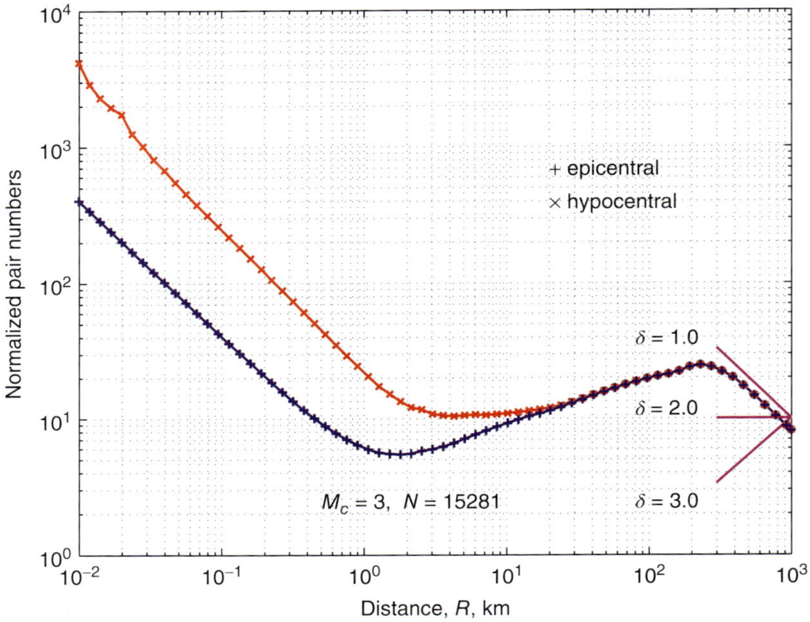

Fig. 7.10 Hypocentral and epicentral spatial moment curves for the CIT catalog 1932–2001. Same polygon as in Fig. 2.5. The upper curve is for the hypocentral moment and the lower for the epicentral. The values of dimensions shown at the right-hand diagram part are for the hypocentral moment, for the epicentral curve the δ_e-value is less by 1.0, i.e., $\delta_e = \delta_h - 1$. *Source:* Kagan (2007a), Fig. 14.

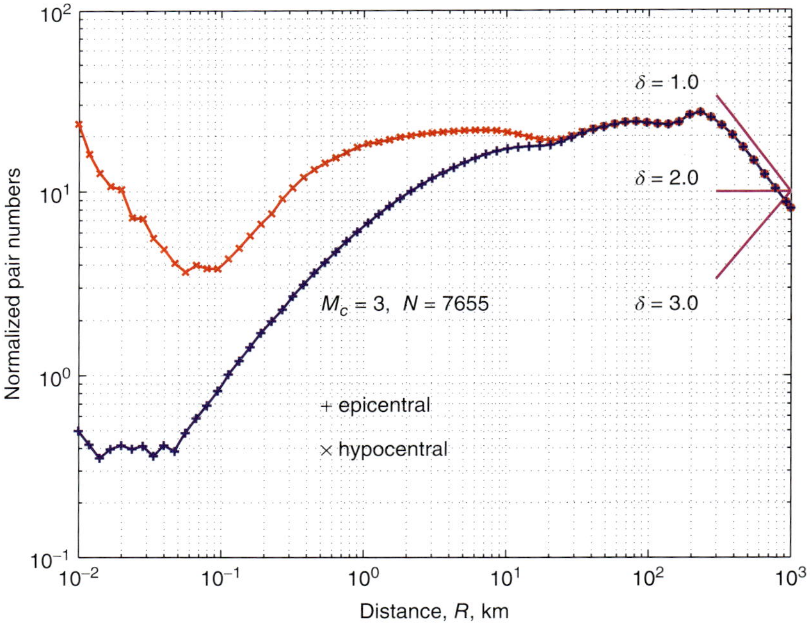

Fig. 7.11 Hypocentral and epicentral spatial moment curves for the CIT catalog 1975–2001 ($m \geq 3.0$). Same polygon as in Fig. 2.5. The upper curve is for the hypocentral moment and the lower for the epicentral. The values of dimensions shown at the right-hand diagram part are for the hypocentral moment, for the epicentral moment $\delta_e = \delta_h - 1$. *Source:* Kagan (2007a), Fig. 15.

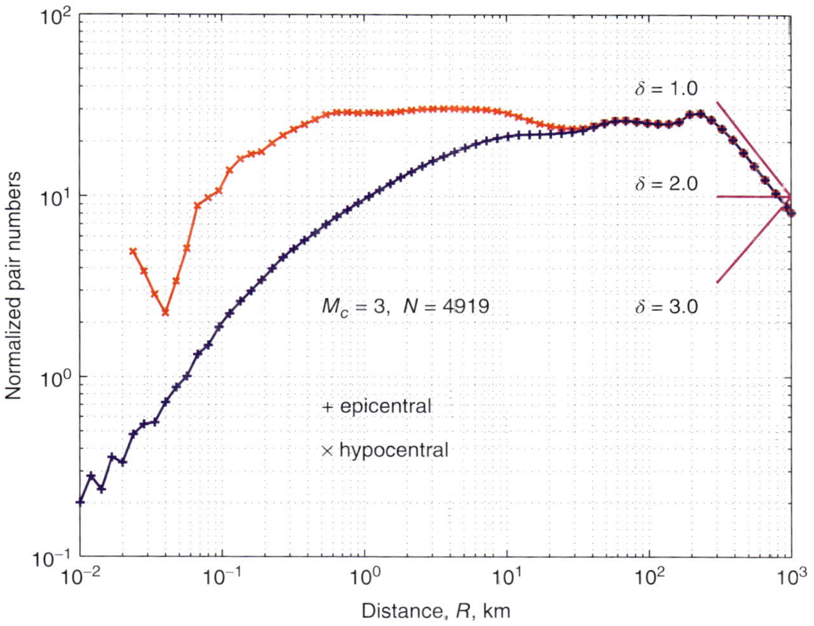

Fig. 7.12 Hypocentral and epicentral spatial moment curves for the southern California relocated earthquake catalog 1975–1997 (Richards-Dinger and Shearer 2000). Same polygon as in Fig. 2.5. Magnitude threshold is $m_t \geq 3.0$. The upper curve is for the hypocentral moment and the lower for the epicentral. The values of dimensions shown at the right-hand diagram part are for the hypocentral moment, for the epicentral moment $\delta_e = \delta_h - 1$. *Source:* Kagan (2007a), Fig. 16.

scale-invariant region. Whereas the time span in both diagrams (Figs. 7.11 and 7.12) is approximately the same, the hypocentral moment in the latter plot extends as a power-law from about 0.5 km up to 200 km. The difference between the epicentral and hypocentral curves is larger for Fig. 7.12. This is due to the higher accuracy of hypocentral solutions. If the vertical errors were comparable to the thickness of the seismogenic zone, the curves would be almost identical for distances comparable to the thickness (Kagan and Knopoff 1980).

Figure 7.13 displays similar spatial distribution curves to the waveform cross-correlation catalog by Hauksson and Shearer (2005). The clusters of earthquakes in the catalogs have been cross-correlated to obtain relative accuracy on the order of tens of meters. Two distance intervals can be seen in the curves: 0.2–20 km and 20–200 km. Apparently the catalog has two earthquake populations: one corresponding to the events in the cross-correlation clusters and the other to the inter-cluster distances. The small distance part ($R = 0.01 - 0.2$ km) of the hypocentral curve indicates that the earthquake spatial distribution behavior is controlled by location errors.

In Figs. 7.10–7.13, location accuracy generally increases over time. In later catalogs it improves as one employs more sophisticated interpretation of seismograms (see Table 2.1). We see that the range of statistical scale-invariant behavior is shifted towards smaller distances. In a complete CIT catalog (Fig. 7.10) which includes early location results, the implied average error is on the order of 3–5 km. In the later part of the catalog (Fig. 7.11), the hypocentral moment's scale-invariant part starts at about 1 km. For the newer catalogs (Figs. 7.12 and 7.13) the curves' scale-invariant behavior extends to 0.5 and 0.2 km, respectively.

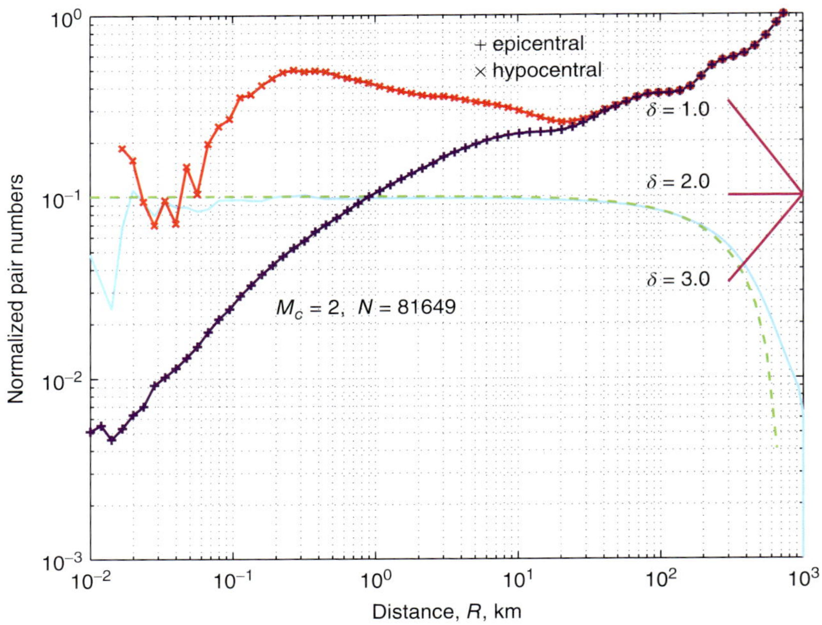

Fig. 7.13 Hypocentral and epicentral spatial moment curves for the southern California waveform cross-correlation catalog 1984–2002, $m_L \geq 2.0$ (Hauksson and Shearer 2005). Same events as in the polygon in Fig. 2.5. The upper curve is for the hypocentral moment and the lower for the epicentral. Here we also plot two curves, demonstrating the boundary effects due to the limited spatial size of the catalog (see Section 7.2.4). The cyan solid curve is obtained by simulation for the polygon in Fig. 2.5. The green dashed line is calculated (Eq. 7.25) for a circle of radius 734.5 km (the maximum distance in the box). Magenta lines at the right show a slope of the curves corresponding to the integer values of the correlation dimension δ_h for the hypocentral moment, for the epicentral moment $\delta_e = \delta_h - 1$. *Source:* Kagan (2007a), Fig. 13.

On the other hand, we see that the later catalogs exhibit more fluctuations at larger distances, due probably to aftershock clusters of a few large earthquakes. This cluster influence and boundary effects make determining the correlation dimension uncertain for large distances in local catalogs. Only global catalogs would allow us to make such evaluation.

7.3.2 Global PDE catalog

In Fig. 7.14 we display epicentral and hypocentral moments for earthquakes in the PDE catalog at three depth intervals. The curves are normalized, so that the horizontal line corresponds to a self-similar distribution with $\delta_h = 2.0$. The curves below the horizontal line have $\delta_h \geq 2.0$ (the fractal dimension is equal to the tangent of the slope angle of the curve plus 2.0). Since the epicentral moments are defined in 2-D, the horizontal line corresponds to $\delta_e = 1.0$. As in Kagan and Knopoff (1980), epicentral moments yield a higher value of the exponent for distance ranges less than or comparable to the thickness of the appropriate layer (Subsection 7.2.3). From simple geometrical arguments, the hypocentral curves are the preferred data input to calculate the fractal dimension (Kagan 1991, 2007a).

Comparing Fig. 7.7 and Figs. 7.10–7.13 we conclude that self-similarity of earthquake geometry is established up to the scale length of 0.1 km and less. Since the equations of elasticity lack any intrinsic scale, we expect that the property of self-similarity can be extended for the brittle fracture of disordered materials

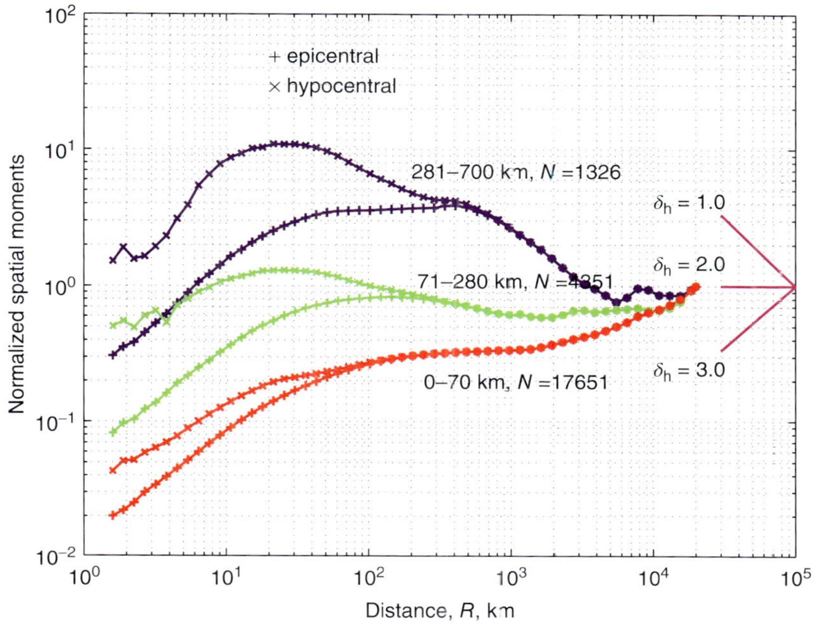

Fig. 7.14 Hypocentral and epicentral spatial moment curves for various depth intervals. The PDE catalog (1965–2003) with $m_b \geq 5.3$ is used. In each of the coupled curves the upper curve is for the hypocentral moment and the lower curve for the epicentral. The two blue upper curves are for the depth interval 281–700 km, the green middle ones are for the depth interval 71–280 km, and the red lower ones are for the depth interval 0–70 km. Solid lines at the right show a slope of the curves corresponding to the integer values of the hypocentral correlation dimension of δ_h. For the epicentral moment $\delta_e = \delta_h - 1$. *Source:* Kagan (2007a), Fig. 18.

(rocks) up to the scale of a few millimeters: the size of rock grains (see Subsection 5.3.5). The upper cutoff for scale-invariance in Fig. 7.14 (around 2000 km) is likely connected to the size of major tectonic plates. The δ values in Fig. 7.14 demonstrate that the dimension decreases as the depth increases.

Taking into account possible errors and biases, we conclude that the hypocentral fractal dimension for shallow seismicity asymptotically approaches 2.20 ± 0.05 for a catalog time span of decades and perhaps centuries. The value of the hypocentral correlation dimension declines to 1.8–1.9 for intermediate events (depth interval 71–280 km) and to 1.5–1.6 for deeper ones.

7.4 Conclusion

Kagan (2007a, see also earlier sections) indicates that evaluating the fractal dimension for earthquake spatial patterns is difficult and prone to many errors and biases. This may explain the contrast with two other classical scale-invariant, universal exponents of earthquake distribution: unlike the G-R relation (Bird and Kagan 2004; Kagan 2010b; Chapter 5) and Omori's law (Kagan 1991b; Kagan and Houston 2005; Kagan 2011; Chapter 6), the properties and value of the correlation dimension are not yet firmly established.

In most studies of earthquake spatial distribution, errors of location and other errors have not been properly considered (Kagan 2007a; this chapter). This might explain the high values of fractal dimensions reported in many publications and the great variability of these values. Such findings may reflect not physical and geometrical properties of earthquake fracture, but rather various location and projection errors peculiar to the catalogs studied.

8 Focal mechanism orientation and source complexity

This chapter is based on the results of several our papers, especially Kagan (1991c, 1992b,c, 1994a,b, 2006, 2009a, 2013). The last two papers (2009a, 2013) are particularly important for this study.

In this chapter we analyze analytically and statistically two tensor fields that are considered to govern earthquake occurrence: stress in the interior of the Earth and earthquake focal mechanisms. This problem is complex because stress and focal mechanisms are tensor-valued variables, thus their analysis requires the application of sophisticated mathematical and statistical tools, many of which are not yet fully developed.

In Section 8.1 we present a general description of the complex stress field and show that stress distribution follows the Cauchy law or even one of the more complicated stable distributions. The tensor stress in the Earth interior is generally not measurable, but the Cauchy stress distribution results in a specific relation between earthquake focal mechanisms that we refer to as the rotational Cauchy distribution. Later in the chapter we show that the focal mechanisms follow this law.

The major part of the chapter (Sections 8.3–8.7) is dedicated to the analysis of earthquake focal mechanism inter-relation: we describe general and statistical properties of the seismic moment tensor (Section 8.3), establish a method for the analysis based on the quaternion technique (Section 8.4), and then apply quaternions for the statistical analysis of earthquake catalogs.

The important property of the focal mechanism is its symmetry, a property which influences the statistical features of any collection of focal mechanisms. Section 8.4 describes our classification of the mechanism symmetry and dependence of the double-couple (*DC*) orientation on its symmetry. Statistical analysis of symmetrical objects has long been a subject of crystallographic texture investigations. Section 8.5 describes the application of crystallographic methods to focal mechanism analysis. Section 8.6 considers theoretical statistical

Earthquakes: Models, Statistics, Testable Forecasts, First Edition. Yan Y. Kagan.

distributions appropriate to *DC* orientation approximation. In Section 8.7 we describe statistical distributions for the rotation of the double-couple mechanisms as well as the results of statistical analysis of the focal mechanism distribution in the GCMT catalog.

Finally, Section 8.8 is dedicated to several projects aimed at a better understanding of the complex structure of the earthquake fault zone. Subsection 8.8.1 starts with the higher-rank seismic moment tensors, proposed by Backus (1977a,b), and applies the tensor techniques, developed in the gravitational wave theory, in order to obtain a more complete description of complex point sources both in the infinite elastic space and in a half-space. In Subsection 8.8.2 we attempt to apply the methods of vector fields analysis developed in statistical hydrodynamics with the goal of describing the moment tensor field of the complex structure of the earthquake fault zone. Due to the mathematical complexity of both problems, considered in Section 8.8, we were not able to proceed far in their solutions. However, we believe that both methods have a significant potential for understanding the earthquake occurrence, which makes them highly relevant for future investigations.

Detailed modern earthquake catalogs with estimates of the focal mechanism and/or seismic moment tensor were compiled beginning in the 1970s. Several extensive catalog datasets are available at present. Frohlich and Davis (1999) and Kagan (2003) discuss the properties of global catalogs and their accuracy. The GCMT catalog includes seismic moment centroid times and locations as well as estimates of the seismic moment tensor components (see Section 2.3). Each tensor is constrained to have a zero trace (first invariant), hence there is no isotropic component (Subsection 8.3.1). Double-couple solutions, or solutions with the tensor determinant equal to zero, are supplied as well. Almost all earthquake parameters are accompanied by internal estimates of error.

8.1 Random stress tensor and seismic moment tensor

8.1.1 Challenges in stress studies

The aim of earthquake seismology is to rigorously describe the tensor stress field which triggers earthquakes. Until now, extensive attempts to study stress fields have been concentrated on stress tensor properties at particular points, especially at hypocentral locations of potential future earthquakes (e.g., Kagan 1994a,b; Harris 1998; Stein 1999; Steacy *et al.* 2005). Chapter 7 results indicate that the earthquake spatial distribution is fractal, with the lower distance limit of fractality in the range of a few meters. The earthquake size distribution (Chapter 5) can be extended in lower limit to ruptures of a few cm. This suggests that the stress field must also be scale-invariant, representing an extremely complicated matrix with critical conditions for earthquake rupture satisfied in close to an infinite number of points. This would correspond to a practically infinite number of micro-earthquake occurrences, if one extrapolates the G-R law for earthquake size distribution (Section 5.2) toward earthquakes of size zero.

While it is apparent that earthquakes are triggered everywhere in seismic regions, the question remains unsolved why small earthquake ruptures develop into giant events which can cause massive destruction. Answering this question adequately will require a detailed description of the 3-D stress field geometry, including its singularities, limit cycles, and possible bifurcations (Gabrielov and Keilis-Borok 1983). This is an extremely difficult and open problem: Gabrielov and Keilis-Borok (*ibid*, p. 489) comment that

> *The [mathematical] problem of the complete description [of the topology of the field ...] has not as yet been solved.*

8.1.2 Cauchy stress distribution

The Cauchy distribution is especially important for representing earthquake geometry. It can be shown by theoretical arguments (Zolotarev and Strunin 1971; Zolotarev 1986, pp. 45-46; Kagan 1990) and by simulations (Kagan 1990) that the stress tensor in the medium with defects follows this distribution. Kagan (1990) argues that the Cauchy distribution of the stress should produce the rotational Cauchy distribution (Eqs. 8.54, 8.55) of earthquake focal mechanisms.

For any point in an elastic medium which is surrounded by point-like defects, the characteristic function for the random stress distribution can be written as (Zolotarev and Strunin 1971; Zolotarev 1986)

$$\log\ \phi(\zeta, \alpha) = \int_0^\infty [\exp(i\zeta\sigma r^{-3}) - 1]\nu(r)r^2 dr, \tag{8.1}$$

where $\nu(r)$ is the density of defects which might depend on r, the distance of the defect from the reference (measurement) point, and σ is the normalized (for $r = 1$) stress Green function of an earthquake; stress decays with distance as r^{-3}. For the uniform 3-D distribution of defects, $\nu = \nu_0$. In this case (8.1) yields the Cauchy stress distribution.

In principle, the above results allow us to find the probability that failure will take place under the condition that some function of the components of the stress tensor will exceed a certain critical level. The simplest case is to compute the probability that any one stress component such as σ_{ij}, for example, exceeds some prescribed value τ_{cr} which can itself be taken to be a random variable. This probability is the cumulative 1-D Cauchy distribution (Zolotarev and Strunin, 1971; Kagan 1990, his Eq. 2.5)

$$F_{ij}(\tau_{cr}) = \text{Prob}\ (\sigma_{ij} \leq \tau_{cr}) = 1/2 + (1/\pi) \times \arctan(\ C_{ij} \times \tau_{cr}), \tag{8.2}$$

where C_{ij} is the normalizing coefficient.

Earthquake spatial distribution, as described in Section 7.3, is fractal. In (8.1) we should substitute the fractal distribution of sources $\nu = \nu_0 r^{\delta - D}$, where $D = 3$ is the Euclidean dimension of the space, and δ is a fractal correlation dimension of earthquake hypocenters. Then (cf. Zolotarev 1986, eq. 1.1.16; Kagan 1994b)

$$\log\ \phi(\zeta, \alpha) = \nu_0 \int_0^\infty [\exp(i\zeta\sigma u) - 1] u^{(\delta/3)-1} du = \nu_0\ \Gamma(-\alpha)\ |\zeta|^\alpha, \tag{8.3}$$

with $\alpha = \delta/3$. The above formula means that if $\delta = 3$ (defects are spatially distributed according to the uniform Poisson law), the resulting distribution is the Cauchy law, whereas for a fractal spatial distribution of earthquakes, $\alpha < 1$ (Kagan 1994b).

8.1.3 Random stress tensors

Analyzing seismic moment and stress tensors has been basic to earthquake seismology. Although tensors are fundamentally important, they have not been sufficiently investigated or interpreted from a statistical point of view in the Earth sciences, with few exceptions. A linear error propagation was first independently proposed to derive the error estimate of the principal stresses and their orientations by Angelier *et al.* (1982) and Soler and van Gelder (1991). The correlation of the invariant quantities of seismic moment tensors was investigated by Kagan and Knopoff (1985a,b). Kagan (1992b,c, 2000) further extended the correlation results on invariant quantities to analyze earthquake catalogs and interpret faulting geometry.

The study of random tensors has its root in nuclear physics (see, e.g., Girko 1990; Mehta 1991) and multivariate statistics (see, e.g., Anderson 1958). For nuclear physics, a simple rotation-invariant distribution has been widely investigated. But in multivariate statistics, only a handful of large sample or asymptotic distribution results involving such distributions are available. These results, despite their significance, cannot be applied directly to the Earth sciences, because the number of tensors derived from the same original source is generally small. More importantly, efforts in nuclear physics and statistical mathematics have largely been focused on the principal invariants, namely, the principal eigenvalues. Very little attention has been paid to random eigendirections, which are equally important in the Earth sciences. Furthermore, the nonlinearity of the mapping onto the eigendirections and eigenvalues has been insufficiently studied. This nonlinearity could strongly affect the estimated eigenvalues and directions if the seismogram noise level is high. Xu (1999, 2002) and Cai *et al.* (2005, and its references) have attempted to develop a probabilistic approach in dealing with random/stochastic tensors in geoscience. The main new results from such studies include exact distributions for the random eigenvalues and eigendirections. They also include accuracy estimates of a higher order and bias computations. More work needs to be done in this study field.

8.2 Geometric complexity of earthquake focal zone and fault systems

Several statistical methods and models can be used to study the geometric complexity of an earthquake fault zone or fault system. Some are based on representing an earthquake as a point source, and the geometric complexity of a source reveals itself in a complex structure of a second-rank seismic moment tensor or higher rank tensors (Subsection 8.8.1). Another method is to investigate the geometric complexity of the fault system as expressed in a set of earthquake locations and their focal mechanisms (Subsection 8.8.2). Due to the difficulty of the problem and insufficient development of the methods, neither of these techniques would provide a complete solution.

Geologic and geophysical studies of earthquake focal zones point to significant complexity in the rupture process (see, for example, King 1983, 1986; Amelung and King 1997). Moreover, almost any large earthquake is now analyzed in detail with its rupture history represented in time-space-focal mechanism maps which usually exhibit an intricate moment release. Although such results suggest that an earthquake focal zone is more complex than the standard planar model for earthquake rupture (Aki and Richards 2002), phenomenological investigations cannot describe the fault patterns appropriate for its quantitative modeling. Such a description needs to be based on statistical treatment of the observed fault geometry.

The geometric complexity of the earthquake rupture process and that of the earthquake fault system are interrelated. Their comparability is the result of a general self-similarity of earthquake occurrence: earthquake rupture propagates over a complex fault pattern. This pattern is then seen in the occurrence of aftershocks and other dependent events. Therefore, we may assume that the fault pattern complexity, when considered for small time intervals, would be similar to the geometric complexity of each earthquake rupture. Fractal features of the spatial distribution of earthquake hypocenters and epicenters (Chapter 7) support this conjecture.

8.2.1 Tensor invariants

The major source of information on the structure of earthquake focal zone are catalogs of focal mechanisms which are represented as seismic moment tensors or as fault-plane solutions for the *DC* source (Okal 2013). Snoke (2003) and Gasperini and Vannucci (2003) consider several equivalent representations for *DC* sources and their properties, and provide mathematical expressions for their mutual transformation. Krieger and Heimann (2012 and its references) review routines for plotting moment tensors and focal mechanisms. Ekström *et al.* (2012, and references therein) discuss their extensive work on evaluating seismic moment tensors for global earthquakes.

A second-rank seismic moment tensor can be represented as a symmetric 3×3 matrix

$$\mathbf{m} = \begin{vmatrix} m_{11} & m_{12} & m_{13} \\ m_{21} & m_{22} & m_{23} \\ m_{31} & m_{32} & m_{33} \end{vmatrix}, \tag{8.4}$$

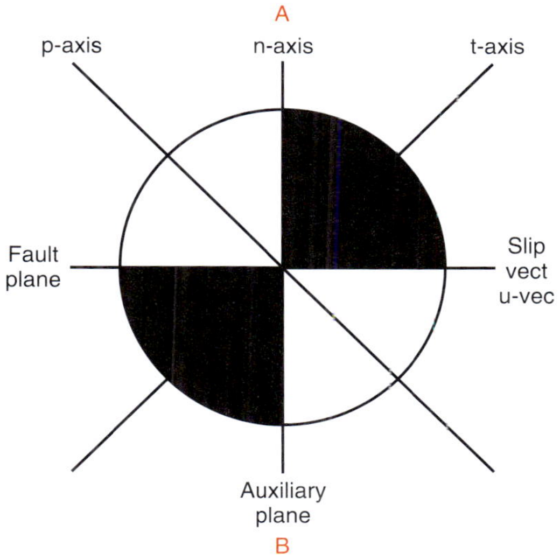

Fig. 8.1 Schematic (beachball) diagram of the *DC* earthquake focal mechanism and its quadrupole radiation patterns. The null (**b**) axis is orthogonal to the **t**- and **p**-axes, or it is located on the intersection of fault and auxiliary planes, that is, perpendicular to the paper sheet in this display. The **n**-axis is normal to the fault-plane; **u** is a slip vector. *Source*: Kagan (2013), Fig. 1.

with $m_{ij} = m_{ji}$, the tensor has six degrees of freedom. The moment tensor is considered to be traceless or deviatoric (Aki and Richards 2002). Hence its number of degrees of freedom is reduced to five.

The eigenvectors of matrix (8.4) are vectors (Fig. 8.1)

$$\begin{aligned} \mathbf{t} &= [1,\ 0,\ 0]; \\ \mathbf{p} &= [0,\ 1,\ 0]; \\ \mathbf{b} &= [0,\ 0,\ 1]. \end{aligned} \tag{8.5}$$

For known eigenvectors **t** and **p**, the *DC* source tensor can be calculated as (Aki and Richards 2002, their Eq. 3.21)

$$m_{ij} = \mu\,[n_i u_j + n_j u_i] = \mu\,[(t_i + p_i)(t_j - p_j) + (t_j + p_j)(t_i - p_i)]/2, \tag{8.6}$$

where μ is a shear modulus, **n** is a normal to a fault plane, and **u** is a slip vector (see Fig. 8.1). Therefore, if we know the orientation of two eigenvectors, the moment components can be calculated.

The invariants of the deviatoric seismic moment tensor **m** can be calculated as (Kagan and Knopoff 1985a)

$$I_1 = \mathrm{Tr}\ [\mathbf{m}] = m_{11} + m_{22} + m_{33} = \lambda_1 + \lambda_2 + \lambda_3 \equiv 0, \tag{8.7}$$

where 'Tr' is a trace of the tensor and λ_i are the eigenvalues of a moment tensor. The second invariant or the norm of the tensor

$$\begin{aligned} I_2 &= -\ (m_{11}m_{22} + m_{11}m_{33} + m_{22}m_{33}) + m_{12}^2 + m_{13}^2 + m_{23}^2 \\ &= -\ (\lambda_1\lambda_2 + \lambda_1\lambda_3 + \lambda_2\lambda_3). \end{aligned} \tag{8.8}$$

For a traceless tensor (8.7)

$$\begin{aligned} I_2 &= m_{12}^2 + m_{13}^2 + m_{23}^2 + (m_{11}^2 + m_{22}^2 + m_{23}^2)/2 \\ &= \frac{1}{2}\sum_{i=1}^{3}\sum_{j=1}^{3} m_{ij}^2 = (\lambda_1^2 + \lambda_2^2 + \lambda_3^2)/2, \end{aligned} \tag{8.9}$$

(Jaeger and Cook 1979, p. 33). The scalar seismic moment is

$$M = \sqrt{I_2}\ . \tag{8.10}$$

To normalize the tensor we divide it by moment M

$$\mathbf{m}' = \mathbf{m}/M. \tag{8.11}$$

In the rest of the section, unless specifically indicated, we use only the normalized moment tensors ($\mathbf{m} = \mathbf{m}'$).

The third invariant is a determinant of a tensor matrix

$$\begin{aligned} I_3 = \text{Det}\,[\mathbf{m}] &= m_{11}m_{22}m_{33} + 2m_{12}m_{13}m_{23} \\ &-(m_{11}m_{23}^2 + m_{22}m_{13}^2 + m_{33}m_{12}^2) = \lambda_1\lambda_2\lambda_3. \end{aligned} \tag{8.12}$$

For a *DC* earthquake source

$$I_3 \equiv 0, \tag{8.13}$$

that is,

$$\min[|\lambda_1|,\ |\lambda_2|,\ |\lambda_3|] \equiv 0. \tag{8.14}$$

Thus, the normalized *DC* moment tensor has 3 degrees of freedom. If all eigenvalues are non-zero, Knopoff and Randall (1970) called the source as 'Compensated Linear Vector Dipole' (CLVD).

8.2.2 CLVD sources and complexity

The CLVD is the simplest measure of earthquake focal zone complexity. Many measurements of focal zone geometry indicate that a planar earthquake fault is only a first approximation; rupture is usually non-planar. However, it is important to know whether the focal zone of a single earthquake or the fault systems of many earthquakes can be represented by a distribution of small dislocations with the *DC* mechanism. If no CLVD component is present in tectonic earthquakes, one degree of freedom for each rupture patch can be excluded with great savings when representing earthquake rupture patterns.

Frohlich *et al.* (1989) and Frohlich (1990) studied the CLVD distribution for earthquakes with the rupture slip along surfaces of revolution and found that a certain geometric slip pattern produces a significant CLVD component. However, such smooth surfaces of revolution are unlikely during a real earthquake rupture. Our results (Kagan 1992a,b,c, 2000, 2007a, see also Chapter 7) rather suggest that both the fault rupture system and focal mechanisms, associated with the rupture, are non-smooth everywhere. They are controlled by fractal distributions.

The Gamma index (Γ) can be used to characterize the CLVD component of a source (Kagan and Knopoff 1985a; Frohlich 1990; Richardson and Jordan 2002; Bailey *et al.* 2009, 2010). The index is

$$\Gamma = \frac{3\sqrt{3}}{2} \times \frac{I_3}{I_2^{3/2}}, \tag{8.15}$$

(see Eqs. 8.8–8.13). For a *DC* source (Eq. 8.15)

$$\Gamma \equiv 0. \tag{8.16}$$

The Γ-index ranges from -1 to 1; $|\Gamma| = 1$ corresponds to a pure CLVD source.

Kagan and Knopoff (1985a) considered the problem of a CLVD Γ-index distribution for a composite source

$$\mathbf{m} = \sum_{i=1}^{N} \mathbf{R}_i \, m^{(i)} \, \mathbf{R}_i^T, \tag{8.17}$$

where $\mathbf{m}^{(i)}$ is a regular moment tensor, $\mathbf{R}_i$ is a random rotation matrix and $\mathbf{R}_i^T$ is its transpose. In a quaternion notation, (8.17) can be expressed as

$$\mathcal{M}(\mathbf{q}) = \sum_{i=1}^{N} \mathcal{M}(\mathbf{q} \times \mathbf{R}^{(i)}), \tag{8.18}$$

where the operator $\mathcal{M}$ is given by Eq. 8.42 and $\mathbf{r}$ is a random rotation quaternion that can be obtained using Marsaglia's (1972) algorithm (see more in Kagan 2005b).

Kagan and Knopoff (1985a, their Fig. 1a) used simulation to show that for the sum of randomly oriented deviatoric sources the Γ-index is distributed uniformly over the interval $[-1, 1]$ for a large number of summands. Even for two *DC*s, the distribution is close to uniform. Similar distribution was obtained by Danilova and Yunga (1990) and by Bailey *et al.* (2009, Figs. C1, C2). It would be interesting to derive an analytical expression for a sum of n randomly rotated deviatoric tensors.

The simplicity of the Γ-index statistical distribution presents a significant benefit in characterizing the CLVD component. Many other measures of non-*DC* properties for an earthquake source have been proposed (see, for example, Julian *et al.* 1998, Eq. 18; Bailey *et al.* 2009), but they lack statistical advantage of the Γ-index.

The appearance of the CLVD component in simulations may be used to explain the non-zero Γ-index value, sometimes obtained for earthquakes with a complex fault zone, that is, an earthquake source comprising several *DC* components of different orientation. However, for tectonic events non-*DC* mechanisms like the CLVD are likely due to various systematic and random errors in determining the mechanism (Frohlich and Davis 1999; Kagan 2003, 2009a). These results suggest that routinely determined CLVD values would not reliably show the deviation of earthquake focal mechanisms from a standard *DC* model.

King (1986) specified two kinds of geometric barriers connected with a change of earthquake failure surfaces: conservative and non-conservative. Kagan (2009a) shows that there are both theoretical and observational arguments suggesting that the structure of a source is complex for tectonic earthquakes but precludes the appearance of a significant CLVD component in the geometric barriers proposed above. It would be very interesting to show theoretically that the configurations considered in the above papers are unique.

8.3 Rotation of double-couple (*DC*) earthquake moment tensor and quaternions

Two general techniques can be employed to study the 3-D rotation: orthonormal rotation matrices and normalized (unit) quaternions. The quaternion method has been used to evaluate rotations in many investigations of earthquake focal mechanisms (see, for example, Kagan 1991c; Frohlich and Davis 1999; Kagan 2009a; Kagan and Jackson 2011). Kagan (2007c) explains how "ordinary" matrices and vectors can be used to obtain 3-D rotation parameters.

Altmann's (1986) book was the first monograph specifically dedicated to 3-D rotations [group *SO*(3)] and quaternions. At present quaternions are widely used to describe rotations in space satellite and airplane dynamics (Kuipers 1999) as well as simulations of virtual reality, robotics and automation (Hanson 2006; Dunn and Parberry 2011). These last three monographs explain quaternions in a more accessible manner. Many journal articles (see references in these monographs and in Kagan 2009a) discuss the practical application of quaternions when analyzing the 3-D rotations.

However, the above publications do not consider the symmetry properties of rotated objects or how symmetry influences orientation analysis. As Kagan (1990, 1991c, 2007c, 2009a) indicated, the techniques considered in those publications cannot be used for the *DC* source orientation studies without major modifications, because of the *DC* symmetry properties, described below in Subsection 8.4.1. The only scientific discipline where symmetry is extensively considered in 3-D rotation analysis is the study of crystallographic texture (Handscomb 1958; Mackenzie 1958; Grimmer 1979; Frank 1988; Heinz and Neumann 1991; Morawiec 2004; Meister and Schaeben 2005; Schaeben 2010). We apply their results for our problems later in Section 8.5.

8.3.1 Quaternions

Kagan (1982, see also Subsection 3.3.2) proposed using normalized quaternions to represent earthquake *DC* focal mechanism orientation. The *DC* focal mechanism, as presented in earthquake catalogs, has a symmetry of a rectangular box (Kagan 1991c, 2005b; Tape and Tape 2012), or orthorhombic symmetry (see Subsection 8.4.1). Therefore, it is necessary to modify the quaternion technique to take into account *DC* symmetry, as will be discussed later in this chapter.

The quaternion $\mathbf{q}$ is defined as

$$\mathbf{q} = q_0 + q_1\mathbf{i} + q_2\mathbf{j} + q_3\mathbf{k}. \tag{8.19}$$

The first quaternion's component q_0 is its scalar part: q_1, q_2, and q_3 are components of a "pure" quaternion. The imaginary units i, $\mathbf{j}$, $\mathbf{k}$ obey the following multiplication rules

$$\begin{aligned}
&\mathbf{i}^2 = \mathbf{j}^2 = \mathbf{k}^2 = -1;\\
&\mathbf{i} \times \mathbf{j} = -\mathbf{j} \times \mathbf{i} = \mathbf{k};\\
&\mathbf{k} \times \mathbf{i} = -\mathbf{i} \times \mathbf{k} = \mathbf{j};\\
&\mathbf{j} \times \mathbf{k} = -\mathbf{k} \times \mathbf{j} = \mathbf{i}.
\end{aligned} \tag{8.20}$$

From (8.20) note that the multiplication of quaternions is not commutative, it depends on the order of multiplicands. Non-commutability is also a property of finite 3-D rotations. Thus, in general

$$\mathbf{q}'' \times \mathbf{q}' \neq \mathbf{q}' \times \mathbf{q}'', \tag{8.21}$$

we need to distinguish between the quaternion right- and left-multiplication.

The conjugate $\mathbf{q}^*$ and inverse $\mathbf{q}^{-1}$ of a quaternion are defined as

$$\mathbf{q}^* = q_0 - q_1\,\mathbf{i} - q_2\,\mathbf{j} - q_3\,\mathbf{k} \quad \text{and} \quad \mathbf{q} \times \mathbf{q}^{-1} = 1. \tag{8.22}$$

The normalized quaternion $\mathbf{q} = [q_0, q_1, q_2, q_3]$ contains four terms which can be interpreted as defining a 3-D sphere (S^3) in 4-D space:

$$q_0^2 + q_1^2 + q_2^2 + q_3^2 = 1. \tag{8.23}$$

Hence the total number of degrees of freedom for the normalized quaternion is 3. Since we practically use only the normalized quaternions, we usually omit the word "normalized." For such quaternion

$$\mathbf{q}^* = \mathbf{q}^{-1}. \tag{8.24}$$

The quaternion can be used to describe a 3-D rotation: in this case the first term in (8.23) represents the angle of the rotation and the following three terms characterize the direction of its axis (Kagan 1991c).

We use quaternions to calculate a rotated vector $\mathcal{R}(\mathbf{v})$ by applying the rules of quaternion multiplication (8.20)

$$\mathcal{R}(\mathbf{v}) = \mathbf{q} \times \mathbf{v} \times \mathbf{q}^{-1}. \tag{8.25}$$

The vector $\mathbf{v} = [q_1, q_2, q_3]$ is represented in Eq. (8.25) as a pure quaternion, that is, its scalar component is zero. In (8.25) the quaternion $\mathbf{q}$ is a rotation operator and the pure quaternion **v** is an operand (Altmann 1986, p. 16). Similarly to (8.25) the whole coordinate system can be rotated (Kuipers 1999).

Quaternion multiplication

$$\mathbf{s} = \mathbf{q} \times \mathbf{r}, \tag{8.26}$$

is used to represent the 3-D rotation of the *DC* earthquake sources.

The above expression can be written in components (Klein 1932, p. 61)

$$\begin{aligned} s_0 &= q_0 r_0 - q_1 r_1 - q_2 r_2 - q_3 r_3; \\ s_1 &= q_1 r_0 + q_0 r_1 \pm q_2 r_3 \mp q_3 r_2; \\ s_2 &= q_2 r_0 + q_0 r_2 \pm q_3 r_1 \mp q_1 r_2; \\ s_3 &= q_3 r_0 + q_0 r_3 \pm q_1 r_2 \mp q_2 r_1, \end{aligned} \tag{8.27}$$

where the upper sign in $\pm$ and $\mp$ is taken for the right-multiplication and the lower sign for the left-multiplication: $\mathbf{s} = \mathbf{r} \times \mathbf{q}$.

Kuipers (1999, p. 133) indicates that the right-multiplication corresponds to the 3-D rotation of an object, whereas the left-multiplication is the rotation of the coordinate system. Distinguishing these multiplications is especially important when considering a sequence of 3-D rotations. The expression

$$\mathbf{q}^{(n)} = \mathbf{q}' \times \mathbf{q}'' \times \ldots \times \mathbf{q}^{(n-1)}, \tag{8.28}$$

is the right-multiplication sequence which we use here. The corresponding rotation is anti-clockwise with the rotation pole located on a 2-D reference unit sphere (Altmann 1986).

3-D rotations for quaternions of opposite signs are equal

$$\mathcal{S}\,[\mathbf{q}] = \mathcal{S}\,[-\mathbf{q}], \tag{8.29}$$

where $\mathcal{S}$ is a transformation operator of a 3-D rotation corresponding to a quaternion **q**. This means that the group *SO*(3) of the 3-D rotations has a two-to-one relation to the quaternions. Altmann (1986, his Ch. 10) describes the complicated topology of rotations due to this representation (see also Weeks 2002, pp. 226–227).

8.3.2 DC moment tensor and quaternions

When applied to the *DC* parametrization, the identity quaternion (zero rotation)

$$\mathbf{1} = [1,\ 0,\ 0,\ 0], \tag{8.30}$$

is identified with the strike-slip *DC* source with plunge angles

$$\alpha_T = \alpha_P = 0^\circ, \quad \text{and} \quad \alpha_B = 90^\circ, \tag{8.31}$$

and azimuths

$$\beta_T = 0^\circ, \quad \text{and} \quad \beta_P = 90^\circ, \tag{8.32}$$

(Kagan 1991c, 2005b). Any other *DC* source corresponds to a quaternion describing the 3-D rotation from the reference *DC* source (Eqs. 8.30–8.32). Thus, the orientation of an arbitrary source may be considered as a rotation from the reference source.

There are several possible representations of rotation in 3-D. Among the commonly used are Euler angles about coordinate axes (Kuipers 1999, his Ch. 4.3) and a rotation by the angle Φ about a rotation axis. The rotation pole is the point where the rotation axis intersects a reference 2-D unit sphere (S^2). We use the latter convention in this chapter since it is more convenient for the quaternion technique. For an arbitrary quaternion, the value of the rotation angle and the spherical coordinates, θ and ϕ, of the rotation pole on a reference sphere are then

$$\begin{aligned} &\Phi = 2 \arccos\,(q_0), \\ &\theta = \arccos\,[q_3/\sin(\Phi/2)], \\ &\phi = \arctan\,(q_2/q_1), \quad \text{if } \phi \leq 0^\circ, \text{ then } \phi = 360^\circ + \phi, \end{aligned} \tag{8.33}$$

where ϕ is an azimuth ($0^\circ \leq \phi \leq 360^\circ$), measured clockwise from North; and θ is a colatitude ($0^\circ \leq \theta \leq 180^\circ$), $\theta = 0^\circ$ corresponds to the vector pointing down.

We use the known correspondence between the orthogonal rotation matrix

$$\mathbf{R} = \begin{vmatrix} t_1 & p_1 & b_1 \\ t_2 & p_2 & b_2 \\ t_3 & p_3 & b_3 \end{vmatrix}, \tag{8.34}$$

and the quaternion (Moran 1975, Eq. 6; Altmann 1986, pp. 52, 162; Kuipers 1999, his Eq. 5.11). The following formula for the rotation matrix can be derived

$$\mathbf{R} = \begin{vmatrix} q_0^2 + q_1^2 - q_2^2 - q_3^2 & 2\,(-q_0q_3 + q_1q_2) & 2\,(q_0q_2 + q_1q_3) \\ 2\,(q_0q_3 + q_1q_2) & q_0^2 - q_1^2 + q_2^2 - q_3^2 & 2\,(-q_0q_1 + q_2q_3) \\ 2\,(-q_0q_2 + q_1q_3) & 2\,(q_0q_1 + q_2q_3) & q_0^2 - q_1^2 - q_2^2 + q_3^2 \end{vmatrix}. \tag{8.35}$$

The above formula can be obtained by applying (8.25) to each of the original **t**, **p**, and **b** vectors. Kagan and Knopoff (1985a, Eq. 5) provide another expression for the rotation matrix using direction cosines of the axes [the first term in the second matrix row should be corrected as $\ell m(1 - \cos\Phi) - m \sin\Phi$].

Comparing (8.34) with (8.35) we derive quaternion components from the rotation matrix direction cosines (Kuipers 1999, p. 169; Hanson 2006, pp. 149–150). For example, if q_0 is not close to zero

$$\begin{aligned} &q_0 = (1/2)\,\sqrt{t_1 + p_2 + b_3 + 1}; \\ &q_1 = (b_2 - p_3)/(4q_0); \\ &q_2 = (t_3 - b_1)/(4q_0); \\ &q_3 = (p_1 - t_2)/(4q_0). \end{aligned} \tag{8.36}$$

Since as many as three of the quaternion components may be close to zero, it is computationally simpler to use the component with a maximum absolute value to calculate the three other components.

The seismic moment tensor for the *DC* in eigenvector coordinates (Eqs. 8.30–8.32) is $\mathbf{m} = \mathrm{diag}\,[1,\ -1,\ 0]$. For the general orientation of an earthquake focal mechanism, the seismic moment tensor (Eq. 8.4) can be calculated from the quaternion as follows (Kagan and Jackson 1994, Eq. A3):

$$\begin{aligned}
m_{11} &= q_1^4 - 6q_1^2q_2^2 - 2q_1^2q_3^2 + 2q_1^2q_0^2 + 8q_1q_2q_3q_0 + q_2^4 \\
&\quad + 2q_2^2q_3^2 - 2q_2^2q_0^2 + q_3^4 - 6q_3^2q_0^2 + q_0^4; \\
m_{12} &= 4(q_1^3q_2 - q_1q_2^3 - q_3^3q_0 + q_3q_0^3); \\
m_{13} &= 2(q_1^3q_3 - 3q_1^2q_2q_0 - 3q_1q_2^2q_3 - q_1q_3^3 \\
&\quad + 3q_1q_3q_0^2 + q_2^3q_0 + 3q_2q_3^2q_0 - q_2q_0^3); \\
m_{22} &= -\ q_1^4 + 6q_1^2q_2^2 - 2q_1^2q_3^2 + 2q_1^2q_0^2 + 8q_1q_2q_3q_0 \\
&\quad -\ q_2^4 + 2q_2^2q_3^2 - 2q_2^2q_0^2 - q_3^4 + 6q_3^2q_0^2 - q_0^4; \\
m_{23} &= 2(q_1^3q_0 + 3q_1^2q_2q_3 - 3q_1q_2^2q_0 + 3q_1q_3^2q_0 \\
&\quad -\ q_1q_0^3 - q_2^3q_3 + q_2q_3^3 - 3q_2q_3q_0^2); \\
m_{33} &= 4(q_1^2q_3^2 - q_1^2q_0^2 - 4q_1q_2q_3q_0 - q_2^2q_3^2 + q_2^2q_0^2).
\end{aligned} \tag{8.37}$$

A more complicated algorithm is needed for the rotation of one *DC* source into another. The methods of quaternion algebra can be used to evaluate the 3-D rotation angle by which any *DC* source can be so transformed (Kagan 1991c). Alternatively, the standard technique of orthogonal matrices can be applied to this calculation (Kagan 2007c).

Given the symmetry of the regular *DC* source or equivalently of the *DC*4 source (Kagan and Knopoff 1985a; Kagan 1990, 1991c; Subsection 8.4.1) the q_0 term in (8.23) can always be presented as the largest positive term in this parameterization. In particular, to obtain the standard *DC* quaternion representation, we right-multiply an arbitrary quaternion $\mathbf{q}$ by one of the elementary quaternions (Kagan 1991c):

$$\begin{aligned}
\mathbf{i} &= [0,\ 1,\ 0, 0]; \\
\mathbf{j} &= [0,\ 0,\ 1,\ 0]; \\
\mathbf{k} &= [0,\ 0,\ 0,\ 1],
\end{aligned} \tag{8.38}$$

if the second, third or fourth term has the largest absolute value, respectively. For example, for the largest second term, q_1

$$\mathbf{q}'' = \mathbf{q}' \times \mathbf{i}. \tag{8.39}$$

If the resulting first term is negative, the sign of all terms should be reversed (see Eq. 8.29).

As the result of multiplication by expressions (8.38) the quaternion $\mathbf{q}$ becomes

$$\begin{aligned} \mathbf{q} \times \mathbf{1} &= [q_0,\ q_1,\ q_2,\ q_3]; \\ \mathbf{q} \times \mathbf{i} &= [q_1, -q_0,\ -q_3,\ q_2]; \\ \mathbf{q} \times \mathbf{j} &= [q_2,\ q_3,\ -q_0,\ -q_1]; \\ \mathbf{q} \times \mathbf{k} &= [q_3, -q_2,\ q_1,\ -q_0]. \end{aligned} \tag{8.40}$$

The transformations (8.29, 8.40) describe an eight-to-one correspondence between an arbitrary normalized quaternion and a quaternion corresponding to a normalized seismic moment tensor. We call this operator $\mathcal{Q}$. Therefore, the quaternion ξ

$$\xi = \mathcal{Q}(\mathbf{q}), \tag{8.41}$$

is an one-to-one quaternion representation of a *DC* focal mechanism.

It is easy to check that all the eight quaternions of the $\mathcal{Q}$ operator (8.41) produce the same moment tensor (8.37). Thus we can write

$$\mathbf{m} = \mathcal{M}(\mathbf{q}) = \mathcal{M}(\xi), \tag{8.42}$$

where $\mathcal{M}(\cdot)$ is an operator (8.37) converting a quaternion into a seismic moment tensor matrix.

In principle, we can use the non-normalized variables. In this case the norm of a quaternion would correspond to that of the tensor (i.e., a scalar seismic moment M for a *DC* source). However, due to the presence of the CLVD component, the general deviatoric tensor has five degrees of freedom even after it has been normalized. Hence it cannot be represented by a regular quaternion.

Similarly to (8.25), rotation of a *DC* requires quaternion multiplication (as shown, for example, in Eq. 8.28). The rotated *DC* source then needs to be converted into a *DC* standard quaternion representation by Eq. 8.41.

Thus, in our representation, an arbitrary quaternion is both a rotation operator and a *DC* source after simple transformations (8.41) have been performed (Kagan 2005b). Although the quaternion does not have the advantage of clearly identifying the *DC* source properties, its benefits are obvious. Multiple rotations of the *DC* source as well as the inverse problem determining the rotation from one source to another are easily computed using the methods of quaternion algebra (Kagan 1991c; Ward 1997; Kuipers 1999; Tape and Tape 2012).

8.4 Focal mechanism symmetry

This section addresses two problems: the symmetry properties of *DC* earthquake sources and how the symmetry influences *DC* rotation angle distribution.

8.4.1 Symmetry of DC source

Depending on the known properties of *DC* earthquake focal mechanism, we consider three types of earthquake source symmetry (Kagan 1990):

1. *DC*1 – double-couple with no symmetry or the identity (*I*) symmetry, if the focal plane and its sides are known.
2. *DC*2 – double-couple with C_2, order 2 cyclic symmetry, i.e., the focal plane is known, but its sides are not.
3. *DC*4 – double-couple with nodal-planes that not distinguishable; it has D_2, order 2 dihedral symmetry, this is the standard representation of focal mechanisms in earthquake catalogs.

These earthquake source symmetries correspond to the following symmetries considered in crystallographic texture analysis (see, for instance, Morawiec 2004): *DC*4 has an orthorhombic symmetry (as in a rectangular right parallelepiped or a rectangular box with unequal sides); *DC*2 has a monoclinic symmetry (as in a 3-D prism with two angles of 90° and one arbitrary angle); *DC*1 has a triclinic, or no symmetry.

Figures 2.1 and 8.1 display the geometry of the *DC* source (Aki and Richards 2002). It represents the quadrupolar "beachball" radiation patterns of earthquakes. The focal plots involve painting on a sphere the sense of the first motion of the far-field primary P-waves: solid (black) for compressional motion and open (white) for dilatational. The two orthogonal nodal planes separating these areas are the fault and the auxiliary planes. During routine determination of focal mechanisms, it is impossible to distinguish between these planes, a property called "nodal-plane ambiguity." The planes' intersection is the null-axis (called also the **b**-axis), the **p**-axis is in the middle of the open lune, and the **t**-axis is in the middle of the closed lune. These three axes are called the "principal axes of an earthquake focal mechanism," and their orientation defines the mechanism.

In principle, if we introduce sign-change or color-change (Lockwood and Macmillan 1978) as one of admissible symmetries, *DC*4 source would have D_4 dihedral symmetry, that is, the **p**- and **t**-axes are not distinguishable. Far-field displacement of seismic waves differs only by a sign in these quadrants. It would be interesting to study the consequences of such a representation.

In seismological literature to make the focal mechanism picture unique, the eigenvectors are pointed down. However, the handedness of the coordinate system formed by the vectors can change as the result of such an assignment. The systems of the opposing handedness cannot be rotated one into another. In most of our considerations, we use the right-handed coordinate system placed at each earthquake centroid.

Figure 8.2 displays four examples of the right-handed coordinate system for a *DC*4 source. The system can be arbitrarily rotated, and the handedness of the system is preserved. The left-handed system can be obtained in this picture if one inverts the direction of any individual axis (vector) or of all three axes. If the direction of two axes is reversed, the handedness of the system is preserved.

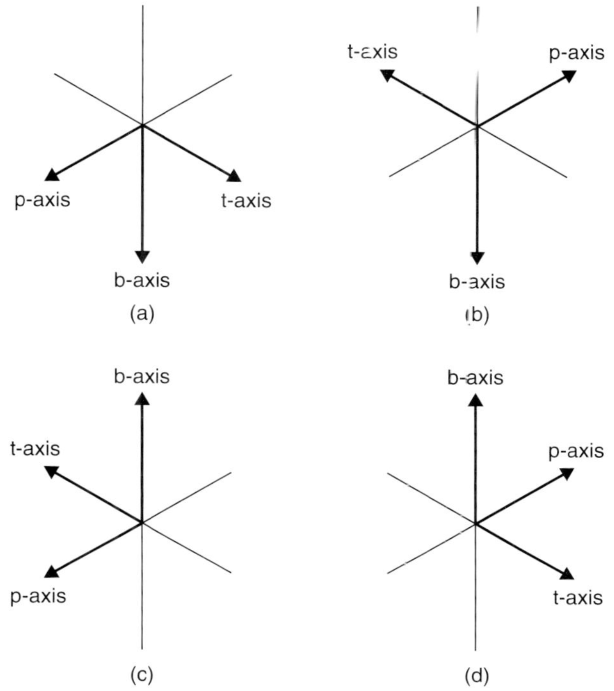

Fig. 8.2 Four schematic diagrams of earthquake focal mechanism with the *DC*4 symmetry. The right-hand coordinate system is used. We show how three vectors can be arranged in a standard representation of a 3-D Cartesian coordinate system. *Source:* Kagan (2013), Fig. 2.

The earthquake focal plane can often be determined by inverting the higher-rank point seismic moment tensors (McGuire *et al.* 2001; Chen *et al.* 2010) or by the aftershock pattern. The face/side (up/down or foot/hanging wall) of a focal plane generally is unknown. In such a case the *DC* focal mechanism has a C_2 symmetry; we call it *DC*2.

Finally, the face or the side of the focal plane can be known, as occurs in geologic studies of earthquake faults; they are shown in Fig. 8.1 by symbols '*A*' and '*B*'. Such a source is called *DC*1. In our simulation of earthquake space-focal mechanism structure, described in Section 3.3.2, the geometrical properties of each micro-dislocation are assumed to be known, thus the source is *DC*1.

Figure 8.3 illustrates the difference between *DC* mechanisms of various symmetries. A vertical cylinder of one material, 'A', shown by the beige color, is rotated counterclockwise in a half-space of another material, 'B'. Suppose that seismic events are registered in two places of their contact, '1' and '2'. If the focal plane is unknown, we would have two *DC*4 sources shown at the right and left ends of the diagram. If the focal planes, which are vertical lines in this plot, are known, the sources are *DC*2, and for the known sides of the fault planes two sources are *DC*1. In the diagram one *DC*4 can be transformed into another by four rotations: the zero (0°) rotation and three rotations by 180° around three principal axes (see Figs. 2.1, 8.1). For *DC*2 two rotations are needed: the zero rotation and one 180° rotation around the **b**-axis. For *DC*1 only the latter rotation is possible.

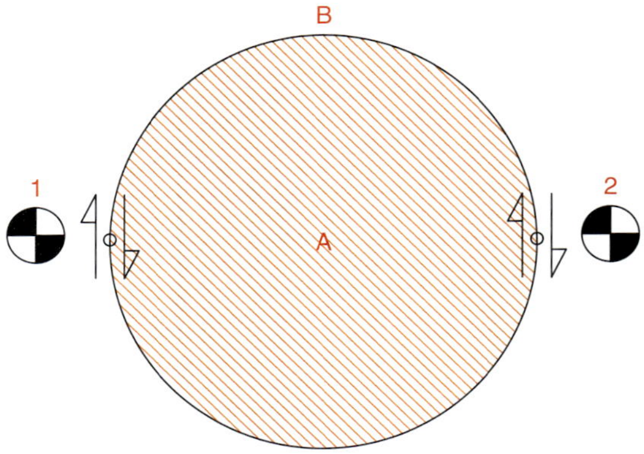

Fig. 8.3 Schematic diagram of earthquake focal mechanisms. Depending on whether a fault plane and its sides are known, the mechanisms could be considered as *DC*4, *DC*2, or *DC*1 (see the text). *Source:* Kagan (2013), Fig. 3.

In Fig. 8.2, any of the configurations appropriately rotated can represent a *DC*1 source, but only pairs (a) and (b) or (c) and (d) correspond to *DC*2 mechanisms. All four diagrams correspond to one *DC*4 source.

8.4.2 DC symmetry and rotation angle

Kagan (1991c) considered the inverse problem of the *DC*4 source rotation: for two earthquake focal mechanisms, to determine all the 3-D rotations by which one mechanism can be rotated into another. The *DC*4 symmetry results in four such rotations with angle Φ range $0° \leq \Phi \leq 180°$ (cf. Fig. 8.2). For most practical purposes, the rotation with a minimum angle Φ_{min} can be selected. Following designation in the crystallographic texture (Morawiec 2004, p. 115), we sometimes refer to general rotation angles as "misorientation angles" and the minimum rotation angle as the "disorientation angle."

The FORTRAN program which determines the 3-D rotation of *DC* sources is available on the Web – ftp://minotaur.ess.ucla.edu/pub/kagan/dcrot.for (see also FORTRAN90 adaptation of the program by P. Bird http://peterbird.name/oldFTP/2003107-esupp/Quaternion.f90.txt). Frohlich and Davis (1999) also discuss the program. Kagan (2007c) supplies simplified algorithms to calculate the angle and other parameters of the 3-D *DC*4 rotations. For the majority of cases these algorithms can be written in a few lines of a computer code.

In the program DCROT.FOR we first compute the orthonormal matrix for each *DC* source and then determine the corresponding normalized quaternion. There is a possibility of losing precision when converting a matrix to a quaternion (Shepperd 1978; Horn 1987). A certain computation technique should be applied to avoid this. We used a similar technique in our programs (Kagan 1991c, the end of the SUBROUTINE QUATFPS) and in DC1ROT.FOR.

How to compute the rotation of a *DC*1 source needed to align it with some reference *DC*1 source? In some cases we need to measure their angle of disorientation or the angular distance. However, it is unlikely that sufficient data would exist on *DC*1 sources for a statistical study of their distribution. Small rotation angles for such focal mechanisms are likely to correspond to $\Phi_{\min}$ for the *DC*4 source, as mentioned above.

To compute the disorientation of a *DC*1 source, we modify our FORTRAN program DCROT.FOR for the *DC*4, discussed earlier. When a fault plane and its faces are known, a focal mechanism would be better specified through a fault plane geometry (Aki and Richards 2002, their Figs. 4.13 and 4.20) with three angles: strike or azimuth (ϕ), dip (δ), and rake (λ). Usually the range of these angles is taken as follows: $0^\circ \leq \phi < 360^\circ$, $0^\circ \leq \delta < 90^\circ$, $-180^\circ \leq \lambda < 180^\circ$. The problem arises when comparing two sources, if the dip (δ) of one focal plane exceeds 90°, so that a foot wall of one mechanism becomes a hanging wall for another source. To simplify the calculations in our program (see below), we extend the δ range to 180°.

If the face/side of a fault plane is unknown, as shown in Fig. 8.3, we need to calculate the second angle of the rotation for the *DC*2 source. As with the *DC*1 source, the data on *DC*2 sources are sparse and likely insufficient for a statistical study, but it would be useful to have a technique to measure their angles of misorientation. An easy way to accomplish this measurement would be in a program for the *DC*1 source to change the strike of the fault plane by 180° and change the rake sign. The modified DC1ROT.FOR program is available at http://jumpy.igpp.ucla.edu/~kagan/dc1rot.for. In this program we use the quaternion technique to determine the rotation angle and the rotation axis parameters to transform one *DC* source into another. Quaternions are used because for the rotation angle Φ close or equal to 180°, the matrix method cannot accurately determine the rotation axis parameters (Kagan 2007c).

If only the rotation angle for the *DC*4 source is needed, then one can use a scalar (dot) product of two quaternions (Hanson 2006, p. 65; Dunn and Parberry 2011, p. 255) to determine the angle:

$$\cos(\Phi/2) = \mathbf{q}^a \cdot \mathbf{q}^b = q_0^a\, q_0^b + q_1^a\, q_1^b + q_2^a\, q_2^b + q_3^a\, q_3^b, \tag{8.43}$$

where $\mathbf{q}^i$ are normalized quaternions for each *DC* source and q_j are the quaternion's components.

8.5 Earthquake focal mechanism and crystallographic texture statistics

Frank (1988) proposes using the Rodrigues vector space to represent 3-D rotation of symmetrical objects. The 3-D Rodrigues vector $\mathbf{r}$ (r_i, $i = 1, 2, 3$) is calculated as

$$\mathbf{r} = \mathbf{n}\tan(\Phi/2) \quad \text{for} \quad 0^\circ \leq \Phi \leq 180^\circ, \tag{8.44}$$

where Φ is the rotation angle and $\mathbf{n}$ is the rotation axis. This representation has an advantage: under any transformation of the Rodrigues map corresponding to a change of the reference orientation, straight lines transform into straight lines, and planes into planes. Each straight line segment in the space corresponds to a rotation around the fixed axis, the length of the segment connecting the origin to the point is equal to $\tan(\Phi/2)$.

For an object with the non-identity symmetry, accepted points in the Rodrigues space lie in a region around the origin, which is called "the fundamental zone of the map." The zone is a polyhedron, bounded by the planes which are orientationally equidistant between the origin and the neighboring equivalent point by a symmetry rotation to the origin. Any points lying outside one of these planes have an equivalent point lying inside the fundamental zone (Frank 1988). For an orthorhombic crystal with three orthogonal axes, the fundamental zone is a cube, with its six faces orthogonal to the axes at a distance from the origin of tan(45°) = 1 (Frank 1988). The cube is surrounded by three neighboring zones, each divided into two at infinity.

Frank (1988), Neumann (1992) and Morawiec and Field (1996) propose using the Rodrigues vector space to display the disorientation of symmetric objects in a fundamental zone as a point in the space. The point coordinates are calculated as follows: the length of a vector **r** in Eq. 8.44 is

$$\zeta = |\mathbf{r}| = \tan(\Phi_{min}/2). \tag{8.45}$$

where Φ_{min} is the minimum rotation angle. Three coordinates of a point in the zone are (see Fig. 8.4)

$$\begin{aligned} x_1 &= \zeta \times \sin(\theta)\ \sin(\phi); \\ x_2 &= \zeta \times \sin(\theta)\ \cos(\phi); \\ x_3 &= \zeta \times \cos(\theta), \end{aligned} \tag{8.46}$$

where θ is the colatitude, and ϕ is the azimuth of the rotation axis. As shown in Fig. 8.4, we identify x_1 with a point on the **p**-axis; similarly x_2 is a point on **t** and x_3 on **b**.

Figure 8.4 shows the fundamental zone for a *DC*4 source (cf. Heinz and Neumann 1991, Fig. 7). It is a cube with corner coordinates $x_1 = \pm 1.0$; $x_2 = \pm 1.0$; $x_3 = \pm 1.0$. Owing to the *DC*4 symmetry and statistical source exchange symmetry (i.e., rotation of s_1 source into s_2 is equivalent to rotation of s_2 source into s_1), an octant of the cube contains full information about the orientation distribution for uniformly random rotation. This octant is called the "MacKenzie cell" (Morawiec and Field 1996, see also their Fig. 1 displaying the cells for the D_3 and D_4 symmetries).

Each point *inside* the cube uniquely corresponds to a certain orientation/rotation with a minimum rotation angle $\Phi_{min} \leq 120°$. The points inside the inscribed sphere of the cube correspond to the rotations with angles $\Phi_{min} \leq 90°$. The other three symmetrically equivalent rotations are situated outside the fundamental zone. For example, the point of the zero rotation is located at the cube center, whereas three other rotation points are at infinity: $x_1 = \pm\infty$; $x_2 = \pm\infty$; $x_3 = \pm\infty$, corresponding to 180° rotations. These points at $\pm\infty$ are equivalent (Altmann 1986). Similarly, for any point inside the cube, three points outside correspond to the rotations with angles $\Phi > \Phi_{min}$.

However, for certain point configurations up to four equal Φ_{min} appear on the cube boundary. To demonstrate the symmetry representation in the

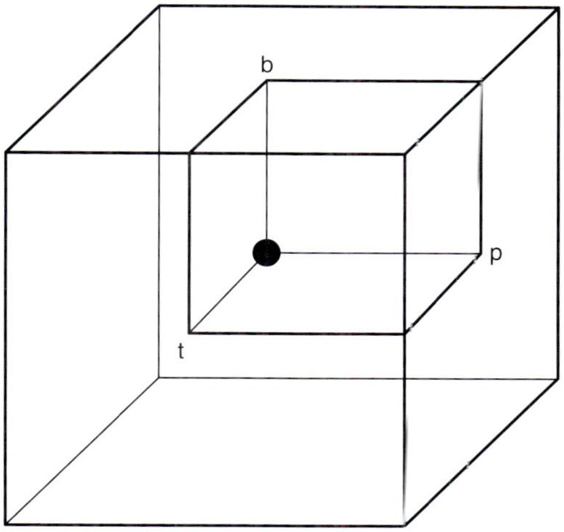

Fig. 8.4 Fundamental zone display for *DC*4 source. The **b**, **p**, and **t**-axes of the source are shown. We identify x_1 with a point on the **p**-axis; similarly x_2 is a point on **t** and x_3 on **b**. The octant of the cube $0 \leq x_1 \leq 1;\ 0 \leq x_2 \leq 1;\ 0 \leq x_3 \leq 1$ is called the "MacKenzie cell." *Source:* Kagan (2013), Fig. 4.

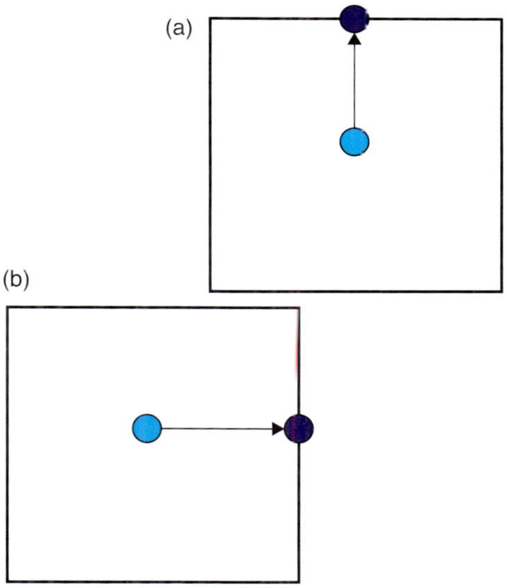

Fig. 8.5 Fundamental zone display for *DC*4 source. Two opposite faces of the fundamental cube (Fig. 8.4) are shown. Colors show two face points corresponding to one source orientation with the angle $\Phi \geq 90°$. The central cyan points correspond to two equivalent rotations $\Phi = 90°$. Similar pairs of points may occur on all 6 cube faces. When a rotation point moves to the edge of one face, an equivalent point on the opposite face also moves to another edge, resulting in rotation with an angle 109.5° (see Fig. 8.6). *Source:* Kagan (2013), Fig. 5.

Rodrigues space and the appearance of these equal minimum rotation angles, we show complicated trajectories of *DC*4 source rotation in the Rodrigues space in Figs. 8.5–8.7.

For example, when a point moving orthogonally from the origin reaches a cube face, it simultaneously appears on the opposite face: two 90° rotations produce the same effect (Fig. 8.5). This means that when we determine the minimum angle Φ_{min} for cyan point rotations shown in Fig. 8.5 using the program developed by Kagan (1991c), we find two equal solutions. The remaining two angles are greater than Φ_{min} (see also the previous section).

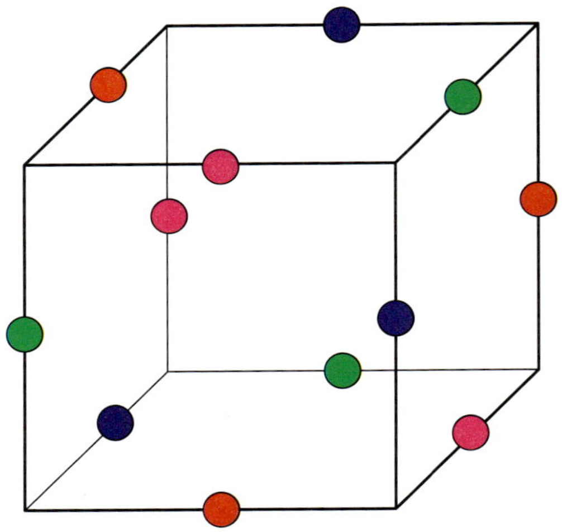

Fig. 8.6 Fundamental zone display for *DC*4 source. Colors show four sets of three edge points. Each set corresponds to one source disorientation with the angle $\Phi \approx 109.47^\circ$ (8.50). Compare to Fig. 8.5 where two edge points are the result of cyan points moving on opposing faces. The third point appears as it moves from outside the fundamental cube to the third edge. Three other sets of points are similarly produced. *Source:* Kagan (2013), Fig. 6.

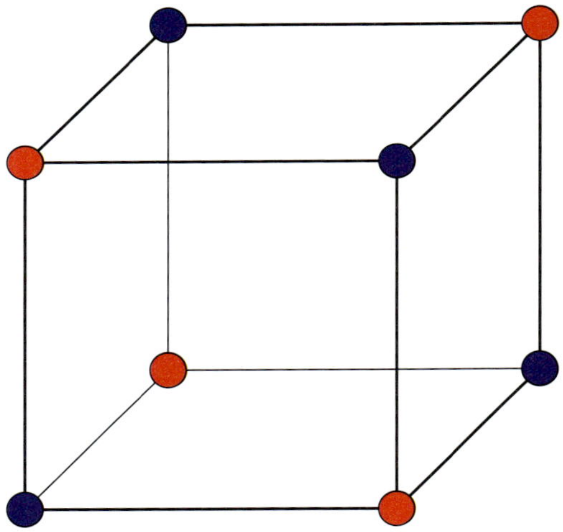

Fig. 8.7 Fundamental zone display for *DC*4 source. Colors show two sets of four vertex points. Each set corresponds to one source orientation with the angle $\Phi = 120^\circ$. *Source:* Kagan (2013), Fig. 7.

If a point on one face moves to an edge, the "identical" point on the opposite face simultaneously moves to another edge until both points reach the middle of the edges. This orientation corresponds to the rotation 109.47° (see Eq. 8.50) and Fig. 8.6 shows that there are three equivalent points at the edges. The third point appears as it moves from the outside of the cube to the third edge. As in Fig. 8.5, this means that three equal angles $\Phi_{\min}$ would be obtained. Finally, when a point is at a vertex, as shown in Fig. 8.7, three other vertices correspond to the same rotation $\Phi = 120^\circ$ (Frank 1988), that is, all four rotation angles are $\Phi_{\min}$.

This arrangement of the orientations for the rotation angles $\Phi \geq 90^\circ$ describes a complex topology for *DC*4 source rotation. This topology involves projective or Möbius transformation (Altmann 1986; Frank 1988). Full analysis of the *DC*4 source orientation, when and if performed, would involve very intricate investigations of rotation angle transformations due to the source symmetry.

The Rodrigues space has no special advantages in displaying the orientation distribution for two other sources: *DC*2 and *DC*1. For *DC*2 source which has C_2 symmetry, the fundamental zone is a layer bounded by two planes perpendicular to the **b**-axis, each at the distance tan(45°) = 1 from 0. For a *DC*1 source, the whole Rodrigues space up to infinity is included in the fundamental zone. In these cases, other spaces are more convenient in displaying the 3-D rotation distribution (Frank 1988; Morawiec and Field 1996). Altmann (1986, pp. 164–176) explains the projective or Möbius topology of rotations in the quaternion parametric ball for non-symmetrical objects.

8.6 Rotation angle distributions

8.6.1 Uniform random rotation of DC sources

The distribution of the uniform random rotation for *DC* sources constitutes a reference for observational distributions of earthquake focal mechanisms where we expect the distributions to be only partially random. These stochastic distributions can be analytically calculated by taking into account the sources' symmetry.

A distribution of the minimum angle $\Phi_{\min}$ for a uniform random rotation of the *DC*4 source was obtained by Kagan (1990, his Eqs. 3.1–3.3), using the results by Handscomb (1958) and Mackenzie (1958) for the random disorientation of two cubes. The PDF is

$$f(\Phi) = (4/\pi)(1 - \cos\Phi) \quad \text{for} \quad 0^\circ \le \Phi \le 90^\circ; \tag{8.47}$$

$$f(\Phi) = (4/\pi)(3\sin\Phi + 2\cos\Phi - 2) \quad \text{for} \quad 90^\circ \le \Phi \le \Phi_S; \tag{8.48}$$

and

$$f(\Phi) = (4/\pi)\left\{3\sin\Phi + 2\cos\Phi - 2 - (6/\pi)\left[2\sin\Phi\,\arccos\left(\frac{1+\cos\Phi}{-2\cos\Phi}\right)^{1/2}\right.\right.$$
$$\left.\left. - (1-\cos\Phi)\,\arccos\frac{1+\cos\Phi}{-2\cos\Phi}\right]\right\} \quad \text{for} \quad \Phi_S \le \Phi \le 120^\circ, \tag{8.49}$$

where

$$\Phi_S = 2\arccos(3^{-1/2}) = \arccos\left(-\frac{1}{3}\right) \approx 109.47^\circ. \tag{8.50}$$

For the *DC*2 source a similar PDF is

$$f(\Phi) = (2/\pi)[1 - \cos(\Phi)] \quad \text{for} \quad 0^\circ \le \Phi \le 90^\circ; \tag{8.51}$$

and

$$f(\Phi) = (2/\pi)\sin(\Phi) \quad \text{for} \quad 90^\circ \le \Phi \le 180^\circ. \tag{8.52}$$

For the $DC1$ source the function is

$$f(\Phi) = (1/\pi)[1 - \cos(\Phi)] \quad \text{for} \quad 0^\circ \leq \Phi \leq 180^\circ. \tag{8.53}$$

Grimmer (1979), also following Handscomb's (1958) and Mackenzie's (1958) results, obtained similar analytic expressions for a completely random rotation of orthorhombic, monoclinic, and triclinic crystals (equivalent in symmetry to the *DC* earthquake source with various restrictions described in Subsection 8.4.1). He listed median angles as well as their mean and standard deviations for all these distributions. Morawiec (1995, 2004, pp. 117–119) derived these distributions by integration in the Rodrigues space.

8.6.2 Non-uniform distributions of random rotations

Two non-uniform rotation angle distributions are useful in analyzing earthquake focal mechanism rotation: the rotational Cauchy law (Kagan 1982, 1992c) and von Mises-Fisher/Bingham rotational distribution (Kagan 1992c, 2000; Schaeben 1996; Mardia and Jupp 2000, pp. 289–292; Morawiec 2004, pp. 88–89).

The Cauchy distribution is especially important for representing earthquake geometry, since it can be shown by theoretical arguments (Zolotarev 1986, pp. 45–46; Kagan 1990, 1994b) and simulations (Kagan 1990) that the stress tensor in a medium with random defects follows this distribution (Subsection 8.1.2). The Cauchy law is a *stable* distribution (Zolotarev 1986). The stable distributions are essential for two reasons (Zaliapin *et al.* 2005; Section 5.4): (a) They are invariant under addition of random variables; (b) Stable distributions have a power-law tail, that is, they are asymptotically scale-invariant.

The PDF of the rotational or angular Cauchy distribution can be written as (Kagan 1982, 1990)

$$\begin{aligned} f(\Phi) &= \frac{2}{\pi}\left[\frac{\kappa A^2(1+A^2)}{(\kappa^2+A^2)^2}\right] \\ &= \frac{4\kappa\ [1-\cos(\Phi)]}{\pi\ [1+\kappa^2+(\kappa^2-1)\ \cos(\Phi)]^2}, \quad \text{for} \quad 0^\circ \leq \Phi \leq 180^\circ, \end{aligned} \tag{8.54}$$

where $A = \tan(\Phi/2)$. The scale parameter κ of the Cauchy distribution represents the degree of *incoherence* or *complexity* in a set of earthquake focal mechanisms. The cumulative rotational Cauchy distribution is

$$F(\Phi) = \frac{2}{\pi}\left[\arctan(A/\kappa) - \frac{A \times \kappa}{A^2+\kappa^2}\right]. \tag{8.55}$$

The Cauchy distribution is assumed to be axisymmetric on the quaternion hypersphere S^3. This means that the rotation axis poles are distributed uniformly over a regular S^2 sphere. For a general case, the axes distribution for earthquake focal mechanisms may need to be specified as non-uniform. In that case certain rotations would be preferred depending on the focal mechanism of a reference event.

Moreover, in our previous investigations the rotation angle and the pole distributions were considered as independent; in reality, these PDFs cannot be factored and should be studied as a joint distribution. However, we have not yet advanced to this stage (see Section 8.7).

The von Mises-Fisher distribution for the 3-D orientation is widely discussed in the literature (Schaeben 1996; Mardia and Jupp 2000; Morawiec 2004). Schaeben (1996) and Morawiec (2004) show that this distribution is essentially equivalent to the Bingham distribution. The simplest von Mises-Fisher distribution is a Gaussian-shaped function concentrated near the zero angle. This distribution can be implemented to model random errors in determining focal mechanisms. The distribution has many forms. However, even the simple axially symmetric expressions, due to the complexity of normalization, represent difficult computations.

For small values of the rotation angle standard error σ_Φ, the von Mises-Fisher-type distribution is equivalent to the rotational Maxwell law used by Kagan and Knopoff (1985a, Eq. A3) and Kagan (1992c, Eq. 12). The latter distribution is obtained by generating a 3-D normally distributed random variable $\mathbf{u}$ (u_1, u_2, u_3) with the standard deviation $\sigma_\mathbf{u}$ $(\sigma_{u_1}, \sigma_{u_2}, \sigma_{u_3})$ and then calculating the unit quaternion

$$q_0 = 1/\sqrt{1 + u_1^2 + u_2^2 + u_3^2},$$
$$q_i = u_i/\sqrt{1 + u_1^2 + u_2^2 + u_3^2}, \quad \text{for} \quad i = 1, 2, 3. \tag{8.56}$$

The 3-D rotation angle is calculated

$$\Phi = 2\arccos(q_0) \approx 2\arccos[1 - (u_1^2 + u_2^2 + u_3^2)/2]$$
$$\approx 2\arcsin\sqrt{u_1^2 + u_2^2 + u_3^2} \approx 2\sqrt{u_1^2 + u_2^2 + u_3^2}. \tag{8.57}$$

The final expression is twice the length of a vector in the 3-D space.

Since components of vector $\mathbf{u}$ are normally distributed, the angle Φ (in degrees) follows the Maxwell law with

$$\sigma_\Phi = 360\ \sigma_u/\pi, \tag{8.58}$$

where we assume that all components of $\sigma_\mathbf{u}$ are equal $(\sigma_{u_1} = \sigma_{u_2} = \sigma_{u_3} = \sigma_u)$. For $0^\circ \leq \Phi \leq 180^\circ$ the Maxwell PDF is (cf. Eq. 7.5)

$$\psi(\Phi) = \sqrt{\frac{2}{\pi}} \times \frac{\Phi^2}{\sigma_\Phi^3} \times \exp\left[-\Phi^2/\left(2\sigma_\Phi^2\right)\right]. \tag{8.59}$$

where a standard error is σ_Φ. The Maxwell cumulative distribution function (CDF) is

$$\Psi(\Phi) = \operatorname{erf}\left(\frac{\Phi}{\sigma_\Phi\sqrt{2}}\right) - \sqrt{\frac{2}{\pi}} \times \frac{\Phi}{\sigma_\Phi} \times \exp\left[-\Phi^2/\left(2\sigma_\Phi^2\right)\right], \tag{8.60}$$

where erf (.) is an error function.

The major problem with these non-uniform random distributions is that they do not consider the symmetry of the rotated object. When rotation angles are small, the distribution is concentrated around the zero angle neighborhood. For the *DC*4 source as shown in Fig. 8.4, almost all distribution density would be inside the fundamental zone. However, for more spread out angle distributions, we should account for cases where the rotation angle exceeds the maximum angles (see, for example, Eqs. 8.47–8.50). Then the distribution would be folded back into the fundamental zone.

Mason and Schuh (2009) propose convolving the angle distribution with appropriate 3-D spherical or 4-D hyperspherical harmonics to obtain a new angle distribution which fits into the fundamental zone. It is not clear whether such calculations can be made analytically. Simulation seems the only practical way to transform both Cauchy and von Mises/Fisher distributions for the D_2 symmetric case (i.e., for maximum Φ_{min} rotation angle 120°). Kagan (1992c, Fig. 3c) displays such distributions.

8.7 Focal mechanisms statistics

8.7.1 Disorientation angle statistics

Since there is no general model of earthquake focal mechanism distribution, we need to study the distribution of mechanisms in earthquake catalogs empirically to infer their properties. How various tectonic and geometrical factors shape the distribution of earthquake sources should be studied as well. Such investigations are difficult because we are dealing with a multidimensional stochastic point process: earthquake size, occurrence time, location, and source parameters serve as potential inputs to the distributions. In Subsection 5.3.3 we used several methods to investigate earthquake focal mechanisms and showed that they have significant complexity.

In this section we are mostly interested in the distributions of rotation between two earthquake focal mechanisms. Even if we fix earthquake time, space, and magnitude interval, the *DC* rotation distribution depends on at least three variables: the rotation angle and two spherical coordinates of a rotation axis pole. Displaying all three degrees of freedom in a distribution presents a difficult problem. Therefore, in our previous investigations we studied partial, marginal distributions. For example, Kagan (1992c, Figs. 6–9; 2009a, Figs. 9–10) obtained various distributions of the rotation angle Φ_{min} between two focal mechanisms. Below we first update our most important results on the distribution of the rotation angle Φ_{min}, and then analyze a three-dimensional distribution of rotation angle and the axes in the Rodrigues space.

Figure 8.8 displays cumulative distributions of the rotation angle Φ_{min} for shallow earthquake pairs with the magnitude threshold $m_t = 5.0$ that are separated by a distance of less than 50 km. We study whether the rotation of focal mechanisms depends on where the second earthquake of a pair is situated with regard to the first event. Thus, we measure the rotation angle for centroids located in

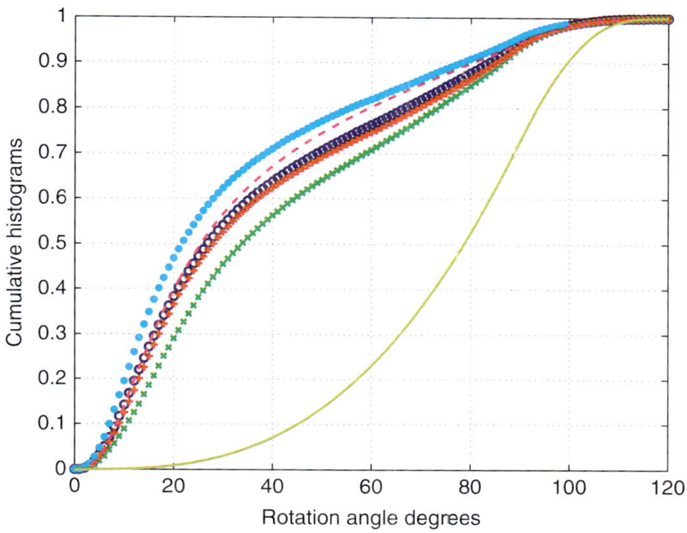

Fig. 8.8 Cumulative distributions of rotation angles for pairs of focal mechanisms of shallow earthquakes (depth 0–70 km) in the GCMT catalog 1977–2011; centroids are separated by distances between 0–50 km, magnitude threshold $m_t = 5.0$. The total number of events is 26,986. Lines from left to right: filled cyan circles are centroids in 30° cones around the **b**-axis; red dashed line is for the Cauchy rotation with $\kappa = 0.1$; blue circles – all centroids; magenta crosses – centroids in 30° cones around the **p**-axis; green x-signs – centroids in 30° cones around the **t**-axis; right yellow solid line is for the random rotation. *Source:* Kagan (2013), Fig. 8.

30° cones around each principal axis of the first event (see curves, marked the **t**-, **p**-, and **b**-axes).

The curves in Fig. 8.8 are narrowly clustered, with about 95% of angles less than 90°, thus situated within an inscribed sphere of the fundamental zone (Fig. 8.4). This pattern can be compared to the uniform rotation (Eqs. 8.47 –8.49) for which 72.7% $(2 - 4/\pi)$ of angles are within 90°. The curves are obviously well approximated by the *DC* rotational Cauchy distribution (Eq. 8.55). This distribution is characterized by a parameter κ; a smaller κ-value corresponds to the rotation angle Φ_{min} concentrated closer to zero. Thus, regardless of spatial orientation, all earthquakes have focal mechanisms similar to a nearby event. Earthquakes in the cone around the **b**-axis correspond to a smaller κ-value than events near the other axes. These results are similar to those shown in Fig. 6 by Kagan (1992c) or Fig. 9 by Kagan (2009a).

Figure 8.9 shows the angle Φ distribution for larger distances (400–500 km) between earthquake pairs. The curve corresponding to fault-planes (the **b**-axis) is clearly separated from the curves connected with the **t**- and **p**-axes. Although the rotation near the **b**-axis is relatively small ($\kappa \approx 0.2$), the earthquakes situated in cones around the **t**- and **p**-axes have focal mechanisms essentially uncorrelated with the primary event: the curves are close to the curve corresponding to a completely random rotation of a *DC*. These variations of the Φ-angle may be caused by a focal mechanism change within one fault system or between different systems or even between different plate boundaries. However, the general properties of the Φ distribution are similar to those observed for the near earthquakes (Fig. 8.8).

The GCMT focal mechanisms are defined on the geographical spherical coordinate system, in their comparison to simplify computations we use the

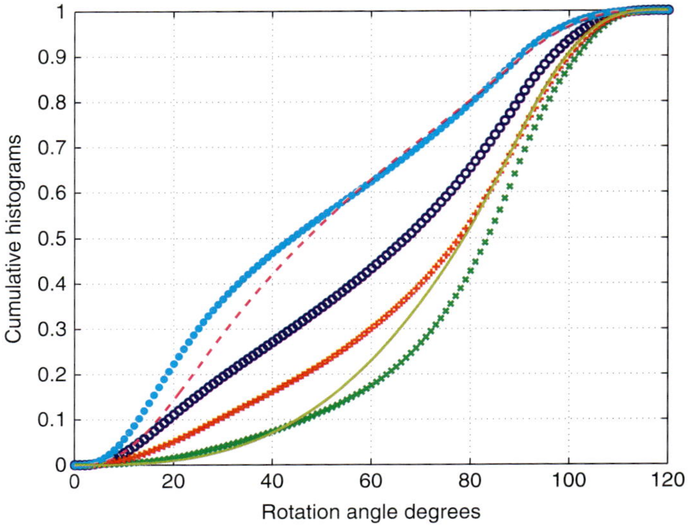

Fig. 8.9 Cumulative distributions of rotation angles for pairs of focal mechanisms of shallow earthquakes. Similar selection criteria and line markings as in Fig. 8.8 are applied, but the centroids are separated by distances between 400–500 km and the red dashed line is for the Cauchy rotation with $\kappa = 0.2$.

Mercator projection (Bugayevskiy and Snyder 1995) to have longitude axes parallel. This projection should introduce a bias, especially near the Earth poles; to reduce the error we limit the distance between the earthquake pairs to 500 km. Fortunately, very few earthquakes occur in polar regions (Fig. 2.2). More complicated calculations need to be carried out to avoid this problem.

Figure 8.10 and Fig. 8.11 show the disorientation angle distribution for the magnitude cutoff $m_t = 5.8$. As may be expected for the higher magnitude, the angles are concentrated closer to zero, and the difference between the curves

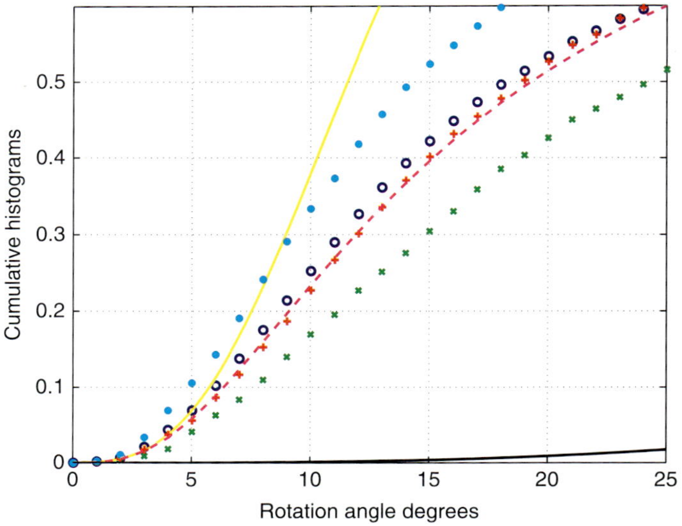

Fig. 8.10 Cumulative distributions of rotation angles for pairs of focal mechanisms of shallow earthquakes. Similar selection criteria and line markings as in Fig. 8.8 are applied, but the magnitude threshold $m_t = 5.8$ and the red dashed line is for the Cauchy rotation with $\kappa = 0.075$. The total number of events is 6,160. The left yellow solid line is for the Maxwell rotation CDF (8.60) with $\sigma_\Phi = 7.5°$; the right solid line (near abscissa axis) is for the random rotation. *Source:* Kagan (2013), Fig. 9.

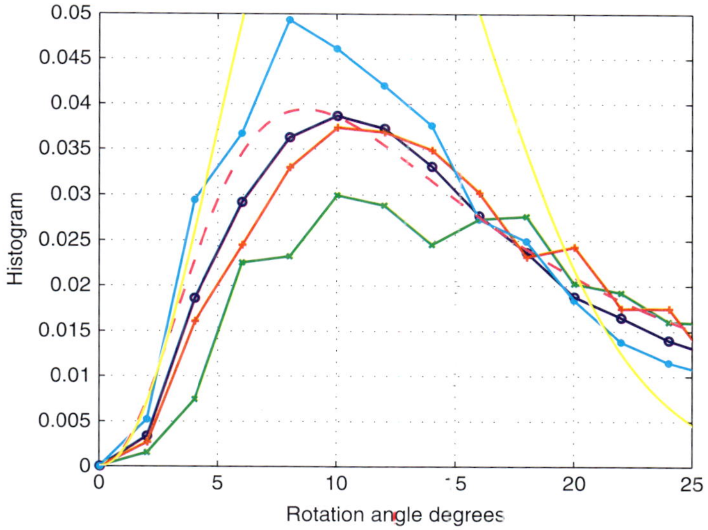

Fig. 8.11 Histograms of rotation angles for pairs of focal mechanisms of shallow earthquakes. Similar selection criteria and line markings as in Fig. 8.10 are applied, but instead of CDFs for the Cauchy and the Maxwell rotation distribution, the PDFs are plotted. *Source:* Kagan (2013), Fig. 10.

corresponding to various cones increases. Maxwell distribution curves are shown to illustrate the possible behavior of the angle distributions near zero. The $\sigma_\Phi = 7.5^\circ$ parameter of the distribution is small, compared to the distribution range ($0^\circ \leq \Phi \leq 120^\circ$). Therefore, the curves are concentrated close to zero; we do not need simulation to consider the curve behavior for large values of Φ_{min}, as is done, for example, in Fig. 3c by Kagan (1992c).

The difference in the distribution curves corresponding to various focal mechanism axes suggests that the Cauchy distribution parameter κ depends upon the geometry of a fault system. Contrary to our assumptions (Eq. 8.54), poles of rotation axes are not uniformly distributed over the S^2 sphere.

Figure 8.12 displays a smoothed map of the average rotation angle Φ dependence on time and distance intervals for a subset GCMT catalog. Well-constrained earthquakes have been selected according to criteria proposed by Kagan (2000). The angle increases with distance between events. The increase with time interval (ΔT) is much less pronounced. For earthquake sequences clustered in time and space, the Φ difference between focal mechanisms is small, on the order 10–15°. These Φ-values are close to the minimum uncertainty in Φ evaluation (Kagan 2003).

8.7.2 Distributions of rotation axes

Mackenzie (1964) derived the distribution of the rotation axes for cubic symmetry. Morawiec (1996) obtained distributions of rotation axes for any symmetric object encountered in crystallographic texture analysis. Using his results we can write down the distribution of rotation axes for the D_2 symmetry: the *DC*4 source.

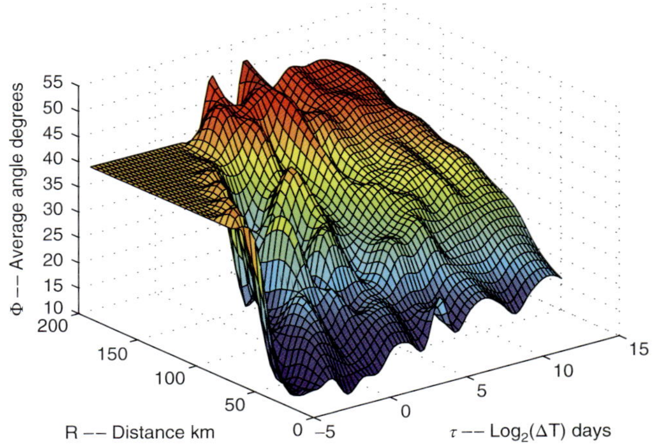

Fig. 8.12 Dependence of the average rotation angle Φ on time difference and distance between two earthquakes for shallow well-constrained events with magnitude $m \geq 5.5$ registered in the time period 1977/1/1–1999/3/31 in the GCMT catalog. Source: Kagan (2000), Fig. 9a.

The Mackenzie cell is shown in Fig. 8.4. We designate the coordinate axes as x_i, $i = 1, 2, 3$, and the distribution depends on distance from the origin. As seen in Fig. 8.4, the distribution should have a 3-fold cyclic symmetry C_3 around the origin or around the cube vertex. Then the PDF for the axes length is

$$p\,(\rho) = \frac{16}{\pi^2}\left[\arctan(\rho) - \rho\,/(1+\rho^2)\right], \tag{8.61}$$

where $\rho = \sqrt{x_1^2 + x_2^2 + x_3^2}$ is the distance from the origin [0, 0, 0] to the walls of the fundamental zone. For small value of ρ, $p(\rho) \propto \rho^3$. The $p\,(\rho)$ density values at $\rho = 1$, $\rho = \sqrt{2}$, and $\rho = \sqrt{3}$ correspond to appropriate values for the D_2 symmetry (Morawiec 1996, Table 2). These ρ-values correspond to the rotation angles $\Phi = 90°$, $109.47°$, and $120.0°$, respectively. For $\Phi \leq 90°$ the rotation axes are distributed uniformly over the S^2 sphere, but they intersect the sphere near the cube vertex close to $\Phi = 120°$.

8.7.3 Rodrigues space statistics and display

A major problem in the orientation visualizing is the high dimensionality of the 3-D rotation space: the orthogonal matrices are characterized by nine values, the seismic moment tensor requires five or six variables, and the normalized quaternion needs four values. The real number of degrees of freedom for a 3-D rotation is three. Thus, in principle, an orientation distribution can be shown in a 3-D diagram.

We obtain a distribution diagram for a set of disorientations. One way to display such a diagram of the fundamental zone is through stereo-pairs (Neumann 1992). Morawiec and Field (1996) display the distribution of disorientation parameters by points in some parallel sections of the fundamental zone. We follow their example.

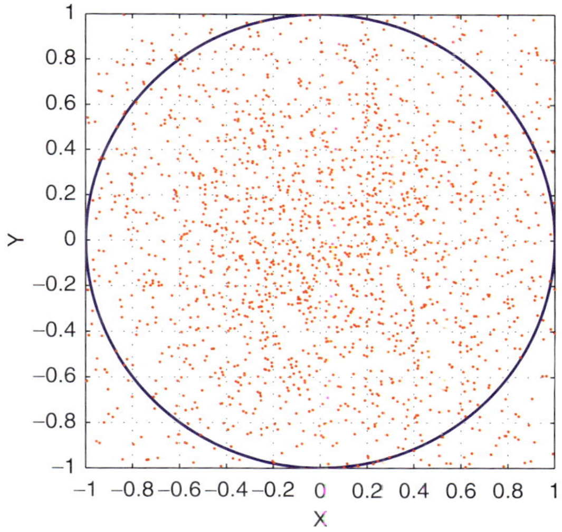

Fig. 8.13 Disorientation distribution in the fundamental zone of the Rodrigues space for randomly rotated *DC*4 sources. The points are shown in the central section of the fundamental zone $-0.05 \leq x_3 \leq 0.05$ (see Fig. 8.4). The total point number is 25,000; 1754 points are in the central section. **X** axis corresponds to x_1, **Y** to x_2. *Source:* Kagan (2013), Fig. 11.

Figure 8.13 shows a distribution for randomly rotated *DC*4 sources in a central section ($-0.05 \leq x_3 \leq 0.05$) of the fundamental zone of the Rodrigues space. For the section of the zone $x_3 = 0$, the points in the inscribed blue circle of the square correspond to the rotations with angles $\Phi_{\min} \leq 90°$, whereas for the square corners of the section the angle $\Phi_{\min} \approx 109.5°$. Fig. 8.14 displays a similar distribution of the earthquake focal mechanism orientation in the GCMT catalog. We show how the second earthquake (s_2) of a pair is rotated in regard to the earlier event (s_1). Since the disorientation of these two events may not be symmetric in time, we display it in a full fundamental zone. As Figs. 8.8–8.11 demonstrate,

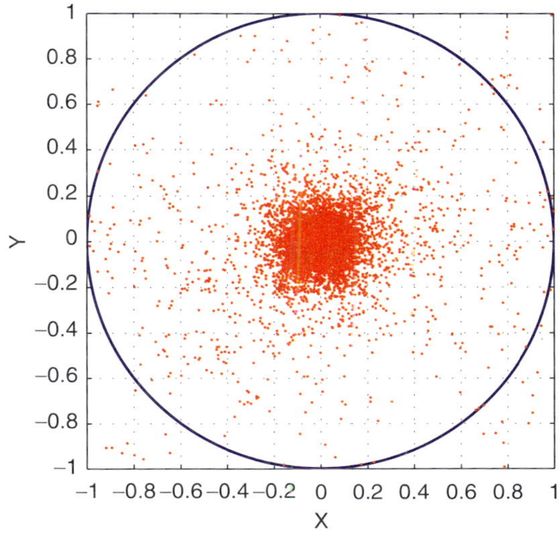

Fig. 8.14 Disorientation distribution in the fundamental zone of the Rodrigues space for shallow earthquakes in the GCMT catalog 1977–2011. Centroids are separated by distances between 0–50 km; magnitude threshold $m_t = 5.8$; the total number of events is 6,160. The points are shown in the central section of the fundamental zone $-0.05 \leq x_3 \leq 0.05$ (see Fig. 8.4). The total number of earthquake pairs is 19,397; 8,725 pair points are in the central section. **X** axis corresponds to x_1, **Y** to x_2. *Source:* Kagan (2013), Fig. 12.

Table 8.1 Properties of disorientation point scatter in the fundamental zone of a *DC*4 source.

#	m_t	N	N_p	N_c	ρ_{pt}	ρ_{bp}	ρ_{bt}	ρ'_{pt}	$\overline{\Phi} \pm \sigma_\Phi$
1	5.6	9,615	43,611	18,381	0.126	0.026	0.013	0.170	31.6 ± 27.5
2	5.8	6,160	19,367	8,725	0.117	0.022	−0.001	0.127	29.9 ± 26.9
3	6.25	2,154	2,741	1,244	0.274	−0.016	−0.026	0.257	28.5 ± 26.1
4	Simul.	25,000	25,000	1754	−0.012	−0.002	0.002	−0.015	75.0 ± 21.0

Notes: The GCMT catalog time interval is 1977–2011. N is the total number of events with magnitude $m \geq m_t$; N_p is the total number of event pairs with centroids separated by less than 50 km; N_c is the number of event pairs in the central section (cf. Fig. 8.14); ρ_{pt}, ρ_{bp}, and ρ_{bt} are the correlation coefficients for all points; ρ'_{pt} is the correlation coefficient for all points within the central section; $\overline{\Phi} \pm \sigma_\Phi$ is the average disorientation angle and its standard deviation. *Source:* Kagan (2013), Table 1.

the distribution of the rotation angles for earthquake sources is strongly concentrated close to $\Phi_{\min} = 0°$. If we compare Fig. 8.13 and Fig. 8.14, this concentration is marked in the fundamental zone display.

Table 8.1 summarizes earthquake focal mechanism disorientation patterns in the fundamental zone of *DC*4. Although the central section $-0.05 \leq x_3 \leq 0.05$ occupies only a small part of the fundamental cube (Fig. 8.14), close to 50% of the pairs are there due to a tight concentration of rotation angles near the zero value. For simulated events in the same central zone, the number N_c is about 7% of the total (see Fig. 8.13).

We also display the correlation coefficients of the point scatter field. Whereas for earthquake focal mechanisms the coefficients ρ_{bp} and ρ_{bt} are close to zero, the ρ_{pt} and ρ'_{pt} coefficients are non-zero, testifying to a certain non-random pattern of focal mechanism rotation. Fig. 11 in Kagan (2009a) also shows that rotation axes are concentrated closer to the **t**-axis. All correlation coefficients for the simulated mechanisms are around zero. A more detailed statistical analysis of this pattern is a task for future work.

The values of the average rotation angle and its standard deviation ($\overline{\Phi} \pm \sigma_\Phi$) show that for larger earthquakes both variables are smaller. This may be caused by a higher accuracy in determining focal mechanism for stronger shocks (Kagan 2003). For simulated focal mechanisms, the $\overline{\Phi} \pm \sigma_\Phi$ values are close to the theoretical estimates for the orthorhombic symmetry – 75.16° ± 20.85° (Grimmer 1979, Table 5).

8.7.4 Summary of results for DC orientation

In this subsection we briefly summarize results for Sections 8.2–8.7, the study of earthquake focal mechanisms orientation. The quantitative study of earthquake focal mechanisms is an important prerequisite for understanding earthquake rupture. Though these investigations began in the mid-1950s, publications have been mostly descriptive until now; relatively little modeling and rigorous statistical analysis have been performed. A major difficulty in analyzing focal mechanisms is both the high dimensionality and non-commutativity of the 3-D rotations as well as the complex topological properties of the rotation space (Section 8.5). This presents a major challenge in analyzing a set of focal mechanisms.

Kagan (1992c, 2000, 2009a) investigated statistical distributions of earthquake focal mechanisms. We found that the disorientation angle is close to zero for spatially close earthquakes, and the angle decreases if the inter-earthquake time interval approaches zero. Kagan (2009a) showed that the CLVD component of focal mechanism tensor is either zero or close to zero for most geometric barriers proposed as common features in an earthquake fault system.

However, the major challenges in describing and understanding the distributions of focal mechanisms still remain. As we see from Figs. 8.8–8.11, the angle distribution is not axially symmetric: in certain directions Φ_{min} is larger than in others. Thus, the distributions proposed to approximate the angle pattern, like the rotational Cauchy distribution (Eqs. 8.54–8.55), need to be made more complex.

The distribution of rotation axes was not investigated as thoroughly as that for disorientation angles. There is still no theoretical model for approximating empirical data, but applying the Rodrigues space may render such analysis more manageable.

However, even these limited results contribute significantly to an understanding of earthquake focal mechanism properties and allow certain quantitative applications for seismic risk evaluation. Kagan and Jackson (2011) explain that the forecasted tensor focal mechanism enables the calculation of an ensemble of far-field low-frequency seismograms for each point of interest on the Earth's surface. Furthermore, the focal mechanism distribution allows us to estimate fault plane orientation for past earthquakes, through which we can identify a preferred rupture direction for future events.

The angle Φ_{min} has also been used to directly compare moment tensors from two different earthquakes (Okal *et al.* 2011). It has also been applied in comparing moment tensors computed for the same events through different techniques (Frohlich and Davis 1999; Pondrelli *et al.* 2007; Yang *et al.* 2012). Such comparisons can help refine the moment tensor evaluation algorithms and lower their computational cost, since they can reveal the relative importance of various assumptions implicit in the algorithms.

Study of focal mechanism rotations carried out in Sections 8.2–8.7 neglected to account for the general geometry of the focal zone and, in particular, for the relation of focal mechanisms to the Earth's surface. Only pairwise *DC* rotations in infinite space have been considered. More complete investigations of source orientations in geographical as well as tectonic reference frames, though much more difficult to accomplish, should also be carried out. Research techniques in crystallographic texture analysis (see, for instance, Morawiec 2004) may be applied in these studies.

8.8 Models for complex earthquake sources

8.8.1 Complex point source solutions

8.8.1.1 Point sources of elastic deformation in homogeneous infinite medium

Higher-rank point seismic moment tensors were introduced by Backus and Mulcahy (1976a,b) and Backus (1977a,b). Silver and Jordan (1983), Silver and

Masuda (1985), Kagan (1987, 1988), McGuire *et al.* (2001), Chen *et al.* (2005) considered various aspects of higher-rank point seismic moment tensors. Kagan (1987) argued that the third-rank seismic moment tensor should show the complexity of the earthquake source, i.e., its difference from the standard planar rupture model. However, with the data available now, inversion results indicate that only the extent and directivity of the planar rupture can be obtained by analyzing higher-rank tensors (Chen *et al.*, 2010, and McGuire *et al.* 2001).

We apply the formalism of gravitation wave theory (Thorne 1980) to the study of complex earthquake sources. We construct a catalog of all possible elementary point sources for static deformation in an elastic solid (Kagan 1987). The familiar *DC*s, CLVDs, centers of compression and dilatation, etc., are all members of the complete catalog. The sources are classified according to the rank of the seismic moment tensor, and according to the weight (or order) of the irreducible tensor representation of the 3-D rotation group. These sources can be classified as belonging to one of three general classes: two spheroidal and one toroidal class. The static excitation functions are calculated for an infinite, homogeneous, isotropic medium for all of these sources. We show that, except for sources belonging to these three general classes, all other sources – which are numerous for the tensors of high rank – are null static sources. These sources do not produce any static displacement outside of the source region. Due to the presence of null sources, an inversion of the static deformation data is non-unique. The expansion of the equivalent-force tensors and the stress glut tensors (or seismic moment tensors) into a set of the symmetric trace-free source tensors is proposed. The sources corresponding to seismic moment tensors of the second, third and fourth ranks are considered in more detail. We identify the third-rank sources with rotational dislocations or disclinations.

8.8.1.2 Point sources of elastic deformation in homogeneous half-space

Kagan (1988) derived expressions for static deformation due to arbitrary sources in elastic homogeneous half-space. The paper had two objectives: (1) to obtain effective formulas for static displacement due to sources of arbitrary order in a half-space (these equations show the limits on non-uniqueness of an inversion based on deformation data); and (2) to extend Mindlin's (1936) interpretation of image forces from that of a single-force to sources of arbitrary order.

After determining the relationship between cylindrical and spherical solid vector harmonics, we arrive at the complete and effective solution for the static deformation due to a multipolar source of arbitrary order in a homogeneous isotropic half-space. Using the correspondence between spherical harmonics and seismic moment tensors obtained earlier (Kagan 1987), we construct a catalog of mirror image sources which yield zero traction on the plane. For each angular order number l the total number of these images is 15 for a general case, of which the first spheroidal source has seven images, the toroidal source has four images, and the second spheroidal source also has four images. We derive the conditions under which image point sources are converted into semi-infinite line nuclei. We analyze several special cases of deformation in a half-space, namely, axisymmetric, generalized Boussinesq-Cerruti, thermoelastic deformations. We present several general and specific models of finite sources of elastic displacement in a half-space and discuss possibilities for inverting complex extended sources of elastic deformation. The non-uniqueness of the inversion can be estimated objectively.

Kagan (1988) considered only theoretical problems of deformation due to arbitrary static sources in a half-space. A derivation of displacement expressions in a final closed form and their application is the problem for future studies. Present inversions of deformation in a half-space are using specific models of deformation source, thus they are not broad enough to allow the construction of general type of sources. The results of this subsection can be applied in rigorous inversion of deformation measurements at the half-space boundary. Such inversions could be useful in the investigation of complex earthquake sources, volcano deformation measurements and in those technical applications where representation of deformation sources in elastic half-space are needed.

8.8.2 Higher-rank correlation tensors

Higher-rank correlation tensors (Kagan and Knopoff 1985b; Kagan 1992b) more completely describe the interrelation between focal mechanisms and their spatial distribution. However, their interpretation is still difficult and the low accuracy of focal mechanism solutions and earthquake locations (Kagan 2003) in available catalogs makes conclusions difficult and uncertain.

The two-point correlation function of the tensor-valued field of seismic moment tensor density can be defined as

$$M_{ijkl}(\mathbf{r},\tau) = < m_{ij}(\mathbf{r}_1, t_1)\ \mathbf{n}_{kl}(\mathbf{r}_2, t_2) >, \tag{8.62}$$

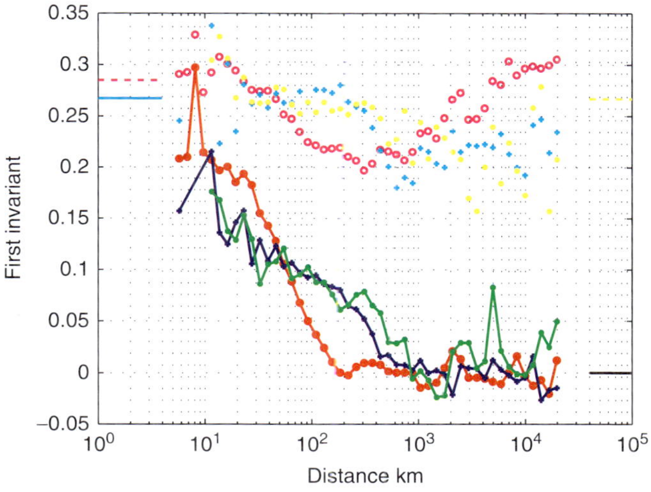

Fig. 8.15 Dependence of invariant J_1 on distance and depth (GCMT catalog, $m \geq 6.0$, 1977–1990). Depth range: red circles – 0–70 km, the number of events N = 6392; blue crosses – 71–280 km, N = 1530; and green stars −281–700 km, N = 636. Solid lines correspond to values of invariant J_1 (8.63), dotted lines correspond to standard deviations for the invariant (see Eq. 8.64), their color is magenta, cyan and yellow, respectively. The short horizontal lines in the left part of the plot correspond to coherent uniformly distributed sources from Table 1 (first column) of Kagan (1992b), the lines in the right side of the plot correspond to incoherent sources from Table 1 (last column).

where **m** and **n** are seismic moment tensors measured at two points $(\mathbf{r}_1, \mathbf{r}_2)$ and two instants of time (t_1, t_2) and $\mathbf{r} = \mathbf{r}_1 - \mathbf{r}_2$; $\tau = t_1 - t_2$. The brackets indicate that a statistical (ensemble) average has been taken.

Because of the symmetry of $\mathbf{m}_{ij}$ and the interchangeability of **m** and **n** in the product (8.62), the fourth-order tensor **M** has the same symmetries as a tensor modulus of elasticity. Without any further simplifications the tensor **M** has 21 degrees of freedom.

Monin and Yaglom (1975, para. 12.4) and others have shown that the correlation tensor **M** can be written in terms of five scalar (invariant) functions for an isotropic medium. For traceless seismic moment tensors (**m**, **n**) there are only three independent scalar functions (Beran 1960, p. 218). We select the three independent quantities to be the three bilinear invariants of **m** and **n** (Lumley 1970, p. 182; Monin and Yaglom 1975):

$$
\begin{aligned}
J_1 &= < (\rho_i \ \mathbf{m}_{ij} \ \rho_j) \times (\rho_k \ \mathbf{n}_{kl} \ \rho_l) >, \\
J_2 &= < (\rho_i \ \mathbf{m}_{ij} \ \mathbf{n}_{jk} \ \rho_k) > \\
J_3 &= < (\mathbf{m}_{ij} \ \mathbf{n}_{ij}) >,
\end{aligned}
\tag{8.63}
$$

where $\rho_i = r_i/|r|$ are the direction cosines of **r**. If $\mathbf{m} = \mathbf{n}$, then $J_3 = -2I_2$, where I_2 is the second invariant of the seismic moment tensor (Kagan and Knopoff 1985a; Eq. 8.8).

We also calculate standard deviations for each of the invariants in (8.63)

$$
\begin{aligned}
\sigma_1^2 &= < [(\rho_i \ \mathbf{m}_{ij} \ \rho_j) \times (\rho_k \ \mathbf{n}_{kl} \ \rho_l)]^2 > -J_1^2, \\
\sigma_2^2 &= < [(\rho_i \ \mathbf{m}_{ij} \ \mathbf{n}_{jk} \ \rho_k)]^2 > -J_2^2 \\
\sigma_3^2 &= < [(\mathbf{m}_{ij} \ \mathbf{n}_{ij})]^2 > -J_3^2.
\end{aligned}
\tag{8.64}
$$

These invariants define the two-point fourth-order correlation tensor (compare Monin and Yaglom 1975, v.2, p. 69).

Figure 8.15 shows the dependence of the invariant J_1 (8.63) and (8.64) on distance and earthquake depth interval for shallow earthquakes in the GCMT

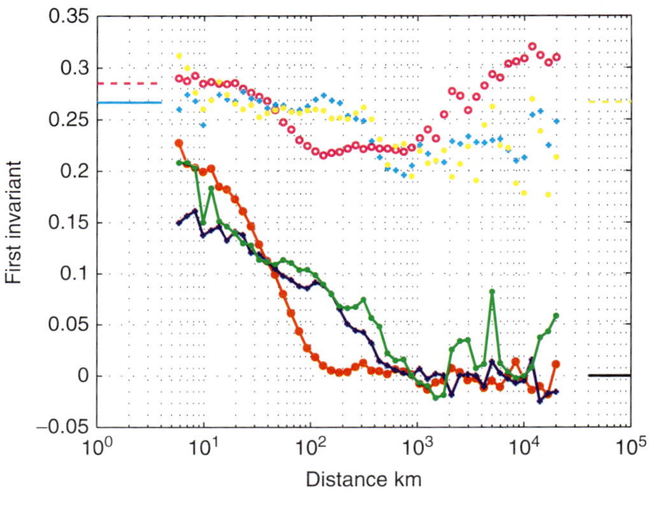

Fig. 8.16 Dependence of invariant J_1 on distance and depth (GCMT catalog, $m \geq 6.0$, 1977–2011). Depth range: red circles – 0–70 km, the number of events N = 29,021; blue crosses – 71–300 km, N = 5226; and green stars –300–700 km, N = 1745. Solid lines correspond to values of invariant J_1 (8.63), dotted lines correspond to standard deviations for the invariant (see Eq. 8.64), their color is magenta, cyan and yellow, respectively. The lines in the left part of the plot correspond to coherent uniformly distributed sources from Table 1 (first column) of Kagan (1992c), the lines in the right side of the plot correspond to incoherent sources from Table 1 (last column).

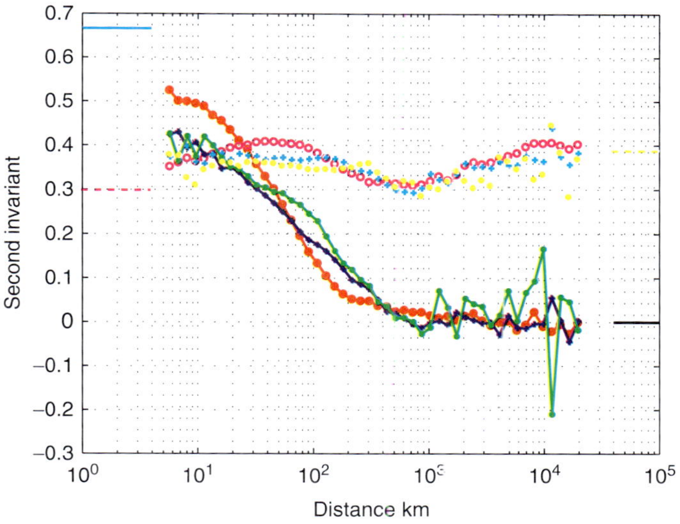

Fig. 8.17
Dependence of invariant J_2 on distance and depth (GCMT catalog, $m \geq 6.0$, 1977–2011). See Fig. 8.16 for notation.

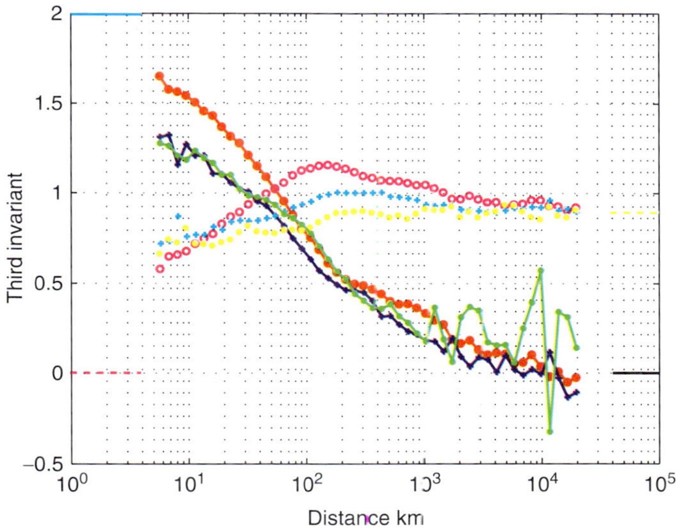

Fig. 8.18
Dependence of invariant J_3 on distance and depth (GCMT catalog, $m \geq 6.0$, 1977–2011). See Fig. 8.16 for notation.

catalog for the same time interval and magnitude limit as Fig. 1a by Kagan (1992b). In Figs. 8.16–8.18 we update all plots of Figs. 1a,b,c by Kagan (1992b) for time period 1977–2011 and slightly changed depth intervals. Updated plots inspection confirms the conclusion we draw from the earlier diagrams (Kagan 1992b, pp. 22–23). Comparing Figs. 8.15 and 8.16 it is obvious that whereas the general behavior of the curves is the same for the longer time interval, random fluctuations are significantly reduced.

The conclusions we can draw from updated plots of the higher-rank correlation function are essentially the same as the conclusions by Kagan (1992b,

p. 33). More theoretical studies and detailed statistical analysis of available earthquake catalogs are needed to explain the properties of earthquake faults and their interaction. The statistical approach developed in this subsection has a significant potential advantage in explaining the complex structure of earthquake focal zone: the fourth-rank correlation tensor (8.62) depends both on focal mechanism and position of earthquake sources. Thus, instead of studying marginal distributions of these features of earthquake occurrence we carried out in Chapter 7 and Sections 8.2–8.7, the entire properties of the zone can be explored. The updated results by Kagan and Knopoff (1985b) and by Kagan (1992b), shown above in Figs. 8.15–8.18, provide a good starting point for a further investigation of higher-rank correlation tensors.

Part III

Testable Forecasts

9 Global earthquake patterns

Several of our publications in condensed form are the basis of this chapter (Kagan and Knopoff 1976; Kagan 1991b; Kagan *et al.* 2010; Kagan 2011). In this and two following chapters we discuss how the results of stochastic model-building (Chapter 3) and statistical analysis of marginal earthquake distributions (Chapters 4–8) can be used to forecast earthquake occurrence and to rigorously test these forecasts. For this purpose we first need to evaluate the parameters of stochastic time-space process models approximating earthquake occurrence.

This analysis has to be performed for local and regional earthquake catalogs as well as for global seismicity. Our previous statistical analysis reported in Chapters 4–8 suggested that there is a certain universality of earthquake occurrence properties over various tectonic zones. Such universality, if shown for time-space properties of earthquake process, would testify to applicability of our stochastic models for forecasting purposes. Kagan and Knopoff (1976, 1977, 1987b) show that the stochastic model parameters are similar for local and worldwide earthquake catalogs.

9.1 Earthquake time-space patterns

In our previous work we have already demonstrated some important variations in seismicity determined by tectonic setting. Bird and Kagan (2004) studied all shallow earthquakes and showed variations in coupling and corner magnitude between subduction, continental collision, continental transforms, and continental rifts (Section 5.2).

Given this proven variability, it would not be surprising if distinct tectonic zones had different earthquake-clustering and/or triggering behaviors. For example, Boettcher and Jordan (2004) and McGuire *et al.* (2005) have already found that earthquakes on oceanic transform faults have more foreshocks and fewer aftershocks. Therefore, our first hypothesis is that significant variations are to be discovered if the empirical clustering parameters in the earthquake branching model (Kagan 1991b) are redetermined in tectonically-defined zones. (Previous studies, which did not distinguish between tectonic settings, presumably gave results dominated by the behavior of subduction zones.) However, in

Earthquakes: Models, Statistics, Testable Forecasts, First Edition. Yan Y. Kagan.

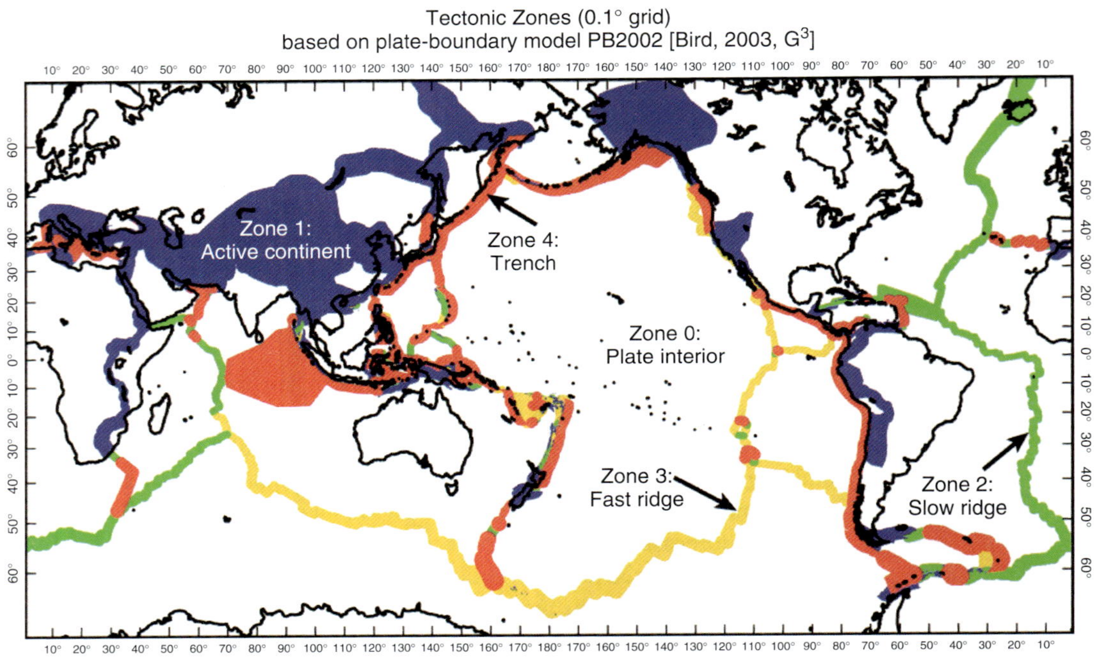

Fig. 9.1 Global tectonic zones according to Bird (2003) and Kagan *et al.* (2010). *Source:* Kagan *et al.* (2010), Fig. 1.

this section we do not divide global seismicity as finely as Bird and Kagan (2004), because their code for classification of earthquakes into subcatalogs by "plate boundary class" is complex and not easy to incorporate into forecasts testable by independent agencies. Instead, we define geographically-contiguous areas of related tectonic style, and define the union of all areas with the same style as a "tectonic zone" (see Fig. 9.1).

These zones are similar in concept to the Flinn-Engdahl zones (Flinn *et al.* 1974; Young *et al.* 1996) which we analyzed in Section 5.6; but these new zones are much larger, and they are based on the detailed classification of the PB2002 model of Bird (2003). The spatial contiguity of each area within each zone is an important advantage:

- Epicenter location errors become relatively unimportant.
- No complex algorithm is needed to decide which actual earthquakes belong to a zone.
- Forecasts and their testing can include earthquakes below the GCMT threshold ($m_t \approx 5.6$) in cases where focal mechanisms are not required.

We start by investigating how different branching models approximate the behavior of global earthquake catalogs, then consider the proposition that different tectonic regimes may have different clustering behaviors. Global catalogs are free of spatial boundary effects and considerably more homogeneous than regional catalogs. Moreover, regional seismicity may be dominated by aftershock sequences of a few strong events, like the *m*7.5 1952 Kern County and the *m*7.3 1992 Landers earthquakes in southern California. Explosions and earthquakes caused by volcanic and geothermal activity are more likely to contaminate earthquake records in local and regional catalogs.

However, it is important to analyze regional catalogs as well. If we see that model parameter values are similar for worldwide and regional catalogs (Kagan 1991b), then we may conclude that the models are relatively robust and therefore suitable for earthquake forecasting even in new regions. It is also important to investigate the various parametrizations of the branching models, especially the spatial and temporal fitting of seismicity patterns (Kagan 1991b) in order to find the best algorithms. Then the results of such a study can be used to develop a technique for calculating and evaluating long- and short-term earthquake rate forecasts in practically any seismically active region of the Earth (see more later in this chapter).

Recently many researchers have applied ETAS branching models (Ogata 1988, 1998; Console and Murru 2001; Console *et al.* 2003b; Ogata *et al.* 2003; Zhuang *et al.* 2005; Ogata and Zhuang 2006; Harte 2013) to Japan, California, Italy, and New Zealand. Lombardi and Marzocchi (2007) and Marzocchi and Lombardi (2008) applied the ETAS model to two global catalogs of large earthquakes. The thorough comparison of ETAS results for local and global catalogs has not yet been made (see more in Section 9.5.2).

9.2 Defining global tectonic zones

Our goal is to divide the Earth surface into a modest number of zones of different tectonic style, defined by objective criteria, which are known from previous research (e.g., Bird and Kagan 2004) to contain interesting variations in seismicity parameters such as seismic coupling and/or corner magnitude. It is reasonable to anticipate that these zones might have different time/space branching behaviors, and our present definition of zones is designed with consideration for practical matters that permit such testing to be conducted relatively easily. Zones are defined here as surface areas into which epicenters (and/or epicentroids) of shallow earthquakes may fall. Precise depth of the earthquake is not a criterion because it is usually poorly known (within the 0–70 km depth range) unless there is local station control. Focal mechanisms are not a criterion because they are not always available for the smaller earthquakes ($m < 5.6$) which make up large portions of the aftershock swarms we wish to include in this study.

The number of zones we have defined is small (five, see Fig. 9.1) so that sufficient earthquakes fall into each zone within a few decades, allowing for reliable optimizations of branching models. We allow one tectonic zone to be the union of many non-contiguous patches. The preceding choices make it impossible to separate strike-slip from normal-faulting earthquakes on mid-ocean spreading ridges, or in continental rift zones. Also, along many trenches it is impossible to separate subduction-related earthquakes from back-arc-spreading earthquakes. Therefore, the tectonic zones are not equivalent to the seven or eight plate-boundary classes of Bird (2003) or Bird and Kagan (2004; Section 5.2), and they have been given new, distinct names that could not be confused with plate-boundary classes.

We propose the following short list of tectonic zones (with identifying integers for compact representation in computer files and plots):

- 4) Trench (including incipient subduction, and earthquakes in outer rise or upper plate);
- 3) Fast-spreading ridges (oceanic crust, spreading rate $\geq$40 mm/a; includes transforms);
- 2) Slow-spreading ridges (oceanic crust, spreading rate <40 mm/a, includes transforms);
- 1) Active continent (including continental parts of all orogens of PB2002 (Bird 2003), plus continental plate boundaries of PB2002); and
- 0) Plate-interior (the rest of the Earth's surface).

Fast-spreading vs. slow-spreading ridges are divided at 40 mm/a because their properties are significantly different (Bird and Kagan 2004). Kagan *et al.* (2010) in their Appendix describe details of zone definition implementation.

Our program assigns tectonic zone integers to grid points rather than drawing curves along boundaries of tectonic zones. It loops through latitudes (in 0.1° steps) and also loops through longitudes (in 0.1° steps), and creates a grid of integers to identify the tectonic zone at each point.

The output is a relatively compact representation (1801 latitudes × 3601 longitudes × 1 to 2 bytes = 6 to 12 Mb). The result of the global tectonic zones classification is shown in Fig. 9.1 and is available at http://bemlar.ism.ac.jp/wiki/index.php/Bird%27s_Zones.

9.3 Corner magnitudes in the tectonic zones

Corner moment and β-slope are the two parameters of the TGR model for the cumulative frequency/moment relation (see Eq. 5.5 and Section 5.2). Bird and Kagan (2004) reported corner magnitudes for seven types of plate boundaries, and these ranged from a low of $5.86^{+0.19}_{-0.16}$ (ranges are 95%-confidence) for normal-faulting earthquakes on Ocean Spreading Ridges (OSR) to a high of $9.58^{+0.48}_{-0.46}$ for shallow events in Subduction zones (SUB), see Fig. 5.3. The spread of corner magnitudes among tectonic zones (Fig. 9.1, Table 9.1) is less, because each tectonic zone (except zone 0) merges the seismicity of different plate-boundary types as defined by Bird and Kagan (2004). At the low end, the corner magnitude for zones 2 and 3 (Slow- and Fast-spreading ridges) is higher than that of OSR/normal because of the inclusion of OTF seismicity. At the high end, the corner magnitude of zone 4 (Trenches) is less than that of SUB because of dilution by Oceanic Convergent Boundary (OCB) and oceanic convergent orogen seismicity, and that of some overlying back-arc OSRs.

To determine the corner magnitudes of the tectonic zones, we divided the shallow ($\leq$70 km depth) events with $m_W \geq 5.63$ (scalar seismic moment $M = 10^{17.45}$ Nm or $M = 2.818 \times 10^{17}$ Nm, see Eq. 2.2) from the GCMT catalog (1982–2008/3/31) into five zone subcatalogs using the gridded zone assignments

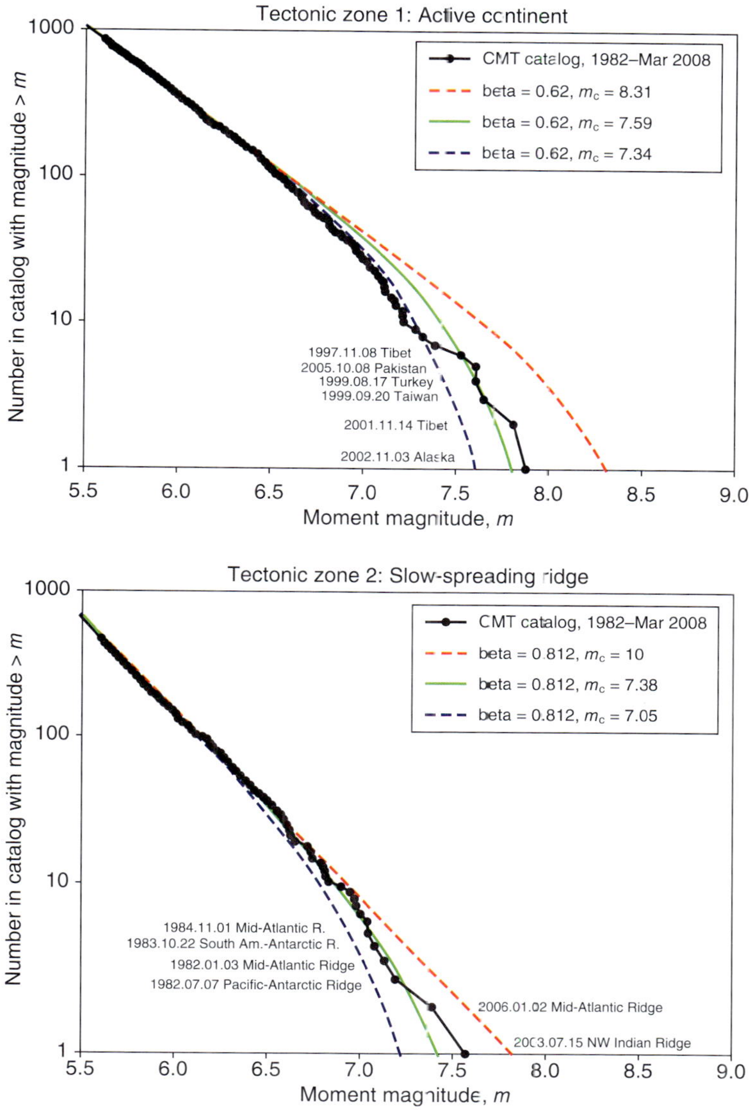

Fig. 9.2 Comparison of TGR model for the frequency / magnitude distribution to actual cumulative distribution. Earthquakes with $m > 7.5$ are identified. In each frequency / magnitude plot, three TGR model distribution curves are shown. Each has the optimal spectral slope β from Fig. 9.3. The three variant curves show the minimum, best-estimate, and maximum corner magnitude (except that a value of $m_c = 10$ is substituted for an unbounded corner magnitude in the lower plot). (a) Upper plot, tectonic zone 1: Active continent. (b) Lower plot, tectonic zone 2: Slow-spreading ridges. *Source:* Kagan *et al.* (2010), Fig. 2.

of Fig. 9.1. In Fig. 9.2 we compare the TGR model (5.5) to actual cumulative distribution for two tectonic zones: "Active continent" (1) and "Slow-spreading ridges" (2). Then each subcatalog was analyzed with program BetaCorner.f90 (Bird and Kagan 2004) which contours the likelihood surface in 2-D (β, M_c) space to determine both maximum-likelihood estimates and 95%-confidence ranges. In Fig. 9.3 we display two maps of the likelihood surface for the same tectonic zones as in Fig. 9.2. For the "Active continent," the corner moment (M_c) estimate is constrained, whereas for the "Slow-spreading ridges," the upper limit is ∞.

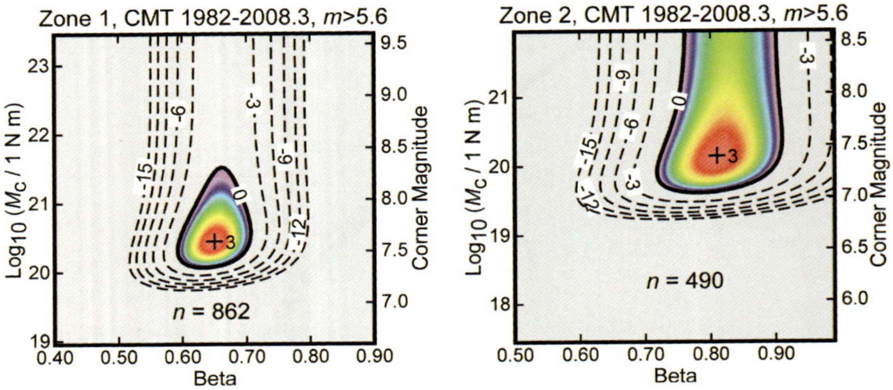

Fig. 9.3 Likelihood map surface for the TGR model in the 2-D parameter for β and corner moment M_c or corner magnitude m_c (left and right ordinates). Color-scale (red high, blue low) and cross for optimal value are used inside the contour which is 3.0 natural-log-units below the peak; this area corresponds to the 95%-confidence range. Dashed contours with interval of 3.0 natural-log-units are used outside. (a) Left, tectonic zone 1: Active continent. (b) Right, tectonic zone 2: Slow-spreading ridges. Note that the 95%-confidence area (zero-contour) is just barely closed on the high-magnitude side for the "Active continent"; ranges for the "Slow-spreading ridges" and some other zones are open on the high side. *Source:* Kagan *et al.* (2010), Fig. 3.

Table 9.1
Parameters of the TGR distribution.

	Tectonic zone	N	N/N_{All}	β	m_c
4.	Trenches	4234	0.648	$0.639^{+0.024}_{-0.024}$	$8.78^{+\infty}_{-0.35}$
1.	Active continent	862	0.132	$0.647^{+0.060}_{-0.055}$	$7.62^{+0.72}_{-0.25}$
0.	Plate-interior	251	0.038	$0.639^{+0.105}_{-0.101}$	$8.18^{+\infty}_{-0.53}$
2.	Slow-spreading ridges	490	0.075	$0.812^{+0.098}_{-0.094}$	$7.41^{+\infty}_{-0.33}$
3.	Fast-spreading ridges	701	0.107	$0.767^{+0.089}_{-0.095}$	$6.82^{+0.30}_{-0.18}$
	Global total (All)	6538	1.0	$0.671^{+0.021}_{-0.021}$	$8.76^{+\infty}_{-0.34}$

Notes: Parameters of the TGR frequency/magnitude relations of the tectonic zones, with 95%-confidence ranges; GCMT catalog, $M \geq 10^{17.45} = 2.818 \times 10^{17}$ Nm ($m \geq 5.63$); N is the number of earthquakes. Compare to Table 5.2. *Source:* Kagan *et al.* (2010), Table 1.

The results for all zones are shown in Table 9.1. The β-value is essentially the same for all five zones. Perhaps the "slow-spreading ridges" may have a statistically significant higher β-value, but that could be caused by a mixture of different earthquake populations. More exhaustive analysis (Bird and Kagan 2004; see Fig. 5.2) suggests that the hypothesis of the universality for β-value ($\beta \approx 0.63$) cannot be rejected based on the present data. The results of the corner moment evaluation are also consistent with previous analysis (Section 5.3).

9.4 Critical branching model (CBM) of earthquake occurrence

9.4.1 *Branching models*

Here we discuss the CBM model which describes earthquake clustering both at the local and global scales. The model was first proposed and applied to the central California earthquake record by Kagan and Knopoff (1987b). The assumptions we make in constructing this model of earthquake occurrence and its application to regional and global seismicity have been summarized in Kagan (1991b), and Kagan and Jackson (2000, 2011, 2012). The ETAS model (Ogata 1988, 1998; Ogata and Zhuang 2006) is essentially similar in design to the CBM and was initially applied to Japanese seismicity. The main difference between these two models lies in the parametrization of the influence functions of dependent events and normalization of these functions. Section 9.5.2 discusses certain potential drawbacks of the ETAS model because of its parametrization of seismicity.

Both the CBM and ETAS models belong to a class of branching point process models known in the statistical literature as *Hawkes* or *self-exciting* point processes (Hawkes 1971, see also Section 3.1 and Fig. 3.1c). For a temporal Hawkes process, the conditional rate of events at time t, given information $\mathcal{H}_t$ on all events prior to time t, can be written

$$\Lambda(t|\mathcal{H}_t) = \nu + \sum_{i:t_i<t} g(t - t_i) \,, \tag{9.1}$$

where $\nu > 0$, is the background rate, $g(u) \geq 0$ is the *triggering function* which describes the event activity induced by a prior event, and $\int_0^\infty g(u)du < 1$ in order to ensure stationarity (Hawkes 1971).

In CBM and ETAS models, seismicity is approximated by a Poisson cluster process, in which clusters or sequences of "parent" earthquakes are statistically independent, though individual earthquakes in a cluster (offspring) are triggered (Section 2.6 and Fig. 3.1c). The parent events are assumed to form a spatially inhomogeneous Poisson point process with a constant temporal rate. The major assumption regarding the events relationships within a cluster is that the interdependence of earthquakes is closely approximated by a stochastic magnitude-space critical branching process which develops in time. Under the *branching* assumption there is a sole trigger for any given dependent event. As shown below, the space-time distribution of the interrelated earthquake sources within a sequence is controlled by simple relations justified by analyzing available statistical data on seismicity. Chapter 4 considers the earthquake number distributions resulting in the application of branching property to earthquake occurrence.

In both models (ETAS and CBM) the branching property include triggering by foreshocks and aftershocks as well as mainshocks. As was mentioned earlier, traditional nomenclature uses the term "mainshock" for the largest earthquake in a cluster, "foreshock" for events within the cluster before the mainshock, and "aftershock" for events within the cluster after the mainshock. Clearly the labeling can only be applied after the cluster is complete and identified by some rule.

We prefer to distinguish between "spontaneous" events, which are the first in their cluster, and mainshocks, which are the largest. They may be the same, but some mainshocks follow a spontaneous foreshock. In practice, the time assigned to a cluster is that of its mainshock, but logically it should be that of its spontaneous event. A common measure of success of a clustering model is that the identified mainshocks should be temporally uniform. However, this presents practical problems for the following reasons. One is that clusters may be independent of one another but jointly affected by some external process such as distant earthquakes or viscoelastic stress variations. The second reason is that the spontaneous earthquakes might be below the magnitude threshold of the catalog. One advantage of global clustering models is that clustering by large distant earthquakes can be included explicitly.

9.4.2 Earthquake clusters – independent events

In the CBM model, the conditional rate $\lambda(t, \mathbf{x}, M)$ at which earthquakes are expected to occur at time t, location $\mathbf{x}$, and scalar seismic moment M, given the history of previous seismicity, is given by (Kagan and Jackson 2011, Eq. 18)

$$\lambda\ (t, \mathbf{x}, M) = \nu\ \phi\ (\mathbf{x}, M) + \sum_i\ \psi(t - t_i,\ \mathbf{x} - \mathbf{x}_i,\ M|M_i), \tag{9.2}$$

where ν is the rate per time unit of the Poisson occurrence of independent (spontaneous) earthquakes in the observed spatial region with scalar seismic moment M above the catalog's moment threshold M_t; function $\phi(\mathbf{x}, M)$ governs their space-seismic moment distribution; $\psi^{(M)}\ (t - u,\ \mathbf{x} - \mathbf{y}, M)$ is the conditional density of succeeding events at time t and location $\mathbf{x}$, if preceding earthquakes have occurred at times t_i in locations with coordinates $\mathbf{x}_i$. We subdivide the spatial coordinates $\mathbf{x}$ into $\mathbf{s} \times z$, where $\mathbf{s}$ are surface coordinates, and z is depth. If the duration of the catalog is T, then νT is the number of independent events, and $\nu T/n$ is the fraction of independent events (n is the total number of earthquakes in a catalog).

9.4.3 Dependent events

We assume that the rate density resulting from a given event *within* a cluster may be broken down into a product of its marginal distributions, that is, the conditional rate density of the j-th shock dependent on the i-th shock ($j > i$ and $t_j > t_i$) with seismic moment M_i modelled as (Kagan and Jackson 2000)

$$\psi\ (\Delta t, \rho, M_j|M_i) = \psi_{\Delta t}(\Delta t) \times \psi_\rho(\rho) \times \psi_M(M_i) \times \phi_M(M_j), \tag{9.3}$$

where $\Delta t = t_j - t_i$ and ρ is the horizontal distance between the i-th and the j-th epicentroids (or epicenters), $\psi_{\Delta t}$, ψ_ρ, and ψ_M are the marginal temporal, spatial, and moment densities, and are detailed below. The seismic moment PDF ϕ_M is given by Eq. 5.6.

The total time-dependent rate density is a sum of effects from all previous earthquakes,

$$\Psi\ (t_j, \mathbf{x}_j, M_j) = \sum_{i<j}\ \psi\ (\Delta t, \rho, M_j \mid M_i). \tag{9.4}$$

The function ψ decays rapidly with time and distance so that only neighboring events substantially contribute to the sum, although the range of strong earthquakes is much longer than that of weak events (see the next seven equations).

A power-law relation is used for the probability density of time intervals between earthquakes within a cluster:

$$\psi_{\Delta t}(\Delta t) = \theta t_M^{\theta} (\Delta t)^{-1-\theta}, \tag{9.5}$$

for $\Delta t \geq t_M$, in accord with Omori-Utsu's law (Utsu 1961). The parameter θ is an "earthquake memory" factor, and t_M approximates the coda duration time of an earthquake with seismic moment M_i. We assume for the coda duration

$$t_M = t_r \, M_i^{1/3}, \tag{9.6}$$

where t_r is the standard (constant) coda duration time, taken here as $t_r = 0.0035$ days (about 5 min.) for an earthquake with the reference seismic moment $M_r = 10^{15}$ Nm, corresponding to $m_r = 4.0$ (Kagan 1991b; cf. Ekström *et al.* 2012, Eq. 1). The probability of the next dependent shock occurring in the time interval (t_1, t_2), for $0 < t_1 < t_2$, given an event of a cluster occurring at time 0, can then be calculated simply as

$$\text{Prob } (t_1 < t < t_2) = t_M^{\theta}(t_1^{-\theta} - t_2^{-\theta}) \, . \tag{9.7}$$

The non-normalized function $\psi_M(M_i)$ which indicates the number of triggered shocks generated on average by an earthquake with seismic moment M_i, is assumed to obey

$$\psi_M(M_i) = \mu \times \left(\frac{M_i}{M_t} \right)^{\delta_a}, \tag{9.8}$$

(Kagan 1991b). To make the distribution parameter δ_a of triggered events comparable to the b-values, in Tables 9.2–9.5 below we use a parameter

$$a = \delta_a \times 1.5. \tag{9.9}$$

Kagan (1991b, p. 142) discusses the δ_a-parameter (called δ over there) and its relation to other measurements of size distribution of triggered events in more detail.

We approximate by a Rayleigh distribution (compare Eq. 7.13) the probability density of the horizontal distance (ρ) between two earthquake epicentroids in a cluster:

$$\psi_\rho(\rho) = \frac{\rho}{\sigma_\rho^2} \times \exp \left[-\rho^2/(2\sigma_\rho^2) \right], \tag{9.10}$$

where σ_ρ is a spatial standard deviation.

If errors in earthquake location have a Gaussian distribution, the inter-earthquake distances should be distributed according to the Rayleigh distribution (9.10), where a spatial standard deviation depends on the standard errors of epicentroid (epicenter) determination and on the seismic moment M_i of the triggering event according to the relation

$$\sigma_\rho^2 = \epsilon_\rho^2 + s_r^2\,(M_i/M_r)^{2/3}, \tag{9.11}$$

(Kagan 1991b). Here ϵ_ρ is the standard error in epicentroid determination and s_r is a characteristic size (a standard deviation) for a focal zone of an earthquake with the reference seismic moment $M_r = 10^{15}$ Nm, $M_r = 4.0$ (cf. Eq. 9.6).

The spatial kernel for short-term forecast differs substantially from that of the long-term forecast (Eqs. 10.1). The reason is that nearby earthquakes in a long-term forecast usually belong to separate clusters. During a long time interval many different earthquake clusters may accumulate in the same region. In contrast, for a short-term forecast we specifically search for earthquake pairs which are closely connected in space and time. The parameters in Eqs. 7.13 and 9.11 characterize the spread of dependent earthquakes (usually aftershocks) in the neighborhood of a parent event. The distance between dependent earthquake pairs is controlled by the focal zone size (s_r) of the first event and epicentral location errors (ϵ_ρ). The ability to determine these parameters is very important in the verification of both the CBM model and the likelihood inversion of catalogs as considered below in the discussion of Tables 9.2–9.5.

In Fig. 9.4 we display a few examples of the Rayleigh distribution density (7.13, 9.10) for several magnitudes. In our calculations we used the values of s_r and ϵ_ρ obtained in Table 9.3 for the PDE catalog: $s_r = 0.38$ km and $\epsilon_\rho = 9.5$ km. The curves for *m*5 and *m*6 earthquakes are very similar, because the location error "overwhelms" the mainshock focal size. Only for an *m*8 event are the aftershock

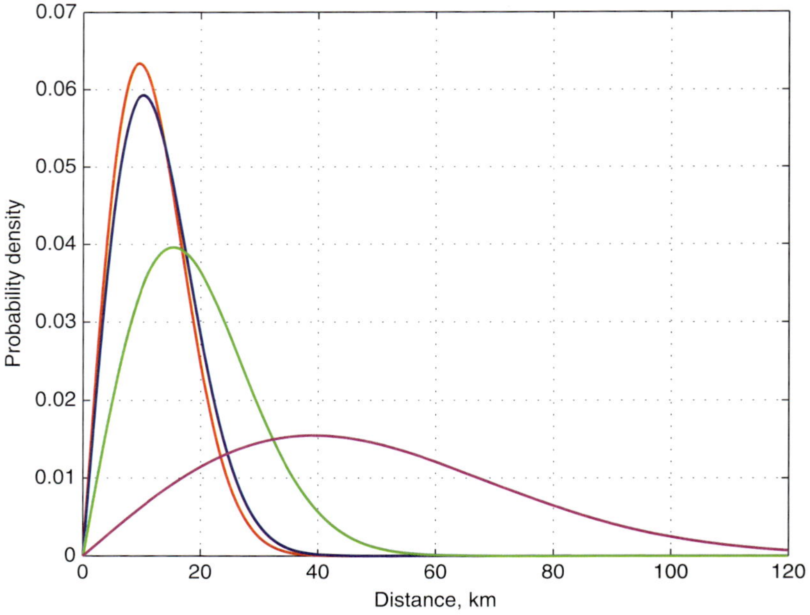

Fig. 9.4 Distance PDFs for mainshocks of different magnitudes: The red line is for the *m*5 event, the blue line is for the *m*6 event, the green line is for the *m*7 event, the magenta line is for the *m*8 event. *Source:* Kagan and Jackson (2011), Fig. 14.

distances mostly controlled by the size of the focal zone (the last term in Eq. 9.11). This aftershock zone size for $m8$ mainshock can be compared to that obtained in Section 5.5: $\sigma \approx 38$ km vs. $\sigma \approx 57$ km. The first value should be smaller since in the inversions of this chapter the zone size for immediate, first generation aftershocks is determined, whereas in Fig. 5.15 the total size of the zone for all aftershocks is evaluated. Moreover, random and systematic errors are better controlled in Fig. 5.15, thus its results are more rigorous.

9.4.4 Stochastic branching processes and temporal dependence

During a likelihood search the θ-values have been restricted within the interval $0.1 \leq \theta \leq 1.0$; smaller or larger estimates are inadmissible because of simple considerations (Kagan 1991b; Kagan *et al.* 2010). These and similar constraints during the likelihood inversion make estimating parameter errors and the correlation between these estimates a difficult and unreliable procedure. Wang *et al.* (2010) carry out the statistical error analysis for the ETAS model. They find significant correlations between statistical ETAS parameter estimates.

Therefore, we investigate a correlation between estimates for two parameters θ and μ of the CBM, these parameters are most important for earthquake temporal interaction (Eqs. 9.5, 9.8). In Fig. 9.5, most θ-estimates are within the interval $0.1 \leq \theta \leq 0.5$ and are negatively correlated with the estimates of the μ coefficient. We also show the correlation coefficients of the parameter estimates for various sets of analyzed subcatalogs. These correlation coefficients also indicate a negative correlation of these estimates.

Several determinations of the time decay exponent are carried out for the ETAS model. Ogata (1998, Tables 2 and 3) obtained the p-values (equivalent to our $1 + \theta$ coefficient), of the order 1.03–1.14. Ogata *et al.* (2003, Tables 1 and 2)

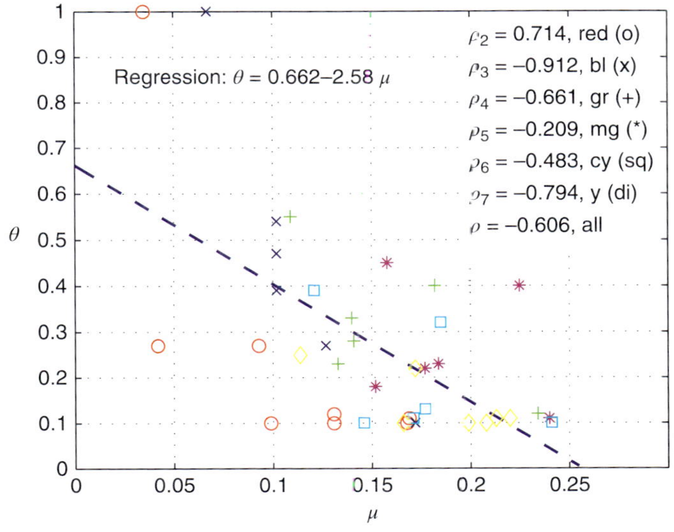

Fig. 9.5 Correlation between maximum likelihood estimates of μ and θ parameters in several earthquake catalogs (Kagan *et al.* 2010). The subscripts i at correlation coefficients (ρ_i) point to the table number in Kagan *et al.* (2010). Tables 9.2, 9.3, 9.4, and 9.5 correspond to subscripts 2, 4, 6, 7, respectively. *Source:* Kagan (2011), Fig. 11.

estimate the p-values to vary within 1.05–1.18. Ogata and Zhuang's (2006, Tables 2 and 3) values are 1.02–1.05. Zhuang *et al.* (2005) obtained p-values of 1.14–1.15. Helmstetter *et al.* (2006, Table 1) used a branching model similar to the ETAS and calculated $p = 1.18–1.20$.

To illustrate the fit of the temporal distribution of dependent earthquakes, average numbers of aftershocks following 15 GCMT $m \geq 8.0$ earthquakes are displayed in Fig. 9.6 (similar to Fig. 13 in Kagan 2004 or Fig. 13 in Kagan and Jackson 2011). We use the time period 1977–2003, so that all large earthquakes are approximately the same size ($8.0 \leq m \leq 8.45$, that is, excluding post-2004 very large earthquakes. Since the GCMT catalog has relatively few dependent events, we selected aftershocks from the PDE catalog. The aftershock rate in the diagram is approximately constant above the estimate of the coda duration t_M. For the logarithmic intervals, this corresponds to the standard form of the Omori law (Eq. 6.1): the aftershock interval number n_a is proportional to $1/t$. For the smaller time intervals, the aftershock numbers decline when compared to the Omori law prediction ($1/t$). This decline is caused by several factors, the interference of mainshock coda waves being the most influential (Kagan 2004; Section 6.2.1). The decline is faster for weaker events (Kagan 2004).

For theoretical estimates in Fig. 9.6, we used the results obtained during the likelihood function search (Table 9.3) for the full PDE catalog, $m_t = 5.0$. The parameters values are: the branching coefficient $\mu = 0.141$, the parent productivity exponent $a_0 = 0.63$ ($a_0 = \delta_a \times 1.5$, see Eq. 9.8), and the time decay exponent $\theta = 0.28$, that is, in the middle of the scatterplot shown in Fig. 9.5. Theoretical estimates in Fig. 9.6 seem to be reasonably good at forecasting time intervals on the order of one day. For larger intervals, the expected aftershock numbers decrease as $n_a \sim (\Delta t)^{-1.15}$: this is stronger than the regular Omori law (Eq. 6.1) would predict. As we suggested in Section 6.4.1, Omori's law assumes that all

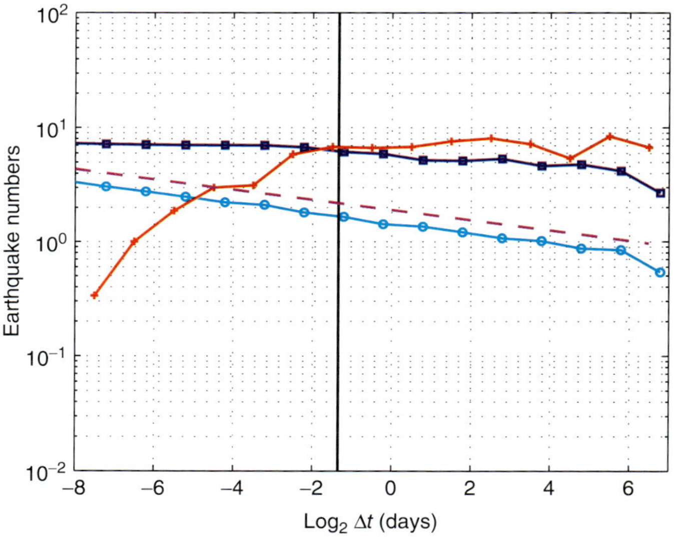

Fig. 9.6 The average aftershock numbers n_a from the PDE catalog in logarithmic time intervals following $m \geq 8.0$ GCMT earthquakes during 1977–2003. The red line with plus signs shows $m_b \geq 4.9$ aftershocks; the red dashed line is the theoretical estimate for the first generation aftershocks. Two curves display the results of earthquake catalog simulation: the cyan line with circles shows the first generation aftershocks, the blue line with squares indicates total aftershock numbers. The black vertical line is the estimate for the coda duration t_M (9.6) for an earthquake $m = 8.1$. *Source:* Kagan (2011), Fig. 12.

aftershocks are direct consequences of a mainshock, whereas a branching model regards any dependent earthquake as a possible progenitor of later events. Thus, later aftershocks are the combined offspring of a mainshock and all consequent earthquakes. With an increase in time, the difference between Omori's law and the CBM predictions would increase as well. Marsan and Lengliné (2008) show that in California catalogs the aftershock rates for direct and secondary triggering differ by a factor from 1.5 to 10. By numerical simulations of the ETAS model, Felzer *et al.* (2002) and Helmstetter and Sornette (2003) estimated that a substantial majority of aftershocks are indirectly triggered by the mainshock.

However, the simulations based on the ETAS model are not without problems. Sornette and Werner (2005, Fig. 2) argue that since events below the magnitude threshold are not included in the earthquake catalogs, in many simulations, the earthquake stochastic interrelation pattern is not fully reproduced. An earthquake which seems to be an independent event may be an offspring of a quake that is below the threshold, similarly an apparent first generation aftershock may also depend on a preceding unobserved smaller event (as in Fig. 3.1c). All earthquake connections above the magnitude threshold are accounted for in the models based on branching along the magnitude/moment axis (Fig. 3.1b). Therefore, in Fig. 9.6 we simulate the aftershock distribution by the programs based on the branching along the magnitude axis.

The p-parameter in Omori's law is often assumed to be 1.0. This p-value implies that the total aftershock number approaches infinity as the duration of the aftershock sequence increases. Figs. 6.10 and 6.11, however, suggest that in the presence of tectonic loading, the time exponent value should increase at longer time periods. Statistical determination of the exponent can usually be made for only short periods; thus, we do not yet have a good estimate of the p-value for the tail of an aftershock sequence.

9.5 Likelihood analysis of catalogs

9.5.1 Statistical analysis results

We applied the likelihood technique to several global and regional earthquake catalogs for the tectonic zones shown in Fig. 9.1. Using the branching model (CBM) of Section 9.4 (see also Section 10.3), we obtain a likelihood function for the earthquake process. We compare the likelihood function for the branching process with a process based on the spatial inhomogeneous Poisson model of seismicity. By searching for the maximum of the likelihood ratio of two models or of their log-likelihood difference, we obtain the statistical estimates of the model parameters.

Two methods can be used to search for the likelihood function maximum: one is the Newton-Raphson optimization technique (needs derivatives of the likelihood function); the other is the simplex algorithm (Press *et al.* 1992, p. 402) in which no derivatives are needed. The simplex method was applied by Helmstetter *et al.* (2006). We used a modification of the Newton-Raphson technique in most of our studies (Kagan 1991b; Kagan and Jackson 2000). We could not apply available standard packages of the Newton-Raphson technique because many of the parameters of our model have either physical constraints (for example, being non-negative) or constraints based on prior geophysical results (Kagan *et al.* 2010). The likelihood equations are rather bulky (Kagan 1991b).

In some calculations we censor the immediate aftershocks from the catalogs (see Eq. 9.6). This censoring which is similar to the removal of earthquakes below the magnitude threshold (see Eq. 4.3), introduces a Heaviside function in the time history of earthquake occurrence, which significantly complicates the calculation of the derivatives. This explains also why we did not apply a triangular window to remove aftershocks in our computations (Eq. 6.4; Kagan 2004; Helmstetter *et al.* 2006). It would be even more difficult to obtain the likelihood function derivatives in such a case. (Eq. 9.6 defines a rectangular-in-time window for the aftershock removal.) Thus, in our calculations the same likelihood-optimization program was applied to two catalog types: one in its original form and the other with all of the early aftershocks which are closer to the previous shock than t_M being deleted (Kagan *et al.* 2010).

In Tables 9.2–9.5 we display some of the likelihood search results (see more in Kagan *et al.* 2010). The tables show the branching model parameter values for four catalogs, with varying magnitude thresholds and for different tectonic zones (Fig. 9.1). Table 9.5 analyzes only one region, California/Nevada belonging to the tectonic zone "Active continent."

Figure 9.1 shows that the tectonic zones, described in Section 9.2, contain many small regions spaced out at large distances throughout the zone. Most of the parameters which depend on the earthquake location separation (like λ, μ, θ) can be expected to be significantly different for these zones compared to the full catalog ("All" column in the tables). However, such a variation is not observed, most likely because these parameters are defined for a close distance interaction between earthquakes.

Earthquake number ratios (N/N_{All}) in the tectonic zones for the GCMT and the PDE catalogs in Tables 9.2–9.4 show some differences, that are especially large for the "Fast spreading ridges" – 10.8% for the GCMT catalog vs. 6.1% for $m_t = 5.0$ PDE catalog. Several factors may influence this variation: the change in the β and the corner moment values (Table 9.1), various systematic effects in the determination of m_b magnitude in the PDE catalog, as well as others (Kagan 2003).

The *b*-values in Tables 9.2–9.5 were computed using the Aki/Utsu formula (Utsu 1999). These values for all the subdivisions of the catalogs are close to 1.0, the *b*-value can be converted to β (see Table 9.1) as $\beta = b/1.5$ (Section 5.2). The only significant deviation from this *b* universality is the values for oceanic ridges. The increased *b*-value for these earthquakes is caused by a relatively low value of their corner magnitude (Table 9.1) which biases the estimates (cf. Fig. 9.2) toward higher slope values (see Section 5.3).

The parameters s_r and ϵ_ρ (see Eq. 9.11) characterize the spread of dependent earthquakes (usually aftershocks) in the neighborhood of a "parent" event. The estimates of the horizontal location accuracy (ϵ_ρ) are in good agreement with the independent evaluations of these uncertainties (Kagan 2003): for the GCMT catalog the accuracy is considered to be of the order 20 km and for the PDE catalog on average it is close to 10 km. The location uncertainties for the PDE catalog depend on the magnitude threshold, m_t. Kagan *et al.* (2010, Tables 3 and 5) provide additional tables for the PDE catalog with $m_t = 5.3$ and $m_t = 4.7$. As one would expect, for a larger magnitude threshold ($m_t = 5.3$) the accuracy is higher (from 8 km to 12 km) than for the lower magnitude ($m_t = 4.7$), where ϵ_p varies from 12 km to 16 km. The lowest ϵ_p-values in the PDE catalog are for zone 1 (Active continent), because most of seismographic stations are located on continents, so location accuracy is higher.

The vertical error ϵ_h in Tables 9.2–9.5 is not informative, since many shallow events are assigned 10 or 15 km depth in the GCMT catalog and 33 km in the PDE catalog. We kept this parameter in our likelihood procedure since it is useful for the analysis of deep and intermediate earthquakes, where relative accuracy of the depth location is higher. This parameter use is also appropriate for those local catalogs which have a low vertical error.

Table 9.2 Parameters for various subdivisions of GCMT catalog, 1982/1/1–2007/3/31, $m \geq 5.55$ or $M \geq 10^{17.325} = 2.113 \times 10^{17}$ Nm.

Zones #		All	Trench (4)	Act-Con. (1)	Inter. (0)	Slow (2)	Fast (3)	All/d	Trench/d (4)
1	2	3	4	5	6	7	8	9	10
1.	N	7720	5022	1004	279	584	831	7471	4845
2.	N/N_{All}	1.0	0.651	0.130	0.036	0.076	0.108	1.0	0.649
3.	m_{max}	9.07	9.07	8.28	8.15	7.67	7.15	9.07	9.07
4.	I/N	1.03	1.18	1.11	0.59	0.30	0.44	0.857	0.997
5.	$\lambda T/N$	0.745	0.690	0.819	0.866	0.935	0.941	0.758	0.707
6.	μ	0.131	0.169	0.093	0.099	0.042	0.035	0.131	0.168
7.	b	1.02	0.98	0.96	1.05	1.28	1.23	1.01	0.98
8.	a	0.64	0.59	0.56	0.45	0.52	0.59	0.62	0.57
9.	θ	0.12	0.11	0.27	*0.1	0.27	*1.0	0.1*	0.1*
10.	s_r	0.30	0.29	0.15*	0.47	0.17	0.27	0.28	0.29
11.	ϵ_p	21.8	22.1	21.6	18.0	19.5	17.7	21.9	22.1
12.	ϵ_h	3.5	4.4	3.0*	5.1	3.0*	3.0*	3.4	4.2

Notes: This table is similar in format to Table 3 in Kagan (1991b). Briefly, N – the earthquake number, m_{max} – maximum magnitude in a subcatalog, I/N – information score per event in bits/eq (1 bit means that the statistical uncertainty can be reduced on average by a factor of 2), $\lambda T/N$ – ratio of spontaneous events to total, μ –branching ratio (more branching–more triggered events), b – *b*-value, a – parent productivity exponent, $\delta_a = a/1.5$ (see Eq. 9.8), θ – temporal exponent, see Eq. 9.5 ($1 + \theta$ is similar to "p" in Omori's law), s_r – focal size for $m4$ earthquake in km (9.11), ϵ_p – horizontal error in km (9.11), ϵ_h – vertical error in km. The values *preceded* by an asterisk (*.) mean that the parameters could not be evaluated and are assigned the upper limit, the values *followed* by asterisk (.*) are assigned the lower limit. */d – means that immediate aftershocks have been removed from a catalog. *Source:* Kagan *et al.* (2010), Table 2.

Table 9.3
Parameters for various subdivisions of PDE catalog, 1968–2006, $m \geq 5.0$.

Note: This table is similar in format to Table 9.2 above. *Source:* Kagan *et al.* (2010), Table 4.

1	2	All 3	Trench (4) 4	Act-Con. (1) 5	Inter. (0) 6	Slow (2) 7	Fast (3) 8
1.	N	45942	29980	7686	2191	3296	2789
2.	N/N_{All}	1.0	0.653	0.167	0.048	0.072	0.061
3.	m_{max}	8.80	8.80	8.50	8.45	7.60	7.30
4.	I/N	1.90	1.99	2.16	1.40	1.28	0.98
5.	$\lambda T/N$	0.680	0.661	0.687	0.716	0.816	0.869
6.	μ	0.141	0.140	0.133	0.234	0.182	0.109
7.	b	1.05	1.04	1.05	1.17	1.17	0.93
8.	a	0.63	0.65	0.64	0.35	0.02	0.18
9.	θ	0.28	0.33	0.23	0.12	0.40	0.55
10.	s_r	0.38	0.37	0.31	0.24	0.15*	0.20
11.	ϵ_ρ	9.5	9.7	7.9	9.8	10.1	12.6
12.	ϵ_h	3.0*	4.5	4.3	3.0*	3.0*	3.0*

Table 9.4
Parameters for various subdivisions of PDE catalog, 1968–2006, close aftershocks removed, $m \geq 5.0$.

Note: This table is similar in format to Table 9.2. *Source:* Kagan *et al.* (2010), Table 5.

1	2	All 3	Trench (4) 4	Act-Con. (1) 5	Inter. (0) 6	Slow (2) 7	Fast (3) 8
1.	N	42925	27648	7205	2127	3217	2728
2.	N/N_{All}	1.0	0.644	0.168	0.050	0.075	0.064
3.	m_{max}	8.80	8.80	8.50	8.45	7.60	7.30
4.	I/N	1.25	1.24	1.50	1.17	1.04	0.76
5.	$\lambda T/N$	0.695	0.674	0.709	0.736	0.831	0.881
6.	μ	0.171	0.177	0.146	0.241	0.185	0.121
7.	b	1.04	1.03	1.04	1.17	1.17	0.93
8.	a	0.57	0.56	0.61	0.27	0.0*	0.0*
9.	θ	0.11	0.13	0.10	0.1*	0.32	0.39
10.	s_r	0.37	0.35	0.32	0.15*	0.15*	0.15*
11.	ϵ_ρ	9.7	9.8	7.8	9.7	10.0	12.5
12.	ϵ_h	3.0*	4.8	4.7	3.0*	3.0*	3.0*

The s_r estimate for the regional catalogs (CIT and ANSS) had to be constrained (Table 9.5). Most likely it is because the location accuracy of these catalogs significantly varies during the long period of observations, starting at 1932. Table 9.5 displays four sets of inversions in which the parameter s_r was constrained at two different values: 0.3 km and 0.15 km. This was done to see whether such a constraint causes significant modifications in the estimates of other parameters. The changes appear to be minor.

The global catalogs are much more recent and the variations of their location accuracy are not that great (Kagan 2003). The s_r values for all the global catalogs (GCMT and PDE) approximately are 0.30–0.35 km (Eq. 9.11). This would correspond to the total size of the focal zone of $m4$ earthquake close to 1 km,

Table 9.5
Parameters for California/Nevada catalogs.

		CIT					ANSS	
	m_t	All 3.0	All/d 3.0	All 3.0	All/d 3.0	All 4.7	All 4.0	All/d 4.0
1	2	3	4	5	6	7	8	9
1.	N	15512	13579	15512	13579	774	4495	3928
2.	m_{max}	7.5	7.5	7.5	7.5	7.5	7.5	7.5
3.	I/N	3.102	2.043	3.105	2.053	1.29	3.263	1.776
4.	$\lambda T/N$	0.475	0.539	0.474	0.539	0.666	0.531	0.586
5.	μ	0.220	0.208	0.213	0.199	0.114	0.172	0.165
6.	b	0.94	0.98	0.94	0.98	0.95	0.95	0.95
7.	a	0.60	0.58	0.61	0.60	0.69	0.66	0.64
8.	θ	0.11	0.10*	0.11	0.10*	0.25	0.22	0.10*
9.	s_r	0.30*	0.30*	0.15*	0.15*	0.30*	0.15*	0.15*
10.	ϵ_ρ	3.6	3.6	3.9	3.7	7.5	6.1	6.3
11.	ϵ_h	3.0*	3.0*	3.0*	3.0*	3.0*	3.0*	3.0*

Notes: This table is similar in format to Table 9.2 above. CIT catalog time limits: 1932–2001. CIT catalog window: Latitude $32.5 - 36.0°N$ Longitude $115.0 - 121.0°W$. ANSS catalog time limits: 1932–2008/05/31. ANSS catalog window: Latitude $31.5 - 43.0°N$ Longitude $113.0 - 125.0°W$.
Source: Kagan *et al.* (2010), Table 7.

the value coinciding with most estimates by other geophysical measurements (Abercrombie 1995; Kagan 2002b). The confirmation of the parameters s_r and ϵ_ρ values, discussed above, makes our results on the statistical analysis of seismicity patterns more credible.

The comparison of the results for the original catalogs with those obtained for the catalogs with immediate aftershocks removed (see Eq. 9.6 and Tables 9.2–9.5 where we mark these catalogs by '*/d' note) shows that the values of most of the parameters are not influenced strongly by this procedure. However, three of the tables' parameter entries do change. The information score per event (I/N) changes significantly because these aftershocks coming close after a parent event with the high conditional rate of occurrence, strongly increase the log-likelihood ratio. The ratio is used in selecting and comparing the best models to approximate an earthquake occurrence. Therefore, in any such testing it is important to know how the immediate aftershocks have been processed.

The ratio of the number of independent events to total $\lambda T/N$ increases since with the removal of the immediate aftershocks fewer triggered earthquakes are left in a catalog. The branching coefficient μ increases apparently because the immediate aftershocks are incompletely sampled in a catalog, and therefore they influence the μ estimate, making it smaller.

The changes due to the removal of close aftershocks are less evident in the GCMT catalog (Table 9.2). This catalog is based on the interpretation of long-period waves and the coda of these waves interferes with the identification of immediate aftershocks. For that reason fewer such aftershocks are present in the catalog (see Fig. 6.1 and its explanation).

Figure 9.7 shows the likelihood search iteration procedure for $m \geq 5$ earthquakes in the PDE catalog (Table 9.3). It demonstrates that the estimates of parameters λ and μ are negatively correlated. The other parameters look less correlated during the search. In principle, to evaluate the parameter

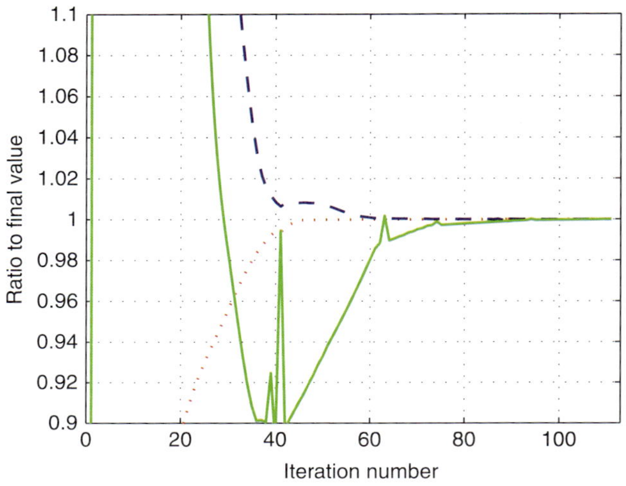

Fig. 9.7 Plot of the log-likelihood function ((the red dotted line), independent event rate λ (the green solid line) and the branching coefficient μ (the blue dashed line). These variables have been normalized to their final values. All shallow earthquakes with $m \geq 5$ in the PDE catalog are processed. *Source:* Kagan *et al.* (2010), Fig. 4.

uncertainties we need to obtain the Hessian matrix (the second partial derivatives of the likelihood function) of the parameter estimates or at least to plot two-dimensional diagrams of the parameter estimates similar to what was done in Fig. 9.3. Such plots have an advantage because if the relation is non-linear or the parameter value needs to be restricted (if, for example, it goes into the negative domain or to infinity, etc.), these displays provide us with more accurate information than the second-derivative matrix.

To assess the statistical significance of the differences in the parameter estimates across the tectonic regimes, we need a reliable way to estimate the multidimensional confidence regions. For the over-determined linear problems with the Gaussian data errors, the confidence regions can be estimated simply from the covariance matrix of the parameter estimates and an estimate of the data uncertainty. For mildly nonlinear over-determined problems, asymptotic methods can yield approximate confidence regions. These methods assume the local linearity of the relationship between the model perturbations and data perturbations, and they use the Hessian matrix to estimate a covariance matrix. In each particular instance the validity of the asymptotic assumption depends on how well the parameters are constrained by observations and prior information, and on the differentiability of the likelihood function (Jackson and Matsu'ura 1985). For catalogs with many earthquakes over a wide magnitude range and with many isolated earthquake clusters, the asymptotic methods might be applicable to branching models. As we explained above, one problem is that positivity and other hard constraints on the parameters, and censoring of immediate aftershocks, introduce discontinuities, thus preventing differentiation of the likelihood function.

We have not yet made a serious study of the confidence regions for the parameters, but indirect evidence suggests that the asymptotic method may not be appropriate, even without aftershock censoring. That evidence comes from: (1) the shape of the likelihood contours for two relevant parameters in Fig. 9.3, which would be elliptical under the assumptions in the asymptotic approach;

and (2) the behavior of the likelihood during convergence of the parameters while searching for the optimum parameter estimates (Fig. 9.7). The likelihood converges relatively slowly, and the tradeoffs between parameters clearly vary through the process.

9.5.2 Comparison of results with the ETAS model

The ETAS model is described in Sections 6.7, 9.1, 9.2, and 9.4. The main difference between the CBM and ETAS models lies in the parametrization of the influence functions of dependent events and normalization of these functions.

Eq. 9.5 is similar to the time influence function used in the ETAS model (Ogata 1998, his Eq. 1.1), where the number of aftershocks is expressed as the Omori-Utsu (Utsu 1961; Eq. 6.2) formula. The p-parameter can be identified with $(1 + \theta)$ in (9.5), whereas the c-parameter is introduced to account for lack and undercount of aftershocks immediately following a strong earthquake (Kagan 2004; Chapter 6).

However, many evaluations of the c-parameter show that it varies strongly, by orders of magnitude, even for mainshocks of similar size (see Chapter 6 and in particular Section 6.2.4). Moreover, the c-estimates vary by an order of magnitude even for the same mainshock when using different data (catalogs). This is in contrast to other seismicity parameters, like the b-value of the G-R relation or the p-value of Omori's law, which hardly (if at all) change (Chapters 5 and 6; Tables 9.1–9.5).

The other problem is that the coefficient c in the present ETAS implementations (as it is in Eq. 6.2) does not seem to be dependent on the magnitude of the preceding event. Eq. 9.6 which defines the aftershocks that are removed from a catalog due to their incomplete sampling, is scaled with magnitude: for a mainshock $m4$ it is about 5 min., whereas for $m8$ earthquake the "dead-time" extends to more than 8 hours. As the above example suggests, a relative temporal decrease of the aftershock numbers modeled by the c-parameter should vary strongly with the magnitude of a mainshock or a preceding event. It can be expected that the model which requires the same short-time behavior for earthquakes of different size may lead to a significant bias in parameter estimation and seismicity forecast.

Using Eq. 6.2 it is assumed that right after a parent event the rate is equal to $1/c$. However, the coda waves of the previous earthquake may hide any aftershocks preventing them from being identified for minutes or hours depending on the event magnitude. Ogata (1983) introduced a coefficient S to describe a practical absence of aftershocks in the initial part of a sequence. His estimate of S is $1/2$ day (p. 120). Therefore, Eq. 6.2 would cause a bias in the evaluation of the branching coefficient.

The estimated temporal parameters for the ETAS model (Tables 1–3 in Ogata 1998; Tables 1–3 in Ogata and Zhuang 2006) demonstrate that the c-value in (6.2) depends on the magnitude threshold. It is smaller for a smaller m_t. This is an additional argument against treating this parameter as having a physical significance.

The p-value is usually slightly larger than 1.0 (ibid.), though for some models the estimate is less than 1.0 – the value that can be rejected on general grounds, since it means that each earthquake has an infinite number of aftershocks. Thus, the p-value ETAS results are in principle consistent, taking into account that $p = 1 + \theta$, with Tables 9.2–9.5.

With regards to the spatial distribution (Ogata 1998, Eqs. 2.2–2.4), there are two problems with the ETAS model: (1) two effects, earthquake location errors and the dimension of the aftershock zone need to be separated in the modeling (cf. Eq. 9.11), since as we discussed above (Section 9.5.1), there are good independent geophysical estimates for these variables. Thus, we can test whether the inversion results are trustworthy. (2) The rupture *area* is used to characterize the source size in the ETAS model. The area is more difficult to measure and is less known than a simple linear size of the focal zone – especially for earthquakes with a near vertical fault-plane due to large vertical location errors for shallow earthquakes. Thus, because of the different parametrization in both models, it is difficult to compare our results for the spatial clustering parameters (s_r and ϵ_ρ in Tables 9.2–9.5) with the corresponding ETAS values.

Eq. 9.8 corresponds to the right-hand part of the ETAS model dependence (cf. Ogata 1998, his Eq. 1.1)

$$\nu_i(t) = n_i(t)\ K_0 \times \exp\ [\alpha\ (m_i - m_t)], \tag{9.12}$$

where K_0 and α are parameters, m_i is a magnitude of a "parent" event, and m_t is a magnitude threshold. Then coefficient a (see Eq. 9.9) should be equal to $\alpha/\log\ (10) \approx \alpha/2.3$. In the above-mentioned publications the most likely α-estimates are between 1.2 and 1.6, that would roughly correspond to a-values in Tables 9.2–9.5. It is unclear how exactly our branching μ-parameter is related to the K_0 parameter, the dimensionality and the normalization of the parameters are different, thus it is difficult to compare their estimates.

Users of the ETAS model do not generally determine the likelihood score for an event. Thus, we cannot compare our information score I/N to the fit of the ETAS model to earthquake occurrence record and we cannot check the model's performance compared to an inhomogeneous spatial Poisson model. Helmstetter *et al.* (2006) performed calculations for the southern California seismicity, using the ANSS catalog and the model they named the ETES model which is largely similar to the ETAS. The parameters values obtained by the authors (ibid., their Table 2) are close to those shown in Table 9.5. The p-value is around 1.2. Their $\alpha \approx 0.4 - 0.8$ can be directly compared to our a-value, since it is an exponent for 10.0. The focal zone size f_d translates to about 0.4 km for an $m4$ earthquake (they fixed the location error at 0.5 km, see their Eq. 19). Their information score $\log_2\ G \approx 3.5$ is also close to our values for a $m_t = 3$ subcatalog (Table 9.5).

9.6 Results of the catalogs' statistical analysis

What differences in the zones' parameters can be seen in Tables 9.1–9.5? As we mentioned at the end of Section 9.3, the β-value is essentially the same for all five zones. Section 9.5.1 b-value measurement results confirm this conclusion.

The results of the corner moment evaluation (Table 9.1) are also consistent with our previous analysis (ibid., see also Fig. 5.3). As we discussed above (Section 9.5.1), the differences in the b-values for various tectonic zones can most likely be attributed to non-linear tapering of the magnitude-frequency relation due to the corner magnitude influence and to various biases and systematic effects in the magnitude determination for the PDE and local catalogs (Kagan 2003).

Can we distinguish the differences in the earthquake occurrence patterns and in particular the earthquake clustering in various tectonic zones? The parameters I/N, $\lambda T/N$, μ, a, and θ need to be compared to see the difference. The θ estimates do not exhibit any consistent pattern. For the rest of the parameters, in all the Tables 9.2–9.5 oceanic ridges seem to stand out compared to continental areas. The ridges exhibit significantly less clustering than the continents: the information score I/N is smaller and the ratio of spontaneous events ($\lambda T/N$) is higher for the oceanic earthquakes.

Two parameters that characterize the degree of branching, that is, μ and the size distribution of descendant events (a) are also different for these zones, though the dissimilarity is not the same for the GCMT (Table 9.2) and the PDE catalogs (Tables 9.3 and 9.4). In the GCMT results the ridges' a-value is similar to the continental (zones 0, 1, and 4) estimates, but the μ-values are significantly smaller. The PDE catalog exhibits an opposite pattern: the a-values drop for the ridges, whereas the μ estimates essentially stay the same. In the PDE dataset the size of smaller earthquakes is characterized by various magnitudes, most often by the m_b magnitude, which exhibit many biases and systematic errors (Kagan 2003). Therefore, it is likely that the size distribution of triggered events may be biased in the PDE catalog. Unfortunately, the magnitude range of the GCMT catalog is too small to see the differences in the branching rate and earthquake clustering due to the variation of the moment threshold. Among the continental zones subduction earthquakes in the GCMT catalog exhibit the strongest clustering pattern, but this feature is not prominent in the PDE catalog.

In summary, we have observed interesting potential variations in branching parameters between tectonic zones, such as a higher fraction of independent events in the spreading-ridge tectonic zones. But we have not been able to document any variations in branching behavior among the tectonic zones with high confidence. At present, the variations in apparent branching obtained at different assumed threshold magnitudes (and/or with different catalogs, and/or with different methods of processing) are comparable in size to the effects we seek. More precise and confident discrimination requires some advance in at least one of three areas: (1) extensions to maximum-likelihood fitting software to provide multi-dimensional confidence bounds on parameters and combinations of related parameters; (2) comparisons of apparent branching parameters obtained from different analysis codes; and/or (3) propagation of apparent differences into earthquake forecasts of global scope which can be scored against future seismicity, to see whether a forecast that discriminates between zones outperforms one that treats all zones as equivalent (Chapter 11).

10 Long- and short-term earthquake forecasting

This chapter discussion follows several of our publications (Kagan 1973; Kagan and Knopoff 1977, 1987b; Jackson and Kagan 1999, Kagan and Jackson 2000, 2012, 2013; Helmstetter *et al.* 2006; Kagan *et al.* 2007). Since the forecasts and their testing are closely connected, many discussions in this chapter and Chapter 11 that follows are intermixed.

10.1 Phenomenological branching models and earthquake occurrence estimation

The fractal dimension of the earthquake process is lower than the embedding dimension: in time (1-D) the fractal dimension is around 0.5 (Chapter 6); the correlation dimension in 3-D space is 2.2 (Section 7.3). This allows us to forecast the spatial and temporal probability of earthquake occurrence. The temporal dimension indicates that earthquake origin times are concentrated on a small subset of total time, determining quantitatively properties of this subset is the basis for the short-term earthquake prediction. Similarly, the long-term forecasts are based on the description of the spatial structure of earthquake occurrence, its dimension points out that seismicity is strongly concentrated in space.

For the phenomenological forecast of seismic activity, we employed both methods shown in Figs. 3.1b,c: (A) branching-in-magnitude and (B) branching-in-time (see Kagan and Knopoff 1977, and Kagan and Knopoff 1987b, respectively). It is not currently clear which technique would be more appropriate for earthquake forecasting. The advantages or drawbacks may depend on catalog properties or on goals of forecasting.

With the first method (A), it is easier to calculate the earthquake rate at the detection threshold and extend the forecasts below the threshold. Forecasting in forward time by (A) method would involve simulating various cluster probabilities (Kagan 1973; Kagan and Knopoff 1977). The second technique (B) is convenient for calculating the earthquake rate at the forward time boundary of an available catalog. This method is currently widely used in earthquake prediction efforts (see citations at the beginning of Chapter 3 and Chapter 9). However, to extend the forecast horizon into the future, simulation is needed (Kagan and Jackson 2000; Helmstetter and Sornette 2004), since we need to

Earthquakes: Models, Statistics, Testable Forecasts, First Edition. Yan Y. Kagan.

consider earthquakes that occur between the end of the catalog data and the prediction time (see more detail in Section 10.3). Similarly, if we want to take into account the influence of past seismicity (as shown in Figs. 3.1b,c, then simulation is needed in both models (A and B).

We have developed a time-independent (long-term) and time-dependent (short-term) earthquake forecast by using several earthquake catalogs (Kagan and Knopoff 1977, 1987b; Kagan and Jackson 1994; Jackson and Kagan 1999; Kagan and Jackson 2000). Since 1999 our aim has been to adapt the clustering model (CBM) to make testable forecasts over large regions of the western Pacific (ibid.). Later we implemented the CBM for long- and short-term regional and worldwide forecasts.

Due to the insufficient capability of the available computers our initial regional forecasts were low-resolution using $0.5^\circ \times 0.5^\circ$ cells. Later we applied high-resolution ($0.1^\circ \times 0.1^\circ$) cells. Several problems need to be addressed in these forecasts: selecting and optimizing smoothing kernels both fixed and adaptive, properly handling high- and low-seismicity areas, forecasts testing, combining in one algorithm long- and short-term prediction, implementing in forecasts the statistical results of Chapters 4–8, and so on. The solutions to these tasks are discussed in this and the following section. The above forecasts are based on approximating the earthquake occurrence by a stochastic point process. In the final section (Section 10.6), we consider forecasts using more detailed models of seismicity.

10.2 Long-term rate density estimates

10.2.1 Low-resolution forecasts

Kagan and Jackson (1994) present long-term earthquake forecasts in several regions using the GCMT catalog (see Section 2.3). They use a spatial smoothing fixed kernel proportional to inverse epicentroid distance $1/r$, except that it was truncated at short and very long distances. In later work Kagan and Jackson (2000; 2011) use a slightly different power-law kernel

$$f(r) = \frac{1}{\pi} \times \frac{1}{r^2 + r_s^2}, \tag{10.1}$$

where r is the epicenter distance, r_s is the scale parameter and $r \leq 1000$ km. The kernel is elongated along the fault-plane, which is estimated from the available focal mechanism solutions. To accomplish this we multiply the kernel by an orientation function $D(\varphi)$ depending on the angle φ between the assumed fault plane of an earthquake and the direction to a map point

$$D(\varphi) = 1 + \Theta \times \cos^2(\varphi). \tag{10.2}$$

The parameter Θ above controls the degree of azimuthal concentration (Kagan and Jackson 1994, Fig. 2). In smoothing we weigh each earthquake according to its moment magnitude

$$w = m/m_t. \tag{10.3}$$

We forecast the focal mechanism of a predicted earthquake as follows: first, we predict the focal mechanism of an earthquake by summing over past events, with a weighting as given above:

$$M_{pq} = \sum_{i=1}^{n} (M_{pq})_i f(r_i) D(\varphi_i) w_i, \tag{10.4}$$

where $(M_{pq})_i$ is the normalized seismic moment tensor of the i-th earthquake in the catalog. Then we calculate the eigenvectors of the sum M_{pq} and assign the eigenvector corresponding to the largest eigenvalue as the T-axis, and the eigenvector as the P-axis corresponding to the smallest eigenvalue of a forecasted event at each location. Thus, though M_{pq} in general is not a double-couple, we take the normalized double-couple component of the tensor as the forecasted mechanism. The angular density function of M_{pq} is then a function of the minimum 3-D rotation angle Φ necessary to take an i-th earthquake focal mechanism into the predicted one. The weighted average rotation angle $\overline{\Phi}$ is also calculated. The angle shows the degree of tectonic complexity at this point. Coordinates of the T- and P-axes as well as $\overline{\Phi}$ are shown in our forecast tables (see an example in Table 10.1 and Fig. 10.14).

An important feature of Kagan and Jackson's (1994) method is a procedure similar to the jack-knife (Silverman 1986) to test the predictive power of the smoothing. It optimizes the kernel parameters choosing those values which, for example, best predict the second part of a catalog (the test or validation period), using a maximum likelihood criterion, from the first part (the training or learning period). We argue that because the seismicity pattern exhibits a long-term clustering (Kagan and Jackson 1991a), such a procedure is better suited than the standard density estimation techniques to predicting the future earthquake rate. We also assume on an *ad hoc* basis (Kagan and Jackson 1994, 2000) that the long-term background rate density is uniform over the whole region and integrates to 1% of the total earthquake rate

$$(\text{Background rate}) = \epsilon \times (\text{Total rate}), \tag{10.5}$$

with the ratio $\epsilon = 0.01$. This parameter should cover "surprise" earthquakes in places far from any event locations in a catalog. A quantitative determination of the ϵ-value would require an extensive study, a task for future work.

Chapter 7 demonstrates that the fractal dimension of earthquake hypocenters, δ, strongly depends on the earthquake catalog time interval. For temporally short catalogs δ is close to zero and approaches the asymptotic value $\delta \approx 2.2$ for catalogs of decades length. In a more intuitive setting, this result signifies that in short time intervals, hypocenters are concentrated in a few point clouds. With increased time, seismicity spreads over a fault system or seismic belt length, eventually occupying a set with the dimension in excess of the 2-D plane ($\delta > 2.0$).

Therefore, if one uses a set of earthquake epicenters in a relatively short catalog to predict the future seismicity rate, the optimal forecast kernel should spread beyond the presently available event points, that is, to be smoother than the standard density estimators (Silverman 1986) would suggest.

The forecasts are expressed as the rate density (that is, the probability per unit area and interval of time). They are updated every day and posted at http://eq.ess.ucla.edu/~kagan/kagan.html. Many FORTRAN programs used in Chapters 9–11 are posted there as well. In Fig. 10.1 we display the long-term forecast map computed for the northwest (NW) Pacific region using the GCMT catalog for 1977–2003 (the training period). Shallow earthquakes in 2004–2006 (the test period) are shown in white. Similar maps for the northwest (NW) and the southwest (SW) Pacific are shown, for instance, in Figs. 8a,b by Kagan and Jackson (2000). The NW-Pacific region boundaries are: latitude limits from 0.25°S to 60.25°N, longitude limits from 109.75°E to 170.25°E. The SW Pacific boundaries are: latitude 0.25°N–60.25°S, longitude 109.75°E–169.75°W.

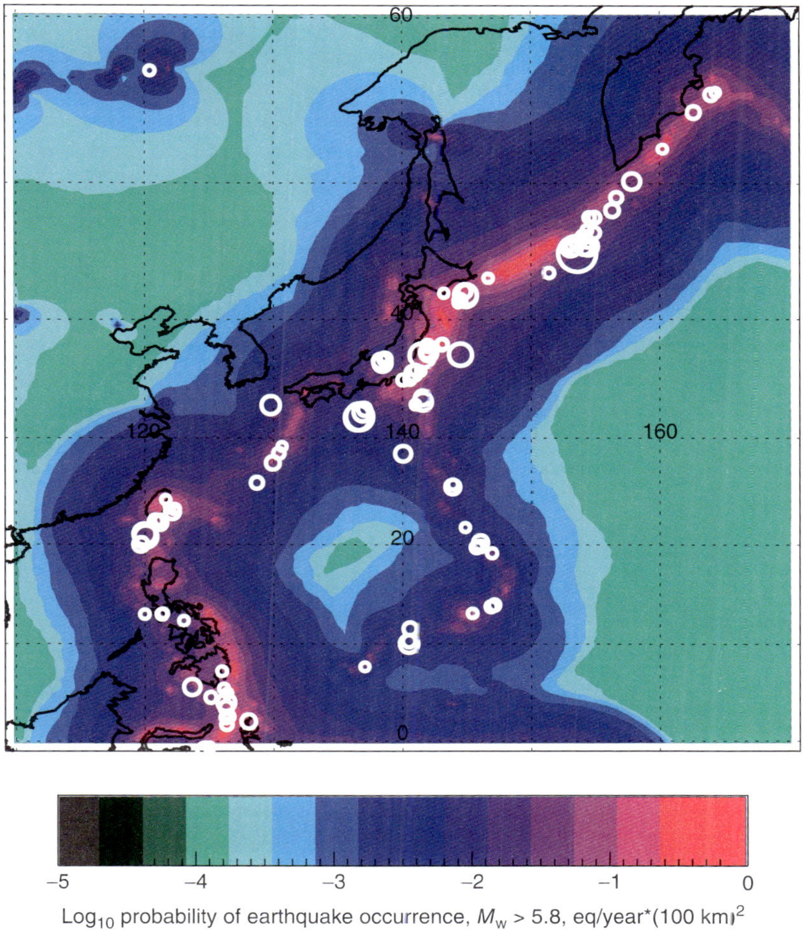

Fig. 10.1 NW-Pacific long-term seismicity forecast. The forecast is calculated at 121 × 121 0.5° grid. Color tones show the rate density of shallow (depth less or equal to 70 km) earthquake occurrence calculated using the GCMT 1977–2003 catalog; 108 earthquakes for 2004–2006 are shown in white. The uniform background rate density ($\epsilon = 0.01$, see Eq. 10.5) can be observed at northwest and southeast corners of the map as grayish-green areas. *Source:* Kagan (2009b), Fig. 1.

For the western Pacific forecasts we use the following values for a smoothing kernel: the spatial scale parameter, r_s is 15 km and 2.5 km for the NW and SW Pacific, respectively. The azimuthal concentration factor Θ in Eq. 10.2 is 100 and 25, respectively. On visual inspection (see Fig. 10.1), the model predicts the spatial distribution of seismic activity reasonably well. We tested this forecast by a Monte Carlo simulation (Kagan and Jackson 1994). Subsection 10.4.2 illustrates implementation of our forecast during the 2011 Tohoku mega-earthquake sequence.

10.2.2 High-resolution global forecasts

We had to solve two problems (Kagan and Jackson 2012) to produce the high-resolution (0.1° cells) whole Earth forecasts: (a) close to the poles we needed to use the exact spherical distance formula (for example, Bullen 1979, Eq. 5, p. 155) which requires about twice the computation time. (b) The size of the grid increased by a factor of 30 compared to quasi-global 0.5° forecast (Kagan and Jackson 2011).

The high-resolution program forecasts earthquakes on a 0.1° grid for the global region $90N-90S$ latitude, the maximum cell size around the equator is about 11 km. We used the PDE earthquake catalog that reports many smaller events ($m \geq 5.0$). The resolution of these forecasts is probably the maximum desirable: the shallow earthquake depth distribution peaks around 10 km, and the PDE catalog location uncertainty is about the same (Kagan *et al.* 2010; see Table 9.3).

The advantage of worldwide forecast testing is that many earthquakes are available. Because these quakes occur over the Globe, there are many practically independent sequences of events; thus seismicity is not dominated by only a few large earthquakes and their aftershocks, as may happen in local and regional earthquake catalogs. Therefore, the testing results are more reliable and reproduceable (Chapter 11). Again the full Earth forecast has lack of boundaries advantage (see the next section).

10.2.3 Smoothing kernel selection

Our previous regional and global forecasts (Jackson and Kagan 1999; Kagan and Jackson 1994, 2000, 2011) were based on fixed kernel smoothing (Eq. 10.1). To take into account the impact of earthquakes outside the boundaries of the region, we allowed events up to 1000 km outside the area to contribute to the rate density inside the forecast boundaries. The additional rate density from outside events is on average balanced by a contribution "leakage" from many "insider" earthquakes close to the boundaries.

Unfortunately, the 1000 km distance limit sometimes causes sharp discontinuities in the smoothed maps as can be seen in Fig. 10.4 around the islands of Hawaii. These discontinuities can be avoided if we use a kernel with the whole Earth support. However, in this case the choice of available kernels is significantly restricted. If we employ a power-law kernel like (10.1), its normalization on a sphere involves the application of cumbersome hypergeometric functions.

Practically, the best simple expression for a spherical surface kernel is the Fisher distribution (Fisher 1953, p. 296; Watson 1983, pp. 14–15; Fisher *et al.* 1987, their Eqs. 4.19–4.23; Mardia and Jupp 2000, their Eq. 9.3.4). These authors propose expressions for the spherical Fisher distribution in a general form. For our purpose we assume that the distribution center is at a pole of a sphere. Then the PDF of the spherical Fisher distribution is

$$\begin{aligned} f(\rho,\eta) &= \frac{\omega}{4\pi \sinh \omega} \exp\ (\omega\ \cos\rho) \times \sin(\rho) \times \phi(\eta) \\ &= \frac{\omega}{2\pi(e^{\omega} - e^{-\omega})} \exp\ (\omega \cos\rho) \times \sin(\rho) \times \phi(\eta), \end{aligned} \tag{10.6}$$

where η is an azimuthal angle, $\phi(\eta)$ is angular azimuthal distribution density, $\rho = r/R$ is the epicenter distance in radians, R is the Earth radius, and ω is a scale parameter.

It is more convenient to consider the Fisher distribution as depending only on distance. Due to a rotational symmetry we take $\phi\ (\eta) = 1/(2\pi)$. For the distance distribution only, $1/(2\pi)$ term in (10.6) can be omitted. Then, the cumulative spherical Fisher distribution function is

$$F(\rho) = \frac{\exp[(\omega\ (\cos\rho) - 1] - 1}{e^{-2\omega} - 1}. \tag{10.7}$$

For $\omega > 100$, Eq. 10.6 can be simplified

$$f(\rho) \approx \omega \exp[\omega\ (\cos\rho - 1)]\ \sin(\rho) = \omega \exp\left[-\omega\ \sin^2(\rho/2)\right] \sin(\rho). \tag{10.8}$$

Because for small distance values $(\sin\rho) \approx \rho$, the above equation suggests that the probability density decays like a Gaussian function. In California forecasts Werner *et al.* (2011) applied the Gaussian as well as the power-law kernel smoothing distributions and found that they yield similar results.

To forecast California seismicity Helmstetter *et al.* (2006, 2007) and Werner *et al.* (2011) employed an adaptive kernel smoothing, where the smoothing distance associated with an earthquake is proportional to the horizontal distance between the event and its nth closest neighbor (Silverman 1986, his Ch. 2.6). Zhuang (2011) applied such variable kernels in the forecasts for the region around Japan. The number of the neighbors, n, is an adjustable parameter, estimated by optimizing the forecasts. However, adaptive smoothing based on the distance between the epicenter and its nth neighbor has a certain disadvantage: if the $(n-1)$th and nth events are separated by a large distance, the kernel width would jump from one value to another. Thus, we propose a method which in effect smoothes the differences in the adaptive kernel width.

To carry out adaptive smoothing based on the Fisher distribution, Kagan and Jackson (2012) follow the advice of Silverman (1986, his Ch. 5.3): we first create an initial weight value estimate for earthquake epicenter location (χ_i) by using Eq. 10.8

$$\chi_i = \sum_{j=1}^{n} f(\rho_{ij}), \tag{10.9}$$

where n is the total number of earthquakes and ρ_{ij} is the distance between two epicenters.

Then the local bandwidth factors (Λ_i) are defined

$$\Lambda_i(\rho) = \omega\ (\chi_i/\chi_g)^{-\alpha}, \tag{10.10}$$

where χ_g is the geometric mean of χ_i

$$\log\ \chi_g = \frac{1}{n}\sum_{i=1}^{n} \log(\chi_i). \tag{10.11}$$

The ω-values in Eqs. 10.9 and 10.10 could be different, but as Silverman (1986, his Ch. 5.3) suggests (also see Section 11.4 below), the parameter value in the initial estimate does not significantly influence the final result. Since the larger ω-values correspond to a narrower kernel (see Eq. 10.8), in actual computations we use an inverse of the $\Lambda_i(\rho)$ function in (10.10). The forecast density at any point $\vec{r}$ is then estimated by

$$\mu(\vec{r}\,) = \sum_{t=1}^{n} \Lambda_i(\vec{r} - \vec{r}_i). \tag{10.12}$$

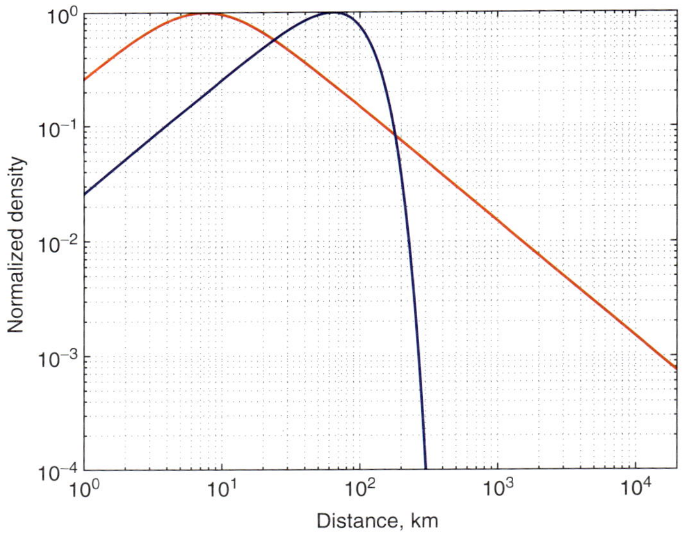

Fig. 10.2 Kernel functions: Red – power-law kernel with fixed r_s = 7.5 km (10.1); blue – Fisher distribution kernels (Eq. 10.8) with ω = 10,000. *Source:* Kagan and Jackson (2011), Fig. 1.

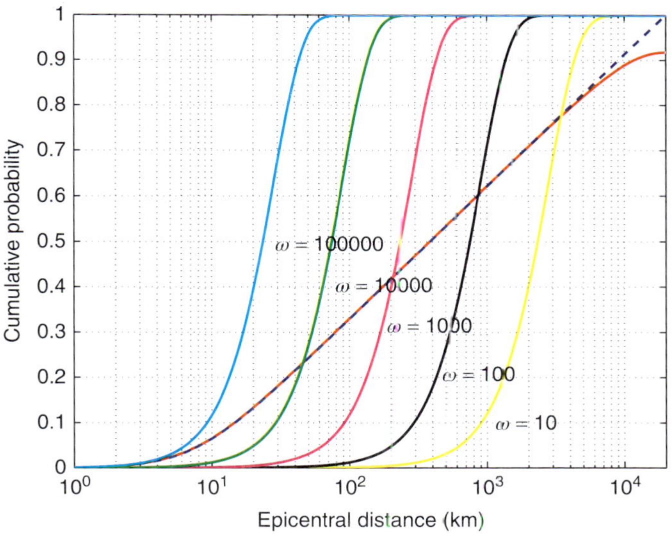

Fig. 10.3 Cumulative kernel functions: Solid and dashed diagonal lines correspond to power-law kernels (Eq. 10.1) with $r_s = 7.5$ km; dashed line integrated over plane surface, solid line integrated over spherical surface. Sigmoidal solid lines from right to left correspond to the Fisher distribution kernels (Eq. 10.7) with $\omega = 10$, 100, 1000, 10,000, and 100,000. *Source:* Kagan and Jackson (2011), Fig. 2.

In Fig. 10.2 we display two kernel examples: the densities for the power-law and for the spherical Fisher distribution. The density maximum for the Fisher law can be calculated by equating the derivative of its PDF (10.6) to zero. For large ω we obtain

$$\cos\rho - \omega \sin^2\rho \approx \text{C}. \tag{10.13}$$

Since for large ω the term $(\cos\rho) \uparrow 1$, the distance for the maximum is

$$\rho_m \approx \arcsin\sqrt{1/\omega}. \tag{10.14}$$

To demonstrate the difference between the kernels, Figure 10.3 shows several examples of the cumulative distribution for various kernels: the Fisher distribution with the change of the ω parameter and two variants of the power-law distribution – linearly normalized for a distance of 20,000 km and normalized on a sphere. The latter kernel normalization was adjusted so that the distributions coincide for smaller distances.

According to Eq. 10.10, the two parameters ω and α control the choice of kernel bandwidth; ω is estimated iteratively from a starting value. But as mentioned above, the final estimate depends very weakly on the starting one. The full spatial model includes one more parameter, ϵ (10.5), a "surprise" factor.

10.2.4 Comparing long-term forecasts

In Fig. 10.4 we show the global long-term forecast made with the PDE catalog using $m \geq 5.0$ earthquakes from 1969 to the present. The fixed power-law

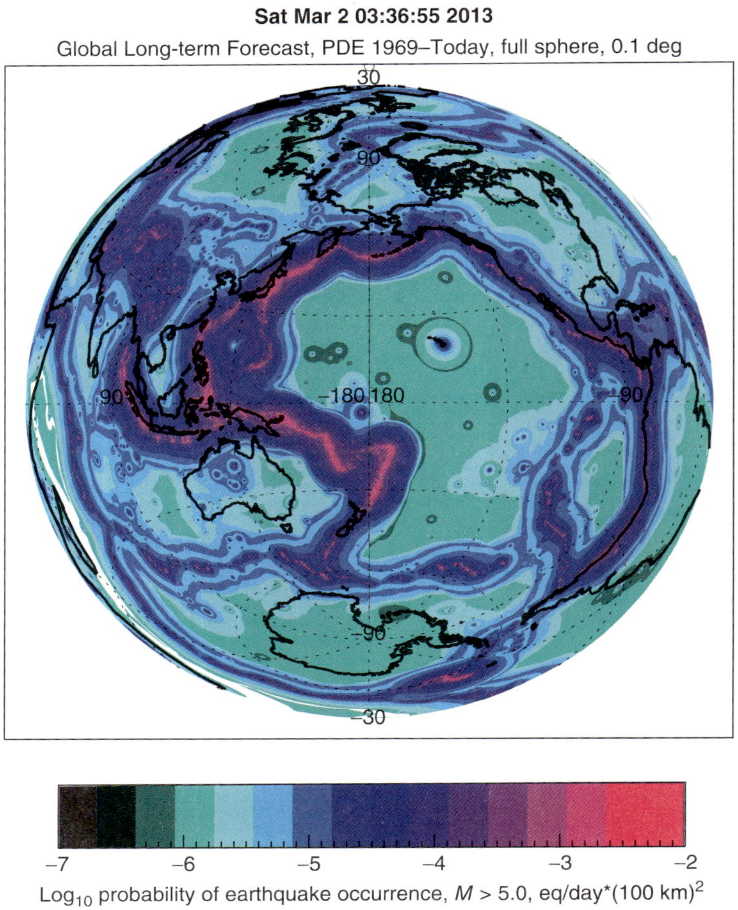

Fig. 10.4 Fixed power law kernel: global long-term earthquake potential based on smoothed seismicity from the PDE catalog 1969 to the present time. Earthquake occurrence is modeled by a time-independent Poisson process. The fixed power-law kernel (10.1) is used on the 0.1° grid. Source: Kagan and Jackson (2011), Fig. 4.

kernel (10.1) with $r_s = 7.5$ km is used on the 0.1° grid. The Lambert projection (Bugayevskiy and Snyder 1995) is applied for display in this and the next two plots. Fig. 10.5 shows the worldwide long-term forecast made using an adaptive Fisher distribution kernel. Comparing this plot with Fig. 10.4 it is obvious that the width of the seismicity peaks is reduced at the subduction zones. This is due to narrower kernels at concentrations of earthquake epicenters. Conversely, seismicity contours for the low activity regions are smoother for the adaptive kernels than for the fixed ones (Kagan and Jackson 2012).

Optimizing the three parameters ω, α, and ϵ is computationally challenging, so we have taken some shortcuts. We used a global low-resolution grid (0.5°cells) to estimate ω and α, and we made only a partial search for the maximum likelihood values of all parameters. The model used in Fig. 10.5 is based on that approximate optimization. We regard it as a respectable model worthy of comparison to others, but the values could probably be improved by further optimization. In Chapter 11 we rigorously test long-term forecasts.

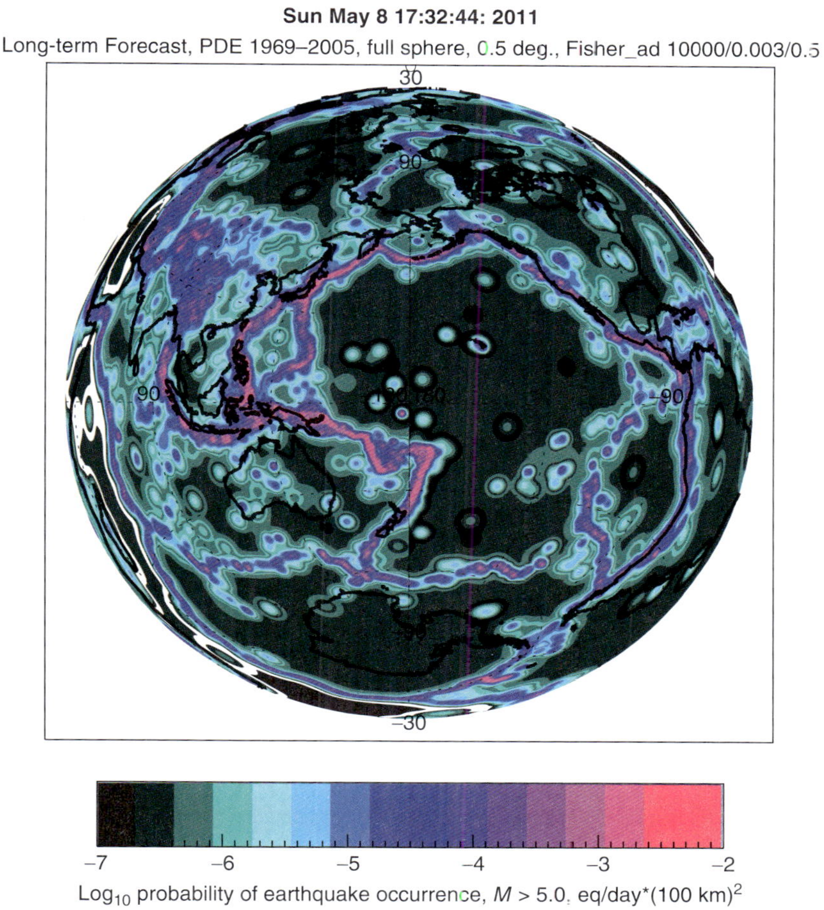

Fig. 10.5 Adaptive Fisher kernel: global long-term earthquake rates based on smoothed seismicity from the PDE catalog 1969 to April 2011. Adaptive smoothing kernel based on the Fisher spherical distribution (Eqs. 10.10–10.12) is used on the 0.5° grid. Values of parameters are: $\omega = 100,000$, $\alpha = 0.5$, and $\epsilon = 0.003$. Earthquake occurrence is modeled by a time-independent Poisson process. *Source:* Kagan and Jackson (2011), Fig. 5.

10.3 Short-term forecasts

The short-term forecast in our work was carried out according to the technique described by Kagan and Jackson (2000, their Section 3.1.2) and Kagan and Jackson (2011, their Section 3.3.3). Examples of the short-term forecast for other regions are shown in several of our papers (ibid., Kagan and Knopoff 1987b; Molchan and Kagan 1992; Jackson and Kagan 1999; Helmstetter *et al.* 2006). More examples of the short-term forecast are displayed in Figs. 10.9–10.12 below. In these short-term forecasts we used the values of parameters obtained during the likelihood function search for the GCMT and PDE catalogs. From Tables 9.2–9.5 we see that the values of the earthquake clustering parameters are similar for the subduction zones (trenches), the active continental zones and the plate interiors. These regions are of special interest for evaluating the seismic hazard. Short-term probabilities computed by these programs can be used in the operational earthquake forecasting discussed by Jordan and Jones (2010), van Stiphout *et al.* (2010), and Jordan *et al.* (2011).

The calculated short-term forecast is valid only for the "next moment" after the end of a catalog (Kagan and Jackson 2000). To extend the forecast further in time, we need to take into account the possibility that other earthquakes could occur in the time interval between the end of a catalog and the forecast time. This accounting can be carried out by the Monte Carlo simulation procedure (Kagan 1973; Kagan and Knopoff 1977; Zhuang 2011): we simulate the first generation of dependent events, which in turn are fed into the algorithm to obtain the second generation, and so on. If the time horizon of the forecast is not long, only a few generations need to be computed for such simulations. For longer time intervals we may simply use our long-term forecast. Helmstetter *et al.* (2006, p. 95) describe a simplified method to extend a forecast to future time. However, because of the computational difficulties in this work we did not follow either of these procedures.

The need for simulation follows from the choice to make and evaluate forecasts on a fixed time interval. Any event in that time interval will change the earthquake rate thereafter, but the model cannot be updated till the end of that time interval. It is worth noting that the fixed time intervals are imposed for practical rather than scientific reasons: public actions cannot always be started or stopped at arbitrary times. Scientific purposes can be well served with forecasts valid only until the next earthquake of a predetermined size. At present, forecasts cannot be practically updated in "real time" because earthquake reports take some time, and final catalogs may take months to be completed. Nevertheless, a forecast model can be retrospectively tested and updated using finalized data for each earthquake in series. Forecasting and testing in this way do not provide information for rapid decisions, but it is as rigorous as a prospective operation as long as the algorithms and parameters are fixed in advance.

Another important problem that we could not solve, is quantitatively determining the effectiveness of the short-term forecast. Chapter 11 discusses the methods to calculate information scores for the long-term forecasts; these methods are based on log-likelihood difference between the forecast and a simple null hypothesis. We also show how to prospectively test the forecasts by using the log-likelihood scores.

A similar scheme could be used for short-term forecasts. Kagan and Knopoff (1987b) tested the short-term forecast by the maximum likelihood technique for a retrospective earthquake forecast on the San Andreas fault. We calculated such short-term log-likelihood scores for the GCMT and the PDE catalogs (Kagan *et al.* 2010; Tables 9.2–9.5). The short-term forecast's efficiency can be measured by its average probability gain per earthquake compared to the temporally uniform Poisson distribution (see, for example, items I/N in Tables 9.2–9.5). For the short-term forecast, the gain G (11.17) is about 3.7 for the PDE catalog relative to the temporally random but spatially localized null hypothesis. For regional catalogs (Table 9.5) with a lower magnitude threshold, the gain increases to $G \approx 9$. One should expect even greater gain values for catalogs obtained with high-frequency seismograms (Subsections 5.3.3, 6.2.4) where aftershocks are detected in coda waves of strong earthquakes. However, these scores may be biased if there is no information about past seismicity, or if there is a finite but inadequate sample. In our calculations of the log-likelihood scores, the earthquake interaction is considered in both time directions. In a prospective forecast only the past record is available, so the score would be smaller.

Kossobokov (2006) determined the effectiveness of our short-term forecast in the west Pacific (Jackson and Kagan 1999; Kagan and Jackson 2000) by using the error diagram (Molchan and Kagan 1992; Kagan 2009b; Molchan 2010). His results indicate that the short-term forecast is significantly "better than random guessing." (However, see Kagan and Jackson's 2006 comment on his paper.)

Figure 10.6 shows the calculated full Earth high-resolution short-term forecast. Red spots delineate places where strong earthquakes have occurred recently, the increased short-term rate can be seen for months and years. Similar short-term maps are shown in the next section (Figs. 10.9–10.12).

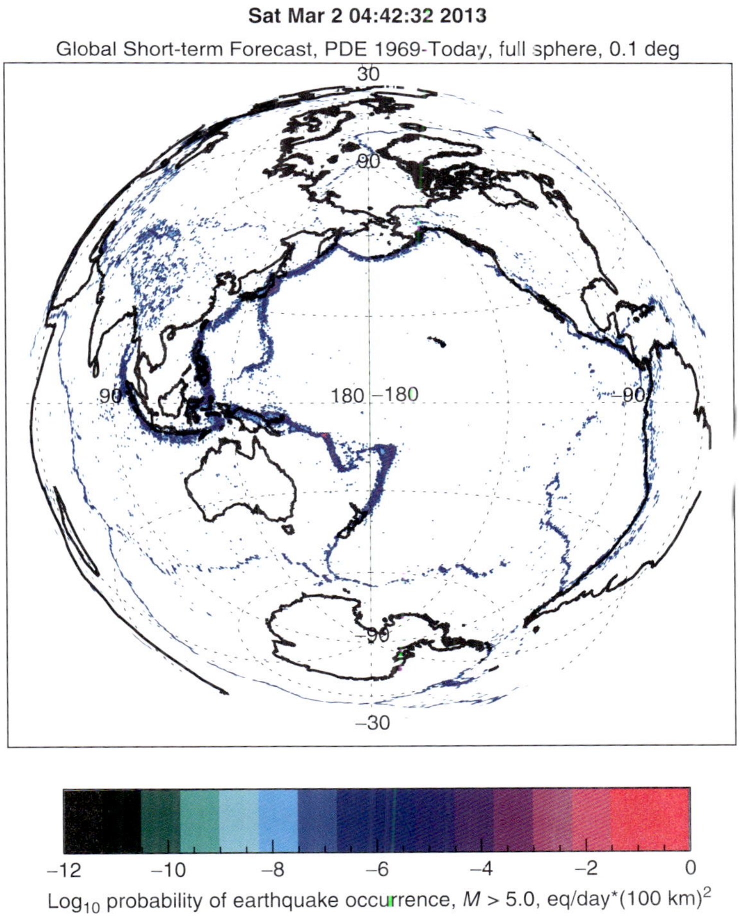

Fig. 10.6 Earthquake short-term potential on 0.1° grid based on smoothed seismicity from the PDE catalog since 1969 to the present time. Earthquake occurrence is modeled by a branching temporal process controlled by Omori's law type dependence. *Source:* Kagan and Jackson (2011), Fig. 10.

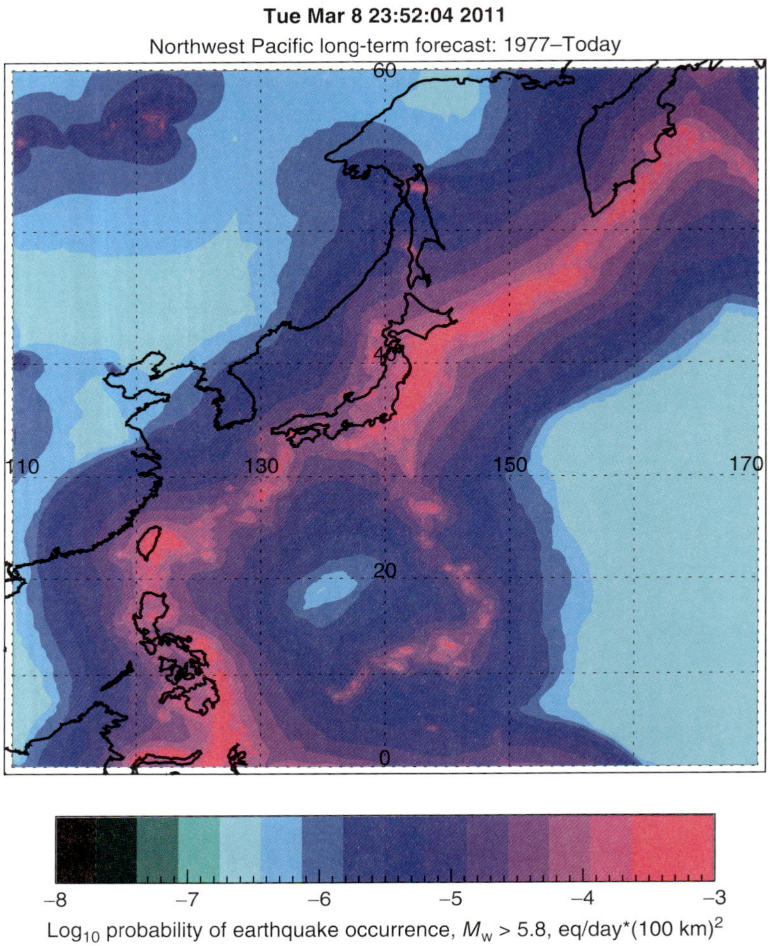

Fig. 10.7 The long-term forecast rate for the north-west Pacific calculated March 8, 2011, before the *m*7.4 Tohoku foreshock. The fixed power-law kernel (10.1) is used on a 0.5° grid.

10.4 Example: earthquake forecasts during the Tohoku sequence

10.4.1 Long- and short-term earthquake forecasts during the Tohoku sequence

Figures 10.7 and 10.8 show long-term forecasts for the north-west Pacific area; one forecast is calculated before the 2011 Tohoku sequence started, the other about one week after the mega-earthquake. There is little difference between these plots – long-term forecasts do not depend strongly on current events. Plate 1 in Kagan and Jackson (1994) and Fig. 8a in Kagan and Jackson (2000) display previous long-term forecasts for the same region. The appearance of both earlier diagrams is similar to Figs. 10.7 and 10.8.

Table 10.1, similar in format to Table 1 by Kagan and Jackson (2000), displays earthquake forecast rates around the epicenter of the *m*7.4 Tohoku foreshock for three time periods – before and after the foreshock as well as after the main-shock. The ratio of the short- to long-term rates (the last column) rises sharply

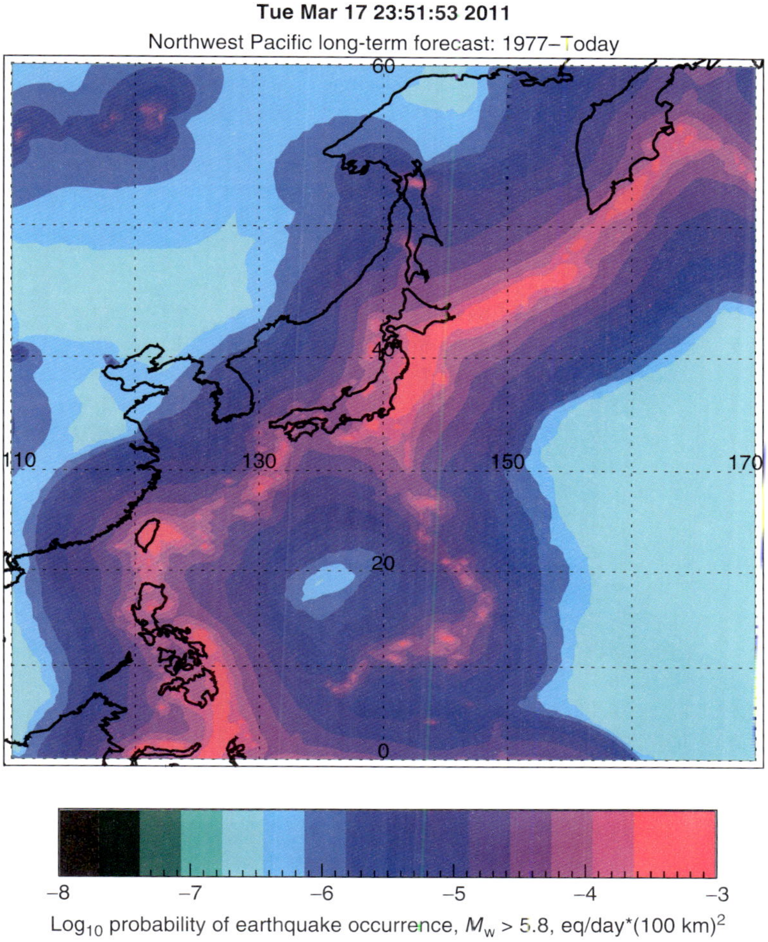

Fig. 10.8 The long-term forecast rate for the NW-Pacific calculated March 17, 2011, after the Tohoku 9.2 mainshock. There is little change compared to Fig. 10.7.

both after the foreshock and after the mainshock. Conclusions similar to those in the previous paragraph can be drawn from Table 10.1: the maximum long-term rates change only by a few tens of a percent. The predicted focal mechanisms (see Eq. 10.4) are also essentially the same for the center of the focal area.

As an illustration, using simple methods we make an approximate estimate of the long-term recurrence rate for large earthquakes in the Tohoku area. In the GCMT catalog, the number of earthquakes with $m \geq 5.8$ in a spherical rectangle 35–40° N, 140–146° E, covering the rupture area of the Tohoku mega-event, is 109 for the years 1977–2010. If we assume that the corner magnitude is well above *m*9.0 (similar to *m*9.6 for subduction zones, see Section 5.6.1), the repeat time for the *m*9 and larger events in this rectangle depends on the assumed *b*-value and is between 300 and 370 years.

Figures 10.9–10.12 display short-term forecasts for this area produced during the initial period of the Tohoku sequence. These forecasts are computed around

Table 10.1 Examples of long- and short-term forecasts during the 2011 Tohoku earthquake sequence.

Latitude	Longitude	LONG-TERM FORECAST						SHORT-TERM	
		Probability $m \geq 5.8$ eq/day*km^2	Focal mechanism					Probability $m \geq 5.8$ eq/day*km^2 Time-dependent	Probability ratio Time-dependent/ independent
			T-axis		*P*-axis		Rotation angle $\bar{\Phi}$ degree		
			Pl	Az	Pl	Az			
			March 8						
141.0	38.5	1.64E-08	81	307	8	107	24.31	5.603E-10	3.423E-02
141.5	38.5	2.31E-08	76	327	11	107	29.62	8.171E-10	3.535E-02
142.0	38.5	1.04E-07	76	0	5	112	31.68	4.414E-09	4.234E-02
142.5	38.5	7.60E-08	63	312	25	113	11.78	2.667E-09	3.510E-02
143.0	38.5	4.98E-08	60	303	29	113	10.25	1.215E-09	2.441E-02
143.5	38.5	3.06E-08	61	303	28	113	13.13	5.691E-10	1.861E-02
144.0	38.5	1.77E-08	61	306	28	114	15.54	1.240E-10	6.993E-03
144.5	38.5	1.21E-08	55	292	35	117	27.55	3.989E-10	3.300E-02
145.0	38.5	2.08E-08	7	275	82	62	27.84	1.238E-09	5.953E-02
			March 10						
141.0	38.5	1.64E-08	81	307	8	107	24.34	5.880E-10	3.583E-02
141.5	38.5	2.32E-08	76	327	11	107	29.56	1.406E-08	0.605
142.0	38.5	1.05E-07	76	360	6	112	31.66	2.482E-06	23.7
142.5	38.5	8.14E-08	63	311	26	113	11.82	7.895E-06	97.0
143.0	38.5	5.87E-08	60	301	30	113	9.76	5.298E-06	90.2
143.5	38.5	3.67E-08	61	303	28	113	11.19	1.868E-06	50.9
144.0	38.5	1.78E-08	61	305	28	114	15.47	2.314E-07	13.0
144.5	38.5	1.21E-08	55	292	35	117	27.45	3.644E-09	0.301
145.0	38.5	2.08E-08	7	275	82	63	27.91	1.241E-09	5.963E-02
			March 11						
141.0	38.5	1.64E-08	81	307	8	107	24.35	4.960E-06	303
141.5	38.5	2.33E-08	76	326	11	107	29.51	9.397E-06	404
142.0	38.5	1.05E-07	75	359	6	112	31.64	4.938E-05	471
142.5	38.5	8.32E-08	63	310	26	113	11.80	4.271E-05	514
143.0	38.5	6.32E-08	59	301	30	113	9.47	3.109E-05	492
143.5	38.5	4.10E-08	61	302	29	113	10.75	1.978E-05	482
144.0	38.5	2.03E-08	60	304	29	114	14.44	9.430E-06	466
144.5	38.5	1.33E-08	55	293	35	117	25.04	5.504E-06	413
145.0	38.5	2.14E-08	9	275	81	74	29.44	6.765E-06	316

midnight Los Angeles time. Figure 10.9, calculated before the *m*7.4 foreshock, shows a few weakly "red" spots in those places where earthquakes had occurred during the previous weeks and months. The short-term rate in these spots is usually of the order of a few percent or a few tens of a percent compared to the long-term rate (see also Table 10.1).

The predicted earthquake rates in the neighborhood of the future Tohoku event increased strongly with the occurrence of a *m*7.4 foreshock (Fig. 10.10).

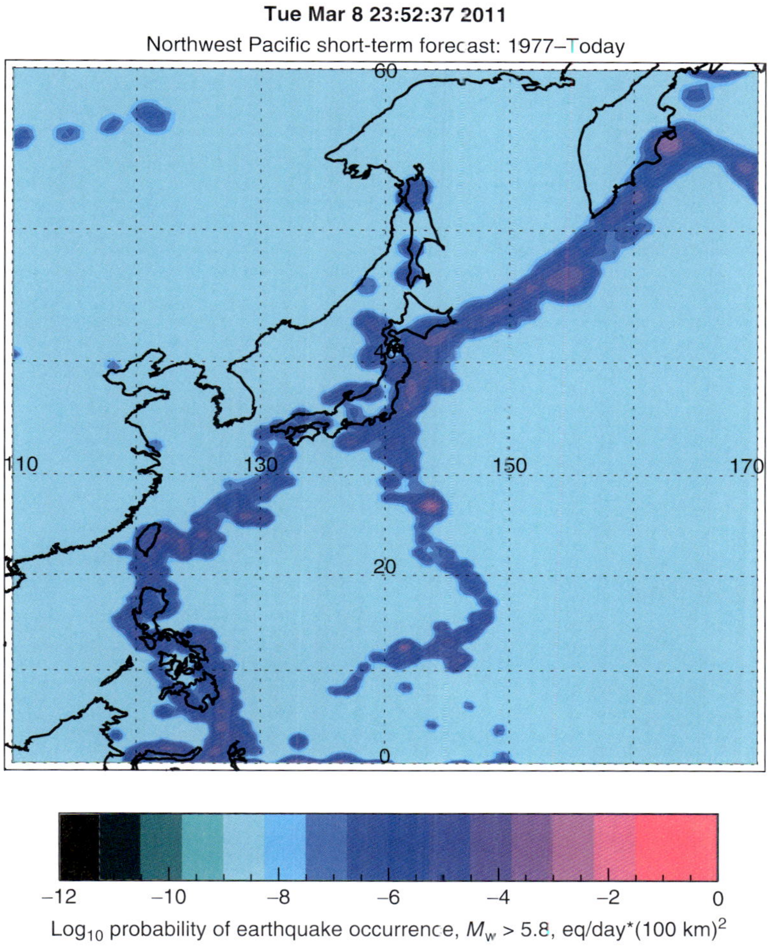

Fig. 10.9 The short-term forecast rate for the NW-Pacific calculated March 8, 2011, before the $m7.4$ foreshock. The fixed power-law kernel (10.1) is used on a 0.5° grid.

As Table 10.1 demonstrates, just before the Tohoku mega-earthquake, the forecasted short-term rate was about 100 times higher than the long-term rate.

The area of significantly increased probability covers the northern part of the Honshu Island following the Tohoku mega-earthquake occurrence (Fig. 10.11). The size of the area hardly decreased one week later (Fig. 10.12), the rate stability is mostly due to aftershocks.

Although only around 5–10% of shallow earthquakes are preceded by foreshocks (Reasenberg 1999), the results shown in Figs. 10.9–10.12 suggest that an operational earthquake forecasting strategy could be developed (see Section 10.3) to take the increase of short-term rates into account.

10.4.2 Long-term earthquake rates for the Tokyo region

As an example of long-term rate calculation we evaluate the rate of occurrence for large earthquakes in the Tokyo area. For comparison we use estimates for

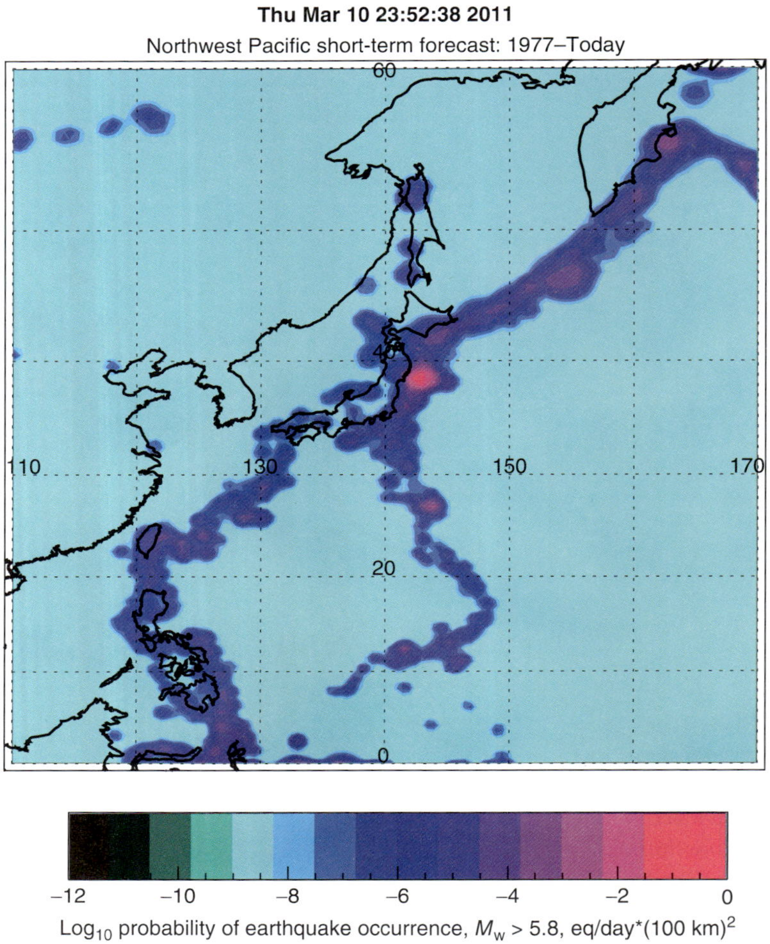

Fig. 10.10 The short-term forecast rate for the NW-Pacific calculated on March 10, 2011, after the *m*7.4 foreshock, just before the Tohoku *m*9.2 mainshock – near the foreshock epicenter the short-term rates are many times higher than the long-term rates.

a repeat of the 1923 Kanto *m*7.9 earthquake made by Stein *et al.* (2006) and by Grunewald and Stein (2006). On the basis of studies of historical seismicity and paleo-seismic data they calculated the probability of an *m*7.9 and greater earthquake in the Tokyo area (window coordinates 139.0–140.5°E, 34.5–36.0°N). Depending on earthquake size distribution, for the Poisson occurrence the 30 years probability is estimated as 7–10%, that is, the recurrence time for such or stronger events is about 300–400 years.

Using forecast tables, calculated during the Tohoku earthquake sequence (see the examples in Table 10.1), we compute the average long-term rate for *m*5.8 and greater events in the window listed above. As we mentioned earlier, the long-term rate was essentially stable during the sequence, increasing by the March 17, 2011 only by 10–15%. Depending on the time the average rate is taken, it is between $1.7 - 2.1 \times 10^{-8}$ eq/(day km^2). For the TGR distribution with the *b*-value 0.96 and the corner magnitude greater than 8.5 (Chapter 5), this would

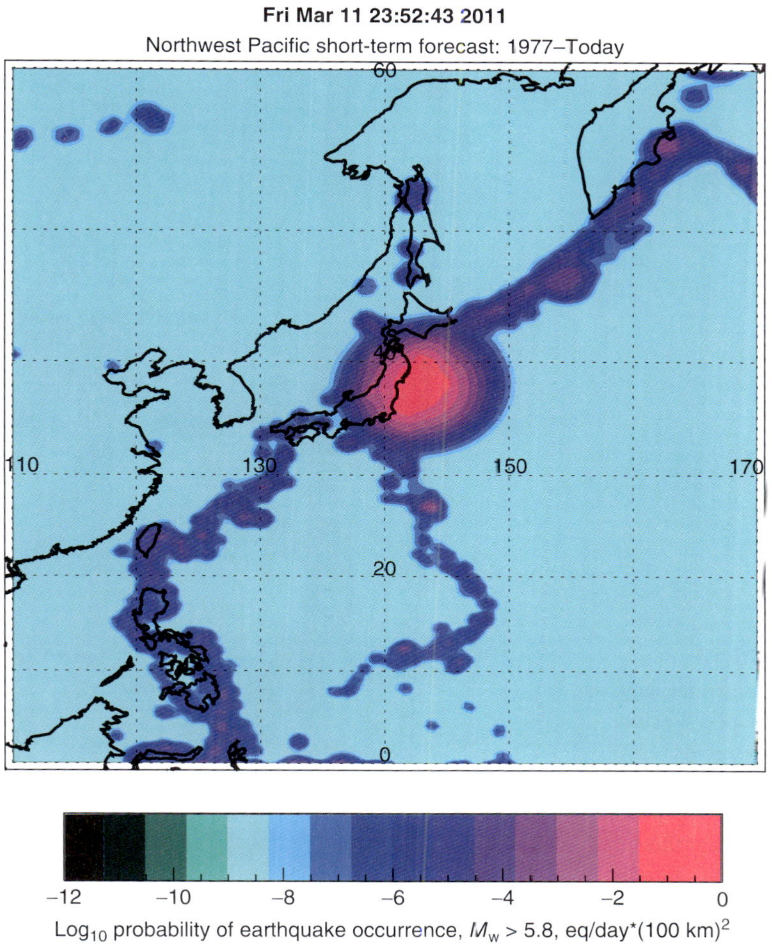

Fig. 10.11 The short-term forecast rate for the NW-Pacific calculated March 11, 2011, immediately after the *m*9.2 mainshock, at its epicenter the short-term rates are about 1000 times higher than the long-term rates.

imply the recurrence interval for $m \geq 7.9$ earthquake of the order 600–750 years, a value greater than the Stein *et al.* (2006) and the Grunewald and Stein (2006) estimate by a factor of two.

Another estimate is based on Bird *et al.*'s (2010) seismic map based on the Global Strain Rate Map. Integrating the map for the window, we obtain the average rate of $m \geq 5.693$ earthquakes 1.7 10^{-19} eq/(m^2 s). Translating it for $m \geq 7.9$ events by assuming that the corner magnitude significantly exceeds *m*7.9, we obtain inter-event time of the order 1100 years.

It is difficult to say why these evaluations are so different. Paleo-seismic and historical magnitude estimates can be strongly biased. Even if they are not, the magnitude uncertainties should be large, likely much greater than 0.25 proposed by Grunewald and Stein (2006). Tinti and Mulargia (1985) calculated that the *a*-value bias (see Eq. 5.2) is proportional to the square of the magnitude standard error. This should cause a significant increase in the estimated rate of occurrence

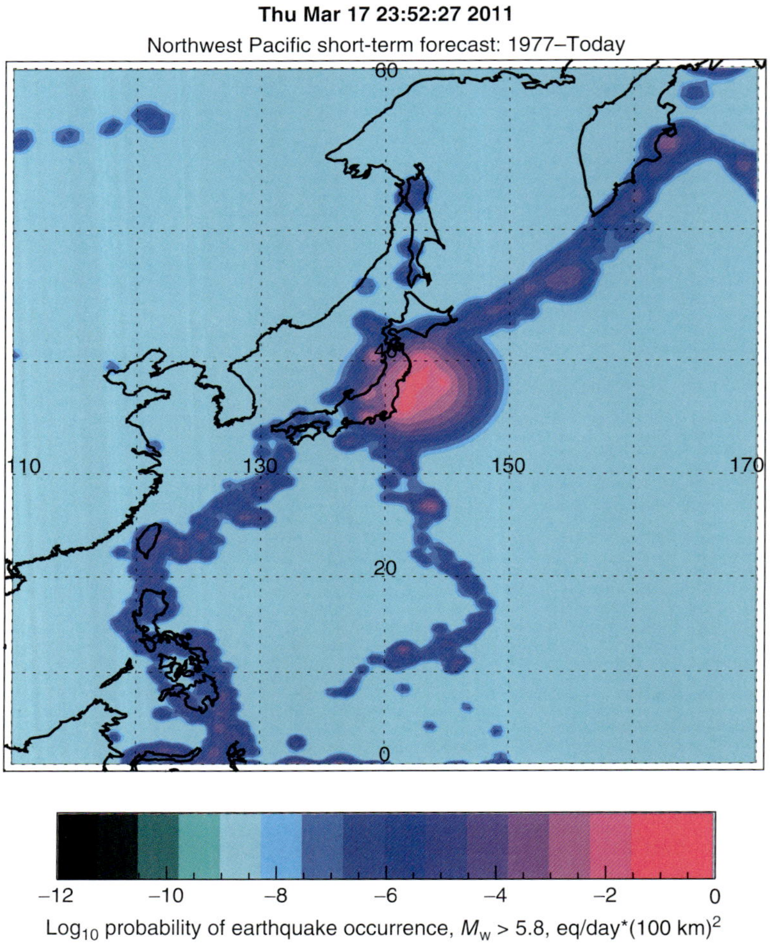

Fig. 10.12 The short-term forecast rate for the NW-Pacific calculated March 17, 2011, about a week after the *m*9.2 mainshock. At its epicenter the short-term rates are significantly higher than the long-term rates.

for large earthquakes and correspondingly a false decrease of the inter-event time. Our long-term forecasts are based on the GCMT catalog, where the magnitude accuracy is high (of the order of 0.1, see Kagan 2003 and Section 5.1), and forecasts are obtained by optimizing a spatial smoothing kernel (Kagan and Jackson 2000, 2012).

10.5 Forecast results and their discussion

The earthquake forecast technique described here continues the work we have pursued since 1977. Our question has been: how can we use the available seismicity record to optimally evaluate future seismic hazard and risk? Because of the fundamentally random character of earthquakes, such a forecast must be statistical.

Several problems must be solved to make such a forecast quantitatively efficient and applicable to the local and global environment. The relevant properties of earthquake occurrence need to be thoroughly investigated, including classical statistical earthquake distributions. How do Omori's law and the G-R relation apply in various regions? We also need to consider in a quantitative, objective manner the spatial distribution of seismicity (Chapter 7). As with any application of measurement results to practical problems, we must have quantitative estimates for the measurement uncertainties and data completeness (Kagan 2003).

Similar to statistical analysis of the time-space earthquake occurrence pattern (Chapter 9), for the long-term model we factor the rate density into the spatial, magnitude, and number distributions, assuming their independence from one another. We estimate the spatial distribution using all earthquakes above the assumed completeness threshold in the catalog. For the magnitude distribution, we assume a TGR distribution (Chapter 5). Any forecast with a given lower magnitude threshold applies as well to larger earthquakes with the maximum (corner) magnitude determined for appropriate tectonic zones. The map of tectonic zones shown in Fig. 9.1 assigns any 0.1° cell to a specific tectonic category. Then we can recalculate any forecast rate using the TGR distribution with the appropriate corner magnitude value (Table 9.1).

Our results (Figs. 5.2 and 5.3, Table 9.1) indicate that whereas the *b*-value of the G-R relation has a universal value 0.93–0.98, the maximum or corner magnitude changes significantly for various tectonic zones, from a high of $m9.6$ for subduction zones to a low of $m5.9$ for oceanic spreading ridge normal faulting earthquakes. Nevertheless, the maximum earthquake size was shown to be the same, at least statistically, for all the studied major subduction zones (Section 5.6). These maximum size determinations, combined with the observed very large ($m \geq 9.0$) events in the other subduction zones before 1997 warn us that such an earthquake could occur in any major subduction zone, including the Sumatra and Tohoku (Japan) regions.

The investigation of earthquake time behavior (Chapter 6) suggests that the Omori's law parameters are also universal: even strong earthquakes are clustered in time instead of being quasi-periodic. We assume that in addition to earthquakes resulting from clustering, spontaneous earthquakes, approximately Poissonian, occur at a rate proportional to the regional strain rate. Thus, as clustered earthquakes decay in frequency according to Omori's law, the total rate becomes asymptotically Poissonian. Therefore, for tectonic plate interiors where the strain rate is low, only Omori's law decay for historical and paleo-earthquakes is usually observed. No transition to the Poisson rate can be unambiguously seen.

England and Jackson (2011) indicate that most earthquake-related casualties occur not at tectonic plate boundaries such as subduction zones but in the continental interiors far from the boundaries. This presents a challenge for seismicity-based earthquake forecasts: if earthquakes are rare, our technique cannot give a good estimate of future earthquake rate with temporally limited earthquake catalogs. However, a forecast based on including the global strain rate data (Bird *et al.* 2010) can significantly improve the seismicity-based forecasts, especially in regions where spatial geodesy and geologic data allow us to evaluate the strain rate accurately. Using the calibration provided by Bird and Kagan (2004) and

Kagan *et al.* (2010), Bird *et al.* (2010) converted the global strain rate to an earthquake occurrence map.

For most of the active continents (the Alpine-Himalayan belt) we have enough earthquake occurrence data and tectonic rate estimates to evaluate the seismic occurrence rate, albeit with some difficulty. However, for plate interiors where the fault deformation rates are less than one mm per year, it is still a problem. In these regions only rough estimates of long-term earthquake rates can be obtained (Section 6.6; Kagan 2011). However, by using our results, an extrapolation of seismic activity in aftershock zones can be obtained.

The forecast method described in this chapter can be considered an important step in the development of seismicity-based earthquake forecasts. The forecast covers the whole surface of the Earth. The spatial resolution of the forecast is unlikely to be significantly improved with the available data, and the adaptive feature of the smoothing kernel produces a nearly optimal adjusted forecast map. However, as discussed in Section 10.3, if a temporarily extended forecast is needed, then future seismic activity needs to be simulated. That may require greater computational resources.

10.6 Earthquake fault propagation modeling and earthquake rate estimation

In the forecasts considered earlier in this chapter earthquakes have been taken as points in space-time. In reality, earthquake ruptures, especially strong ones, are extended over large distances and their propagation time cannot be ignored. Thus, in this section we consider forecasts using more detailed models of seismicity.

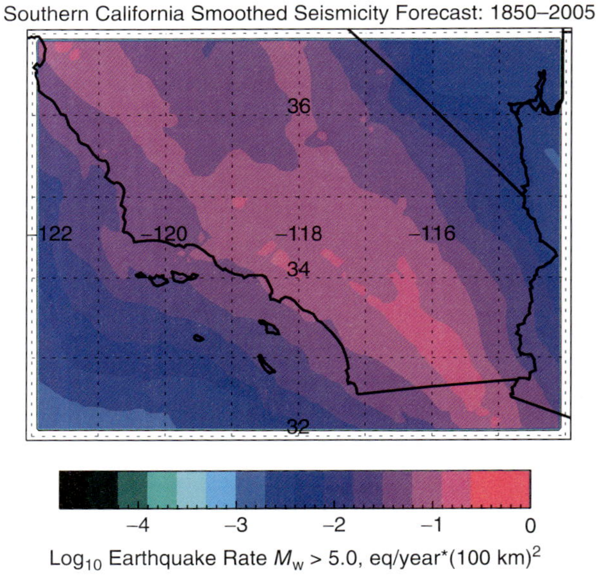

Fig. 10.13 The California-Nevada earthquake potential based on smoothed seismicity. Earthquakes from the RELM catalog since 1850 are used. Earthquake occurrence is modeled by a time-independent (Poisson) process. Color scale tones show the long-term probability of earthquake occurrence. *Source:* Kagan *et al.* (2007), Fig. 1.

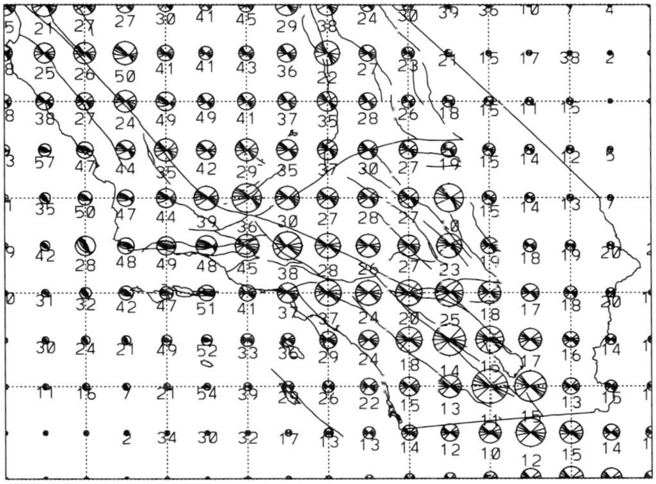

Fig. 10.14
Long-term forecast diagrams of earthquake focal mechanisms in southern California. Lower hemisphere diagrams of focal spheres are shown. Size of the focal mechanism diagram is proportional to forecasted rate of occurrence (Fig. 10.13). Stripes in beachballs are concentrated towards the assumed earthquake fault-plane. The numbers below the diagrams of earthquake focal mechanisms correspond to the weighted average rotation angle $\overline{\Phi}$ (Eq. 10.4). Points without a beachball diagram and rotation angle uncertainty denote places for which data are inadequate to forecast focal mechanism. *Source:* Kagan *et al.* (2007), Fig. 2.

10.6.1 Earthquake extended rupture representation and earthquake rate estimation

Fig. 2.4 shows how moderate and large earthquakes in California can be represented as extended sources. We also demonstrate in Section 5.5 the extent of a focal zone of an earthquake which makes it possible to approximately calculate zone size given its seismic moment.

Figure 10.13 shows the rate density of *m*5 earthquakes forecasted by the smoothed seismicity model (Kagan *et al.* 2006; Kagan *et al.* 2007). In this model all large events ($m \geq 6.5$) are represented as spatially extended sources. We assume that larger earthquakes will have a rate consistent with a TGR magnitude distribution with $b = 0.95$ and $m_c = 8.0$ (Bird and Kagan 2004). Because we assume a spatially uniform magnitude distribution, a rate density map for *m*6 and above, for example, would look the same as this figure, if the color scale were shifted accordingly.

In Fig. 10.14 we extended long-term seismicity forecast to include focal mechanisms as a predicted quantity (see Eq. 10.4 and below it) with a statement of their uncertainty (cf. Figs. 5–6 in Kagan and Jackson 1994; Fig. 2 in Kagan *et al.* 2007). These focal mechanism forecasts have implications for seismic hazard because ground shaking depends on faulting style

10.6.2 Earthquake fault propagation modeling

One way of using the continuum-state branching model of fault propagation (Section 3.3) is to apply it to maps of geologic faults and past earthquakes to predict the propagation of future earthquake ruptures. As we discussed earlier (Section 2.5), geometric compatibility conditions imply that faults must rupture virgin rock. The model which reproduces branching properties of real earthquake faults can be applied to extrapolate fault data. Kagan and Knopoff (1984) made an early attempt to see how such a forecast could be formulated. They extrapolated the known fault traces as the result of model simulation. In such an extrapolation existing faults can propagate in an unruptured medium, branch into new fault orientation, and so on. In principle, if appropriate Green's functions are available, this model can generate a set of seismograms for each synthetic sequence. Such seismograms would incorporate short-term variations of seismic moment release as shown, for example, in Fig. 3.4.

11 Testing long-term earthquake forecasts: Likelihood methods and error diagrams

This chapter is based on the main findings reported in our papers (Kagan 1991b, 2007b, 2009b; Kagan and Jackson 2012) discussing how to test earthquake forecasts. We briefly considered the short-term forecast tests in Section 10.3, in this chapter we concentrate on the long-term forecasts verification where more extensive treatment is possible now.

Kagan and Jackson (1995) proposed three tests to verify forecast models: *N*-test, *L*-test, and *R*-test (see also Kagan and Jackson 2011). The *N*-test compares the number of predicted earthquakes with the actual number of events during the test period. To do so we need to know the statistical distribution of earthquake numbers (Chapter 4). Computations for the *L*-test are discussed in Section 11.2. The *R*-test measures the performance difference of two or more various forecasts, we will not discuss it in this book.

Several problems need to be addressed in these tests: comparing the likelihood and error diagram methods, calculating information scores and probability gains, applying the tests in low- and high-seismicity areas, testing various regional and global catalogs, and implementing the results of Chapters 4–8 in tests.

11.1 Preamble

This chapter continues our analysis of stochastic point process forecast verification (Kagan 2007b, 2009b). There we had discussed two interrelated methods for measuring the effectiveness of earthquake prediction algorithms: the information score based on the likelihood ratio (Kagan 1991b) and the "Error (Molchan) Diagram" (ED). Kagan (2007b) applied these methods to 1-D temporal renewal stochastic processes, but only for very long processes with the number of events approaching infinity. Kagan (2009b) extended the analysis by (1) discussing spatial (not temporal) random processes (fields); (2) considering forecast testing if the number of events is relatively small; and (3) applying newly developed techniques to long-term earthquake forecasts.

Earthquakes: Models, Statistics, Testable Forecasts, First Edition. Yan Y. Kagan.

Two issues are related to the problem of testing point process forecasts:

- Spatial random point fields density evaluation and its prediction is a mature discipline with many publications. Baddeley *et al.* (2005), Baddeley (2007), Daley and Vere-Jones (2003, 2008) provide reviews. As we explain below, the earthquake forecasting problem is different in many respects from regular density evaluation and requires special treatment. However, some results of this chapter can be applied to test a forecast of a random spatial pattern.
- Well-developed application and validation methods exist in weather and climate prediction; their reviews have been recently published by Jolliffe and Stephenson (2003), Palmer and Hagedorn (2006), DelSole and Tippett (2007). These prediction methods and their testing consider continuous processes and fields; however, with necessary modifications some of these methods can be used for stochastic point processes.

11.2 Log-likelihood and information score

In earlier studies Kagan and Jackson (1994, 2000) have tested the long-term forecast by a Monte Carlo simulation. In Fig. 11.1 we show the forecasted earthquake density for the north-west (NW) Pacific with 10 sets of synthetic catalogs, each having 108 events (see Fig. 10.1). Earthquakes are assumed to occur at the centers of the grid cells with the rates defined by the forecast.

Table 11.1 displays a slightly modified small extract of the forecast tables. The values of earthquake rate densities (column 3) or cell rates (column 5) are calculated at the end of year 2003 and they are used as a forecast for 2004–2006 earthquakes. In this table we sorted (ordered) entries by the values of earthquake forecast densities (column 3). To create a synthetic catalog we normalize the cell rates as shown in Table 11.1 (column 5) and simulate a random number uniformly distributed in the interval [0, 1]. The random number corresponding to a particular segment of the cumulative normalized rate curve defines the cell where an event occurs. We obtain one synthetic catalog by repeating this procedure n times.

Some of the grid points are occupied by more than one event. Some of the simulated points occur in areas of low seismicity (compare Fig. 11.1 with Fig. 10.1). As mentioned above (Eq. 10.5), this feature of the forecast is used to prevent surprises, that is, an occurrence of earthquakes in zones where no nearby events happened in 1977–2003.

Table 11.2 summarizes annual earthquake rates for both western Pacific regions. Because events outside the region's boundaries have influence (as explained in Subsection 10.2.3), the rates calculated through the smoothing procedure and evaluated by a direct method (dividing the earthquake numbers by time interval) are close but do not coincide. The difference between the predicted (υ_0) and observed (υ_i) numbers is not statistically significant; for a Poisson process the standard error is close to $\sqrt{\upsilon}$. In Chapter 4 we argue that the earthquake number distribution follows a negative-binomial law which has a higher variance than the Poisson distribution with the same mean.

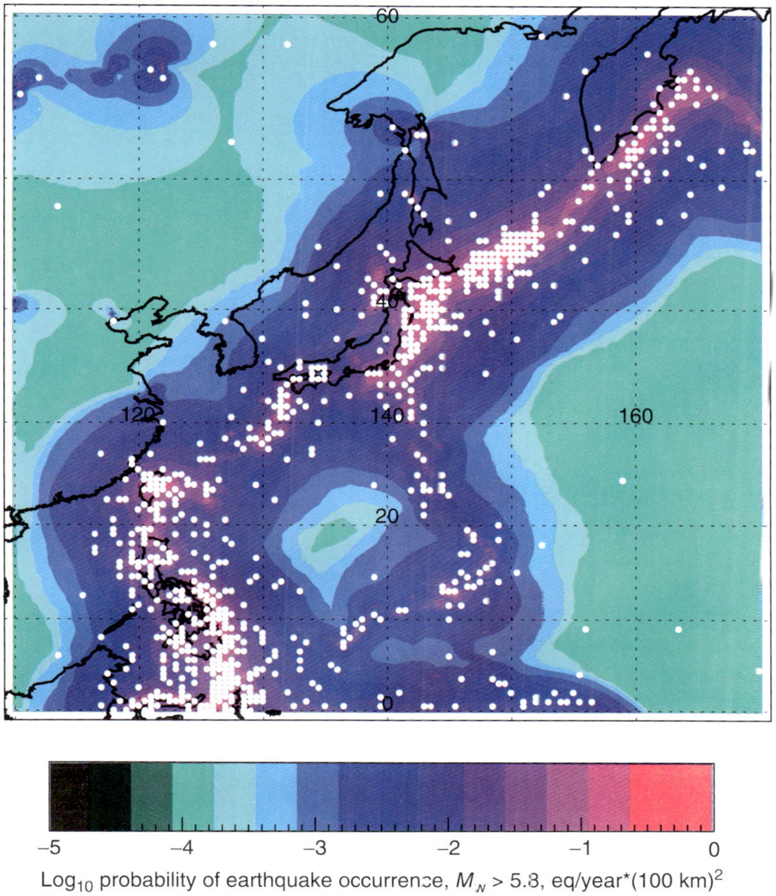

Fig. 11.1 NW-Pacific long-term seismicity forecast. Color tones show the rate density of earthquake occurrence calculated using the GCMT 1977–2003 catalog; 1080 simulated earthquakes for 2004–2006 are shown in white. *Source:* Kagan (2009b), Fig. 2.

The standard method of statistical analysis for a stochastic point process is a likelihood function computation. For an inhomogeneous Poisson process in which n points are observed $(x_1, \ldots, x_n)$ in a region A, the log-likelihood can be written as (Daley and Vere-Jones 2003, their Eq 7.1.2)

$$\log L\,(x_1, \ldots, x_n) = \sum_{i=1}^{n} \log \lambda\,(x_i) - \int_A \lambda\,(x)\,\mathrm{d}x, \tag{11.1}$$

where $\lambda\,(x_i)$ is the process rate (density) at a point x_i.

The log-likelihood of an inhomogeneous Poisson process is normally compared to a similar log-likelihood, L_0, calculated for a Poisson process with constant intensity (ξ) to obtain the log-likelihood ratio (Daley and Vere-Jones 2003, their Ch. 7; Schorlemmer *et al.* 2007)

$$\log (L/L_0) = \ell - \ell_0 = \sum_{i=1}^{n} \log \lambda\,(x_i/\xi) - \int_A [\lambda(x) - \xi]\,\mathrm{d}x. \tag{11.2}$$

Table 11.1
Beginning and end of earthquake rate forecast table for NW Pacific.

Lat. θ	Long. ϕ	EQ Rate (λ_i) EQ/(day×km²)	1°×1° (S_i) km²	Cell Rate (R_i) EQ/[yr×(.5°)²]	EQs 04-06	Probab. Gain (G_i)	EQs 77-03
1	2	3	4	5	6	7	8
7.0	127.0	1.7909E-07	12254	0.2003970	0	68.3700	6
40.5	143.0	1.6302E-07	9388	0.1397471	0	62.2331	1
45.5	151.5	1.5134E-07	8653	0.1195765	0	57.7738	5
24.0	122.0	1.4759E-07	11278	0.1519959	1	56.3445	10
44.5	150.0	1.4496E-07	8805	0.1165462	0	55.3376	8
44.5	149.5	1.4252E-07	8805	0.1145830	0	54.4055	9
12.5	125.5	1.4152E-07	12053	0.1557541	0	54.0252	6
44.0	148.5	1.4150E-07	8881	0.1147490	0	54.0181	8
....	....	....	....	....	....	....	....
48.5	128.0	2.6198E-11	8180	0.0000196	0	0.0100	0
48.5	127.5	2.6189E-11	8180	0.0000196	0	0.0100	0
48.5	127.0	2.6183E-11	8180	0.0000196	0	0.0100	0

Note: This table is based on forecast for the GCMT catalog (see Section 2.3) for 1977–2003 and ordered by descending rate density (column 3, λ). Cells are 0.5° × 0.5°; they form a 121 × 121 grid; EQ – earthquake(s). S_i in column 4 is the area of one square degree. The probability gains (G_i) (see Eq. 11.17) for displayed cells are the ratio of the rate density (λ_i) to the Poisson rate density (Eq. 11.6). *Source:* Kagan (2009b), Table 1.

In our calculations we normalize both rates (λ, ξ) by the observed event number n, hence the integral term in (11.2) is zero.

Kagan and Knopoff (1977, see also Vere-Jones 1998) suggested measuring the performance of the earthquake prediction algorithm by first evaluating the likelihood ratio to test how well a model approximates an earthquake occurrence. In particular, they estimated the information score, $\hat{I}$, per one event by

$$\hat{I} = \frac{\ell - \ell_0}{n} = \frac{1}{n}\sum_{i=1}^{n} \log_2 \frac{\lambda_i}{\xi}, \tag{11.3}$$

where $\ell - \ell_0$ is the log-likelihood ratio (11.2), n is the number of earthquakes in a catalog, $\log_2$ is used to obtain the score measured in the Shannon bits of information, λ_i is the rate of earthquake occurrence according to a stochastic

Table 11.2
Annual earthquake rate (υ_i) estimates.

	Time Interval	Pacific Regions NW	Pacific Regions SW
		υ_i – Annual Rate	
υ_0	77-03 Forecast	35.7159	60.7509
n_1	77-03 EQs	968	1591
υ_1	77-03 EQs	35.8546	58.9304
n_2	04-06 EQs	108	170
υ_2	04-06 EQs	35.9918	56.6537

Note: Actual rate calculations are made with time intervals measured in days (9861 days in 1977–2003 and 1096 days in 2004–2006). For display convenience, we convert the daily rates into annual rates by multiplying them by 365.25.
EQs – earthquakes, n_i – earthquake numbers. See north-west (NW) and south-west (SW) regions' boundaries in Section 10.2.1., *Source:* Kagan (2009b), Table 2.

model, conditioned by the past events:

$$\lambda_i = \text{Prob}\{\text{ an event in}\,(t_i,\, t_i + \Delta)\,|\,I(t_i)\}, \qquad (11.4)$$

where $I(t_i)$ is the past history of the process up to the moment t_i, and ξ is a similar rate for the event occurrence according to the Poisson process with a uniform rate over a region. The interval Δ is infinitesimally small so that the probability and the intensity (rate) are equivalent.

The Poisson process rate can be calculated by normalizing the seismicity level in the forecast regions. Several rates, such as shown in Table 11.2, can be used in the normalization. To make the results comparable to the forecast rate density, we use υ_0 values

$$\xi = \frac{\pi\,\upsilon_0}{180.0 \times [\sin(\theta_u) - \sin(\theta_l)](\phi_u - \phi_l) \times 111.111^2 \times 365.25}, \qquad (11.5)$$

where υ_0 is the annual rate of earthquakes in each region in 1977–2003 (Table 11.2), θ_u and θ_l are the upper and lower latitudes, respectively, ϕ_u and ϕ_l ditto for longitudes. For the NW Pacific and SW Pacific regions the rates measured in eq/(day $\times$ km^2) are

$$\xi_{NW} = 2.6289 \times 10^{-9} \qquad \text{and} \qquad \xi_{SW} = 3.3479 \times 10^{-9}. \qquad (11.6)$$

Below we use ξ without a subscript, since it is usually clear which west Pacific region is discussed.

Several methods can be used to calculate the information score for a set of forecasted events. Using the forecasted rate values (λ_i for cell centers in which earthquakes occurred) we compute I_1 score (success)

$$I_1 = \frac{1}{n_j} \sum_{i=1}^{n_j} \log_2 \frac{\lambda_i}{\xi}, \qquad (11.7)$$

where n_j is the earthquake number in two Pacific regions during 2004–2006 ($j = 1, 2$ for NW or SW, respectively, see Table 11.2). I_1 measures the degree to which the earthquakes in the test period are concentrated in cells where the forecast rate is high.

In Eq. 11.7 and in derivations below, we assume that earthquakes in the cells are independent identically distributed (i.i.d.) events. The assumed independence may be challenged by the clustered nature of earthquake occurrence of which foreshock–mainshock–aftershock sequences are the most clear example (Kagan and Knopoff 1977; Kagan 1991b; Chapter 6). However, given the high magnitude ($m_t = 5.8$) threshold for the GCMT catalog, the clustering is less pronounced. The dependent events on average constitute only about 20% of the total seismic activity (Kagan and Jackson 2000, Eq. 23). Thus, we expect that earthquake statistical inter-dependence will have a relatively small impact. A more complete investigation of this problem should be done in future work.

Table 11.3
Information scores for one event in west Pacific regions.

#	Info Score	Pacific Regions (n_j)	
		NW (108)	SW (170)
1	I_0	2.3645	3.3772
2	I_1	2.3675	3.0506
3	I_2	2.4067	3.2777
4	$< I_3 >$	2.3609	3.3768
5	I_4	3.0970	3.9823
6	$I_0 - I_1$	−0.0030	0.3266
7	$I_0 - I_2$	−0.0422	0.1002
8	σ	2.2102	2.9720
9	η	−0.6574	−0.2264
10	ψ	0.3685	−0.3078
11	σ_n	0.2151	0.2296
12	η_n	−0.0196	0.0078
13	ψ_n	−0.0369	0.0441

Note: For I_0 and I_4 scores see (11.14) and (11.23). The standard error (σ), coefficients of skewness (η) and kurtosis (ψ) (11.19 −11.21) are also shown. The numbers in parentheses are event counts in 2004–2006 for each region, n_j. Variables σ_n, η_n, and ψ_n (higher order moments, rows 11–13) are for the set of n_j events. They are calculated by simulating sets of n_j events and using Eqs. 11.18, 11.20, 11.21. *Source:* Kagan (2009b), Table 3.

As another option, instead of (11.7) we compute the information score for the actual epicenter (centroid) locations (λ_k)

$$I_2 = \frac{1}{n} \sum_{k=1}^{n} \log_2 \frac{\lambda_k}{\xi}. \tag{11.8}$$

As we see from Table 11.3, the values of I_1 and I_2 may be significantly different. The score I_1 depends on the cell grid subdivision and therefore is a less reliable indicator of the forecast effectiveness. On the other hand, its calculation can be accomplished using an available forecast table such as Table 11.1, whereas I_2 requires new extensive computation.

The scores I_1 and I_2 show the statistical behavior of earthquakes that occurred after the forecast was made. In our examples 2004–2006 events are compared to the forecast based on a smoothed seismicity of 1977–2003. We compare these forecasts with a score estimate for an extrapolated earthquake record based on the forecast. One way to calculate such a record is to create synthetic earthquake catalogs (Kagan and Jackson 1994; Fig. 11.1).

In simulated catalogs we generate multiple (for example, $\mathcal{N} = 10,000$) sets of n events (Table 11.2) and calculate the rate for cell centers as the earthquake location (see Fig. 11.1)

$$I_3 = \frac{1}{n} \sum_{l=1}^{n} \log_2 \frac{\lambda_l}{\xi}, \tag{11.9}$$

and

$$< I_3 > = \frac{1}{\mathcal{N}} \sum_{\ell=1}^{\mathcal{N}} (I_3)_\ell . \tag{11.10}$$

Similar to I_1 calculation, this method has an advantage that we do not need to compute the rate densities again (as for I_2), but instead we use the previously computed forecast tables (as shown in Table 11.1) to evaluate the scores. The score $< I_3 >$ is equivalent to (the opposite of) an entropy measure (Daley and Vere-Jones 2003), since the summation (sum of the earthquakes, Eq. 11.10) is similar to

$$I = \frac{1}{n} \sum_{l=1}^{n} \lambda_l \log \lambda_l, \tag{11.11}$$

(sum of the cells) for ensemble average.

The score $< I_3 >$ is a measure of the potential prediction effectiveness as compared to a Poisson process with a uniform rate over a region. Although the score is constructed using a previous earthquake record, its usefulness for forecast purposes needs to be established – it may be argued, for example, that earthquakes "avoid" the locations of past recent strong events (as in the seismic gap model, see Rong *et al.* 2003; Kagan *et al.* 2012). Therefore, we calculate the score difference $I_3 - I_2$ or $I_3 - I_1$ to characterize the forecast properties. In short, if the difference is close to zero, it means that future earthquakes in a statistical sense follow the predictions. In effect, this difference corresponds to the *L*-test – fit of tested events to the forecast.

In Fig. 11.2 we display the log-likelihood function distribution differences for the simulation as shown in Fig. 11.1. We simulate n earthquake locations according to 1977–2003 forecasts for each region. For each synthetic catalog we calculate the log-likelihood function and subtract the log-likelihood function value obtained for earthquakes which occurred during 2004–2006. Thus, we display the histograms of $I_3 - I_2$ (Eqs. 11.8 and 11.9). Kagan and Jackson (1994, see their Fig. 7 and its discussion) show a cumulative version of these distributions and explain the test in more detail.

11.3 Error diagram (ED)

Starting with Molchan's (1990) paper, previous EDs were almost exclusively time-dependent. We apply here the ED to time-independent 2-D spatial earthquake distributions. In some respects, the earthquake spatial pattern is more difficult to analyze than the temporal distribution. In the latter case, we have a reasonable null model (the uniform in time Poisson process) which can be compared to any test model. In the spatial case, the simple model of uniformly distributed seismicity can hardly serve as an initial approximation; even large earthquakes (which often can be well approximated by a Poisson temporal process, see Chapter 6) are strongly clustered in space. This property of seismicity is caused by the fractal nature of earthquake spatial distribution (Chapter 7). Although in our forecasts (Kagan and Jackson 1994, 2000, 2011, 2012) we use a projection of earthquake

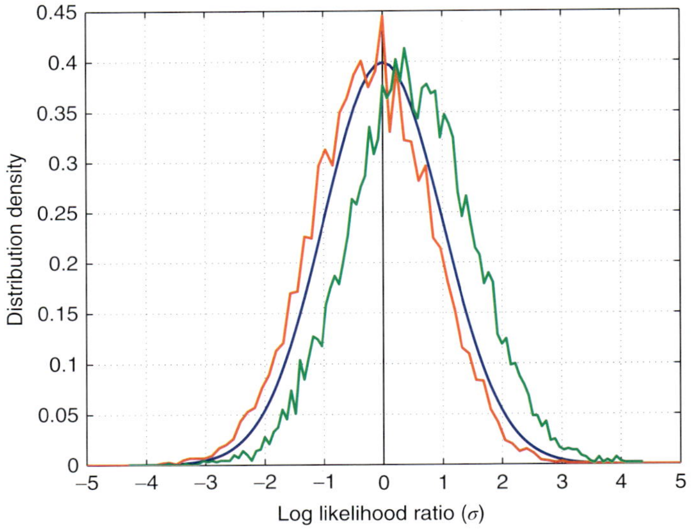

Fig. 11.2 Histograms of the log-likelihood function differences for 2004–2006 simulated earthquakes (see Fig. 11.1). The functions are normalized to have a unit standard deviation. We simulate 10,000 sets of 108 events for the NW Pacific and of 170 events for the SW Pacific. The blue line is the Gaussian curve with a zero mean and unit standard deviation. The red curve corresponds to the simulation distributions for the NW Pacific; the green curve to the SW Pacific. Curves on the right from the Gaussian curve correspond to simulations that are on average worse (smoother) than a real earthquake distribution; curves on the left correspond to simulations that are less smooth than a real earthquake distribution. Both empirical curves demonstrate the reasonable effectiveness of forecasts. *Source:* Kagan (2009b), Fig. 3.

centroids on the Earth's surface which smoothes their spatial distribution (Subsection 7.2.3), the spatial distribution still preserves a self-similar fractal pattern with large parts of the Earth practically aseismic.

Diagrams similar to EDs have been used previously to describe the spatial distribution of seismicity: Rong and Jackson (2002, Fig. 3) as well as Helmstetter *et al.* (2007, Fig. 4) created spatial "Concentration Diagrams" to characterize the agreement (or lack thereof) between the predicted seismicity distribution and future earthquakes. These diagrams plot the fraction of the event success rate (equivalent to $1 - \nu$) versus the normalized area (τ), sorted by probability density (see Section 11.1 or Fig. 11.3). The sorting is largely analogous to "water-level threshold" analysis (Zechar and Jordan 2008). These concentration diagrams can easily be converted to EDs by adding an ascending diagonal and then reflecting the plot in the line ordinate ($\nu = 1/2$).

To construct the error diagram, we sort the cells in decreasing order of reference probability (as in Table 11.1) and compute a Cumulative Distribution Function (CDF) for each hypothesis. A theoretical error diagram is a plot of the complementary CDF ($CCDF = 1 - CDF$) for each hypothesis as a function of the horizontal axis which represents a kind of 'water level,' determined by a threshold cell probability; what is plotted is the probability that an event would be forecast by the reference model. The vertical axis represents the probability that the earthquake would not be forecast according to the model in question (see an example in Fig. 11.3).

The error diagram is related to the *Relative Operating Characteristic* (ROC) (Swets 1973; Mason 2003, pp. 66–76), used in signal detection and weather prediction efforts. In the ROC diagrams the success rate of an event prediction is plotted against the false alarm rate. In principle, ED diagrams can be treated like ROC plots where cells (bins) with events are considered as a success and empty cells as false alarms. However, this interpretation encounters difficulties

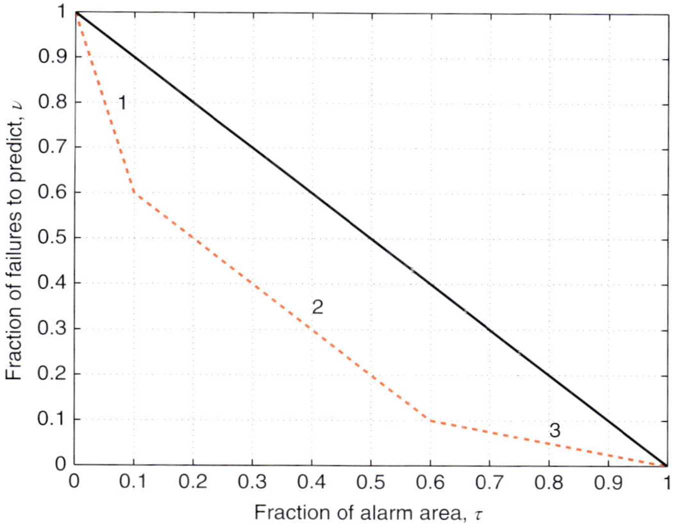

Fig. 11.3 Error diagram (τ, ν) schematic example (see Eq. 11.12). Two ED plots are shown: (1) for the uniform Poisson process, corresponding to the black diagonal of the square, and (2) for a region consisting of three sub-areas with different occurrence rates in each – the red dashed line (see text). *Source:* Kagan (2009b), Fig. 4.

when the cells are not infinitesimally small, so some may contain more than one event. Even for very small cells due to the fractal nature of earthquake spatial distribution (Chapter 7) several events may be located in a single cell. Furthermore, for a point process on a sphere it is difficult to define cells of equal size. A naïve sphere subdivision yields cells of unequal size that are larger at the equator and smaller towards the poles. Below we discuss the techniques for overcoming these difficulties.

Characterizing prediction performance is a major challenge for ED analysis. Since prediction results are represented by a function (curve), it is important to find a simple one-parameter criterion (a functional) that briefly expresses the efficiency value. Several functionals have been proposed as a measure of ED forecast efficiency: the *minimax* strategy (Molchan 1997; Molchan and Keilis-Borok 2008), the *sum of errors* $(\nu + \tau)$ (ibid.), the *area skill score* (Zechar and Jordan 2008), etc. Each of these criteria has some advantages and disadvantages. For example, Kagan and Jackson (2006, p. 36) in their discussion of Kossobokov's (2006) paper, show that two ED trajectories with a very different forecast behavior have the same "sum of errors" value.

In this chapter, as well as in our earlier papers (Kagan 2007b, 2009b), we advocate the use of the log-likelihood to characterize the ED performance. The advantage of the likelihood score is that it is well known in statistics and its properties are discussed in many statistical treatises. Moreover, as Kagan and Jackson (2006, p. 36) argue, the likelihood score yields a better measure of forecast performance in terms of possible earthquake warning strategies and their cost.

11.3.1 Relation between the error diagram and information score

We illustrate the ED by a sketch in Fig. 11.3. For the spatial point distribution, this example is easier to construct and explain than for temporal renewal processes

(Kagan 2007b). Fig. 11.3 shows a theoretical pattern for two ED diagrams. The square's diagonal corresponds to the uniform Poisson distributions of the points in a region, that is, a random guess forecast strategy or *unskilled* forecast. As a test example, we assume that the region consists of three sub-areas, their normalized surface τ_i is 0.1, 0.5, and 0.4 of the total, and the normalized number of events ν_i is 0.4, 0.5, and 0.1, in each zone respectively. Contrary to the temporal one-dimensional point process models (as discussed in Kagan 2007b), for a spatial field these sub-areas do not need to be contiguous, each of them can be subdivided into any number of patches. It is assumed that coordinates of all patches are known. The points in these zones are distributed according to the Poisson spatial process with the density ν_i/τ_i. Then, the information score for such a point distribution can be calculated as (see Eq. 11.3)

$$
\begin{aligned}
I = \sum_{i=1}^{3} \nu_i \log_2 \frac{\nu_i}{\tau_i} &= 0.4 \log_2 4.0 + 0.5 \log_2 1.0 + 0.1 \log_2 0.25 \\
&= 0.8 - 0.2 = 0.6 . \qquad (11.12)
\end{aligned}
$$

For the normalized point Poisson distribution in the ED, the point density is unity. Hence its contribution to the information rate (11.12) is zero.

The information score can be calculated for continuous theoretical concave curves in an error diagram (Kagan 2007b, 2009b; Molchan 2010)

$$
I = \int_0^1 \log_2 \left(- \frac{\partial \nu}{\partial \tau} \right) d\nu . \qquad (11.13)
$$

If the ED consists of several linear segments (as in Fig. 11.3), then Eq. (11.13) converts to a sum

$$
I_0 = \sum_{i=1}^{N} \nu_i \log_2 \left(\frac{\nu_i}{\tau_i} \right) , \qquad (11.14)
$$

where i are cell numbers, N is the total number of grid points, and ν_i and τ_i are the normalized rate of occurrence (R_i) and cell area (S_i):

$$
\nu_i = \frac{R_i}{\sum_{i=1}^{N} R_i} \quad \text{and} \quad \tau_i = \frac{S_i}{\sum_{i=1}^{N} S_i} , \qquad (11.15)
$$

see Table 11.1. In our ED plots, we use as abscissa and ordinate of the diagrams the cumulative fraction of alarm space and failures to predict

$$
\tau = \sum_i \tau_i , \quad \text{and} \quad \nu = \sum_i \nu_i . \qquad (11.16)
$$

The score I_0 (specificity) in Eq. 11.14 contains no forecast earthquake data; rather, it measures the degree to which the forecast rate itself is concentrated relative to the null hypothesis of uniform rate density. The equation can easily

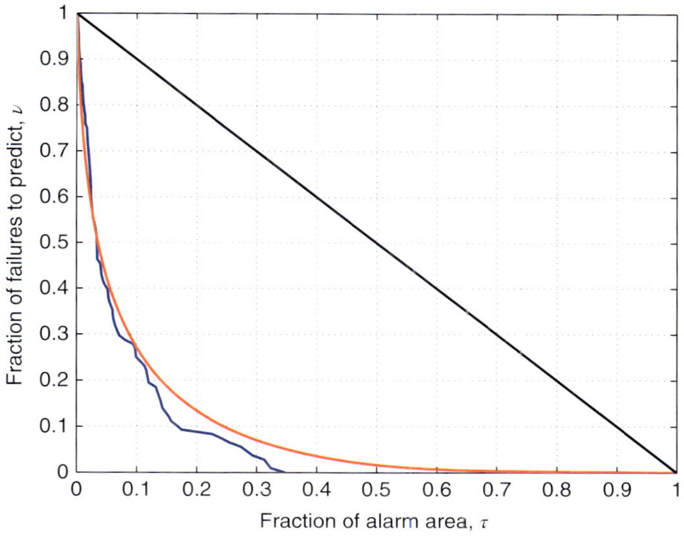

Fig. 11.4 Error diagram (τ, ν) for NW Pacific long-term seismicity forecast calculated using the GCMT 1977–2003 catalog at 121×121 grid. Solid black line – the strategy of random guess, red line – the ordered density for long-term forecast, blue line –earthquakes in 2004–2006. *Source:* Kagan (2009b), Fig. 5.

be generalized to measure the information score relative to any other reference hypothesis.

When calculations of τ are made for a spherical surface (as in Figs. 10.1 and 11.1), the τ_i steps are usually unequal in size, unless a special effort is made to partition a sphere into equal-area cells (see more in Kagan and Jackson 1998). This cell inequality complicates the calculation.

Figures 11.4 and 11.5 show the EDs for both west Pacific regions. The red curves are for the forecast, based on 1977–2003 seismicity, and the blue curves are for the earthquakes which occurred in these regions from 2004 to 2006. Both sets of curves are calculated using the forecast tables like those in the example (Table 11.1). In principle, the calculations such as in (11.15) can be made with unordered cells. The density ordering in Table 11.1 and Figs. 11.4, 11.5 is performed to create the ED diagrams.

The score values I_0 (11.14) in Table 11.3 are calculated using the distributions shown by the red curves in Figs. 11.4 and 11.5. To obtain the average probability gain G or the predictive ratio (Kagan and Knopoff 1977, p. 101) we calculate

$$G = 2^{I_0}. \tag{11.17}$$

Zechar and Jordan (2008, Eq. 9) derived a similar formula for the average probability gain of a forecast; their equation can be converted into Eq. 11.13 taking into account their different normalization.

Equation 11.17 means that for the NW and SW Pacific on average the probability gain (G) is a factor of 5 to 10 ($2^{2.36}$ to $2^{3.38}$) when using the long-term forecast compared to a random guess. Of course, these I_0 values do not fully describe the forecast advantage. The boundaries of both regions have already been selected to contain the maximum number of earthquakes in relatively small

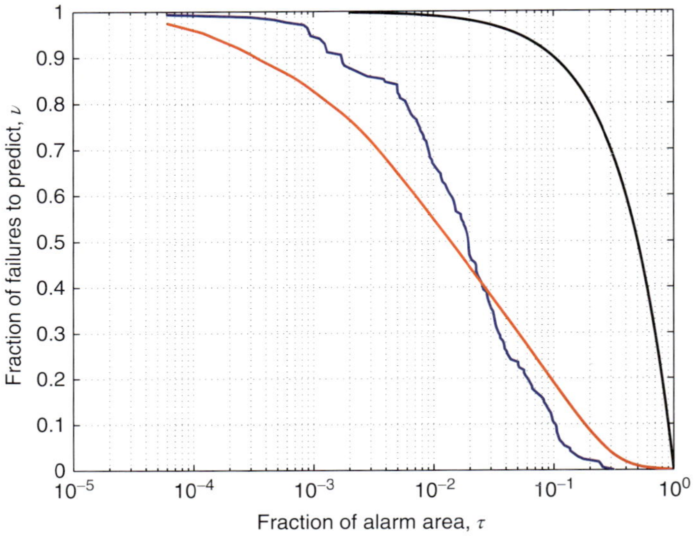

Fig. 11.5 Error diagram (τ, ν) for SW-Pacific long-term seismicity forecast calculated using the GCMT 1977–2003 catalog at 121 × 161 grid. Solid black line – the strategy of random guess, red line – the ordered density for long-term forecast, blue line – earthquakes in 2004–2006. *Source:* Kagan (2009b), Fig. 6.

areas. If we extend any of the regions toward the seismically quiet areas, the information score would significantly increase. The proper measure of long-term forecast performance would extend the forecast method globally, that is, over the whole Earth surface. Limited numerical experiments suggest that depending on the degree of smoothing, the value of ϵ (Eq. 10.5), and other factors, the *G*-value for world-wide seismicity varies from about 10 to 25 (Kagan 2009b).

The above values of the probability gain, *G*, can be compared with similar calculations by Rhoades and Evison (2005, 2006) and Rhoades (2007). These authors calculated *the information rate per earthquake* for a model of smoothed seismicity (PPE), similar to our long-term model. The PPE model was compared to a stationary and spatially uniform Poisson model. The probability gain, computed using the information rate, for New Zealand, Japan, Greece, and California is about 4.5, 1.6, 1.6, and 3.4, respectively. These relatively small gain values are caused by the authors' choice of the regions that include only seismically active areas (see ibid.). Helmstetter *et al.* (2007, Table 1) obtained the *G*-values ranging from 1.2 to 4.8 for different long-term seismicity predictive models in California.

These estimates of the *G*-values depend strongly on catalog properties and methods of calculation. When the accuracy of the earthquake location increases, the gain value should increase as well. Moreover, in our computation we use 2-D epicenter maps, whereas the spatial information score generally should rely on 3-D hypocenter distribution (see more discussion in Section 7.2). Since the spatial distribution of hypocenters is fractal set, the *G*-value in 3-D should approach infinity, as location uncertainty decreases, thus a proper definition of the spatial information score needs to be developed for fractal distributions.

The ED curves for earthquakes (blue lines) in Figs. 11.4, 11.5 are similar to the forecast earthquake curves (red lines). The computation of the likelihood scores

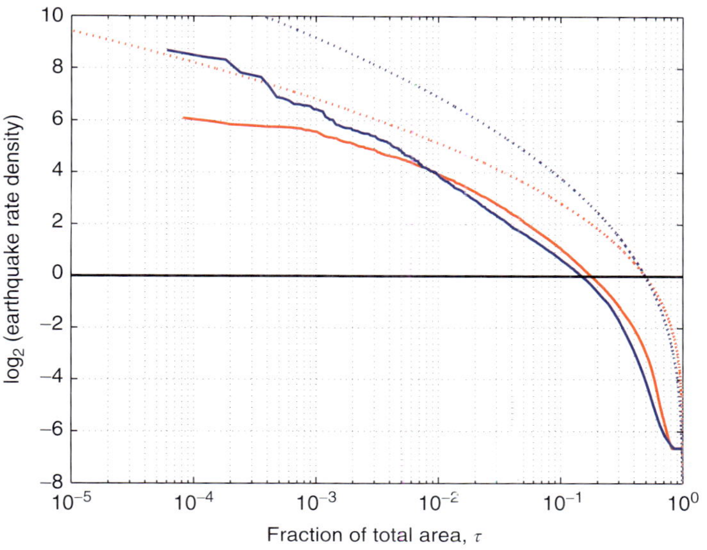

Fig. 11.6 Likelihood function dependence on the fraction of the total area, τ, for west Pacific long-term seismicity forecasts: solid red curve – NW Pacific, blue curve – SW Pacific (see boundaries in Section 10.2.1). Dotted lines are for the Gaussian distribution of log forecast rates: red – NW Pacific, blue – SW Pacific. The thick horizontal black line shows the log-likelihood distribution if all cells are assumed to have a uniform Poisson rate. *Source:* Kagan (2009b), Fig. 7.

(Eqs. 11.7, 11.8) shows that the NW earthquakes have a better score than the forecast, whereas SW events display the opposite behavior (see also Fig. 11.2). The scores using the actual centroid position (I_2) are larger than those for the cell centers (I_1), an anticipated feature. Similarly, Table 11.3 shows that the average scores for synthetics ($< I_3 >$) are very close to those of I_0, which is understandable, since the simulation runs are extensive (see Eqs. 11.9, 11.10).

Figure 11.6 shows the frequency curves for the log-likelihood function of both western Pacific regions. We display $\log_2$ of the normalized rate (see column 5 of Table 11.1) against the normalized cumulative area of the cells (column 4). Curves for both regions exhibit high values of the rate (R_i) concentrated in a relatively small fraction of the area. For example, the rate density exceeding the Poisson rate (ξ, see Eq. 11.5) is observed only at less than 18% of the total area. Low values at the right-hand end of the diagram correspond to the assumed uniform background rate density (Eq. 10.5, see also Figs. 10.1 and 11.1). We also display two curves which are calculated as if the log of forecasted normalized earthquake density was distributed according to the normal law with the standard deviations (σ) as shown in Table 11.3.

The curves in Fig. 11.6 show the dependence of the log-likelihood function for one forecasted event on the fraction of total area τ. The log-likelihood distribution for both regions is highly non-Gaussian: very few cells with low density would be expected for the Gaussian law, whereas in our forecast a significant part of the area has the density equal to the ϵ-value (see Figs. 10.1 and 11.1). When we test the prediction for several events, the distribution seems to be close to the Gaussian distribution (Fig. 11.2). The tests of the forecast effectiveness are much simpler if the distribution is Gaussian or sufficiently close to it.

To measure the difference between the actual log-likelihood distribution and the Gaussian one, we calculate the higher order central moments for the error curve (I_0 of Eq. 11.14 corresponds to the first moment μ_1)

$$\mu_k = \sum_{i=1}^{N} \nu_i \left[\log_2 \left(\frac{\nu_i}{\tau_i} \right) - I_0 \right]^k , \tag{11.18}$$

where $k = 2, 3, 4, \ldots$.

The standard deviation of the log-likelihood for the set of n events is

$$\sigma_n = \sqrt{\mu_2 / n}. \tag{11.19}$$

The coefficient of skewness (Abramowitz and Stegun 1972, p. 928) is

$$\eta = \mu_3 / \mu_2^{3/2}, \tag{11.20}$$

and coefficient of kurtosis (ibid.) is

$$\psi = \mu_4 / \mu_2^2 - 3. \tag{11.21}$$

These coefficients (η and ψ) characterize how the likelihood curve differs from the Gaussian distribution; for the latter law both coefficients should be zero. The *Central Limit Theorem* states that the distribution of the sum of a large number of i.i.d. events with finite second moment (variance) should approach the Gaussian law. If the event number is small, we would need to find an efficient way to numerically approximate the distribution of the sum of i.i.d. random variables.

In Table 11.3 both coefficients are large for one event likelihood curve (see also Fig. 11.6), but for the set of n events they are small: the distribution is close to the Gaussian law as demonstrated in Fig. 11.2. The difference between the score values I_0 to I_2 is less than the standard error value (see Table 11.3). Thus both forecasts can be considered statistically successful.

The differences

$$I_0 - I_1 \quad \text{or} \quad I_0 - I_2, \tag{11.22}$$

show the predictive efficiency of a forecast, that is, whether on average earthquakes in 2004–2006 occurred at the sites listed in the prediction table (see an example in Table 11.1). For this particular time interval, both forecasts are sufficiently good. However, as other examples (Kagan and Jackson 2000, Fig. 9) demonstrate, this is not always the case. The values of differences (negative for the NW-Pacific and positive for the SW Pacific) correspond to those simulations in Fig. 11.2, where we display the distribution of the difference $I_3 - I_2$.

By applying (11.14) to the blue curve of earthquakes in 2004–2006 in Figs. 11.4, 11.5 we evaluate the information score

$$I_4 = \frac{1}{n} \sum_{i=1}^{n} \nu_i \log_2 \left[\frac{\nu_i}{\tau_i} \right], \tag{11.23}$$

(see Table 11.3). The value of I_4 is obviously significantly larger than all the other estimates of the score. Earthquake simulations provide an explanation for this feature (see Fig. 11.9).

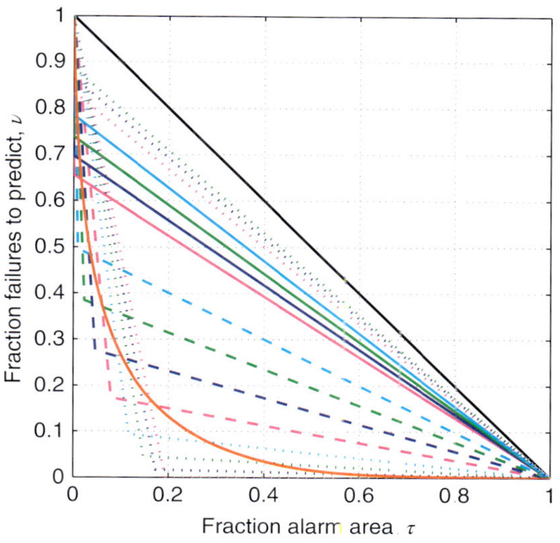

Fig. 11.7 Error diagram (τ, ν) for the NW Pacific long-term seismicity forecast, approximated by two-segment distributions. The solid thick black straight line corresponds to a random guess, the solid thick red line is for the NW forecast. Two-segment lines are for the curves with the information score $I_0 = 2.3645$ bits. The slope D_1 for the right-hand first segment with the corner at $\nu = 0.0, \tau = 0.194$ is $D_1 = -2^{I_0}$. For the next first segments slopes are $D_1 \times 1.1$, $D_1 \times 1.25$, $D_1 \times 1.5$, $D_1 \times 2.0$, $D_1 \times 3.0$, $D_1 \times 5.0$, $D_1 \times 10$, $D_1 \times 50$, $D_1 \times 100$, $D_1 \times 250$, $D_1 \times 1000$, $D_1 \times 10,000$, $D_1 \times 100,000$, and $D_1 \times 1000,000$. *Source:* Kagan (2009b), Fig. 8.

11.3.2 Two-segment error diagrams and information score

Similarly to Fig. 5 in Kagan (2007b), in Fig. 11.7 we display an approximation of the ED for the NW Pacific by several two line segment diagrams with the same value of the information score, I_0. These ED segments correspond to two sub-areas with different point densities. As we discussed above (Eq. 11.12), for a spatial stochastic field these sub-areas may be subdivided into any number of patches with known coordinates.

For the assumed information score I, the contact point of two segments is defined by the equation

$$D_1 \left[\frac{\nu}{\nu - 1 - D_1} \right]^{\nu} = -2^I. \tag{11.24}$$

By solving this equation for any value of the first segment slope D_1 (non-positive by definition), one obtains the ν-value for the contact point of two linear segments and

$$\tau = (1 - \nu)/D_1. \tag{11.25}$$

The first of these curves has the second segment coinciding with the abscissa axis. This means that one can obtain the same information score by concentrating all the points in the $2^{-I_0} = 0.194$ "active" part of the region. However, though the I-value for such a pattern would be 2.36 bits, all points would have the same value of the probability gain. Hence, for such a likelihood value distribution, the variance and higher order moments would be zero: very different from the actual probability gain pattern (Table 11.3). In this two-segment curve we presume that there are no events at the non-active part of the region, thus the ϵ-value (10.5) is zero. Such an arrangement is possible only if the model is assumed to be correct, because if even one event occurs in such a subregion, the value of the likelihood score would immediately be $-\infty$. This is why in the real-life situation, where an earthquake occurrence model is not known exactly, we need to adopt a non-zero ϵ-value.

If we modify the two-segment model to distribute the events with different non-zero densities over both sub-areas, the variance and the other moments would be also non-zero. In Fig. 11.8 we show the dependence of the lower order moments for the likelihood score on the D_1 slope. We calculate the moments for each two-segment model, using a modification of (11.18). For example

$$\sigma = \sqrt{(1-\nu)\left[\log_2\left(\frac{1-\nu}{\tau}\right) - I_0\right]^2 \quad + \quad \nu\left[\log_2\left(\frac{\nu}{1-\tau}\right) - I_0\right]^2}, \qquad (11.26)$$

where ν and τ are defined in (11.24 –11.25) and I_0 for this case can be calculated as

$$I_0 = (1-\nu)\log_2\left(\frac{1-\nu}{\tau}\right) + \nu\log_2\left(\frac{\nu}{1-\tau}\right). \qquad (11.27)$$

the higher order moments (11.18), as well as the skewness (11.20) and the kurtosis (11.21) are computed similarly.

For $D_1 = -2 \times 2^I$ (the dashed magenta line, fifth curve from the bottom) the 2nd, 3rd, and 4th moments correspond roughly to the moments of the forecasted densities. Thus, such a two-segment model would reasonably well approximate the actual event distribution. The contact coordinates of two segments for this curve are: $\nu_5 = 0.1732$ and $\tau_5 = 0.0803$. Therefore, the point pattern having apparently the same lower order moments as the actual earthquake forecast would have about 83% of points concentrated in 8% of the area, that is, the point density is 10.3 times higher than the uniform Poisson rate. The rest of the events would be distributed in 92% of the area and have the rate of 0.19 compared to

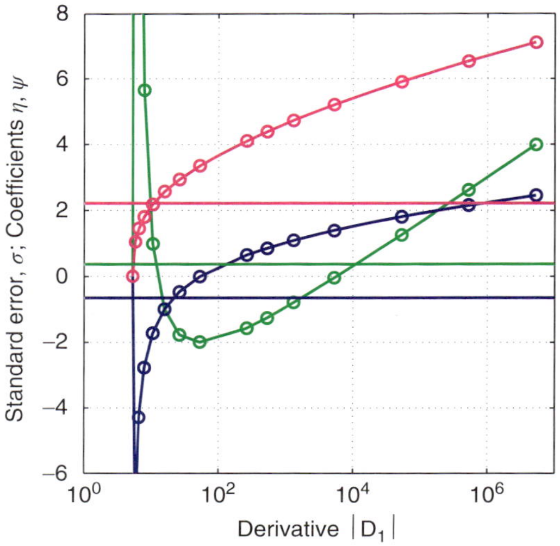

Fig. 11.8 Dependence on slope $|D_1|$ of standard deviation (magenta), coefficients of skewness (blue) and kurtosis (green) for two-segment curves in Fig. 11.7. Horizontal lines are these variables for the red curve in the cited plot. *Source:* Kagan (2009b), Fig. 9.

the uniform Poisson distribution. As we mention in Section 10.2.1, in our Pacific forecasts 0.01 part of the total earthquake rate is spread over the entire region (see Eq. 10.5 and Figs. 10.1 and 11.1).

The equality of two scores, one based on the ED (I_0) and the other on the likelihood ratio ($< I_3 >$), shown in Table 11.3, demonstrates again that the ED is a more complete description of the point process. One can obtain the information score from the ED, but not vice versa: as we see from Fig. 11.7 even for two-segment diagrams there is by their construction an infinite number of ED curves having the same score.

11.3.3 Information score for GCMT and PDE catalogs

ED displays in Figs. 11.4, 11.5 are inconvenient since the most interesting parts of the curves are concentrated near the ν- and τ-axes. The reason for this feature is that seismicity is concentrated in relatively narrow seismic belts having a fractal spatial earthquake distribution. Now we focus on how other curves deviate from the forecasted (red) one. To make these deviations show more prominently, we need to display the curves in a relative abscissa format, using the 1977–2003 forecast density as a template or baseline for the likelihood score calculation.

Figure 11.9 shows several curves in a new format; in effect, we convert the red curve in Fig. 11.4 to the diagonal. This is equivalent to calculating the information scores by using λ_i as a reference density

$$I = \frac{1}{n} \sum_{i=1}^{n} \nu_i \log_2 \frac{\zeta_i}{\lambda_i}, \tag{11.28}$$

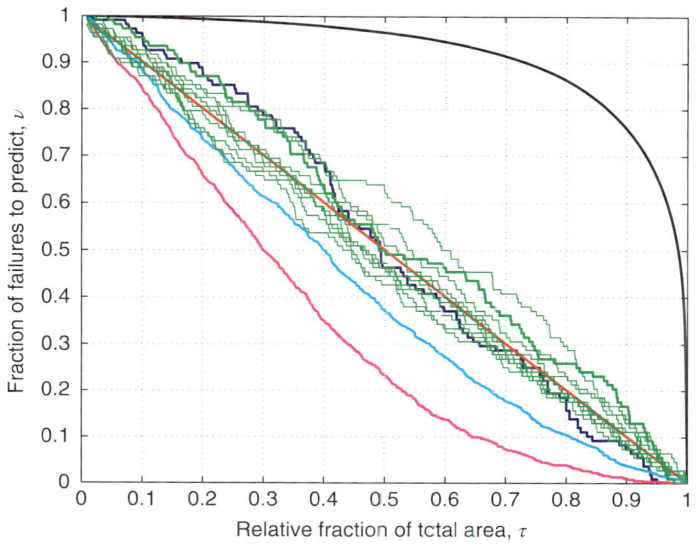

Fig. 11.9 Error diagram (τ, ν) for NW Pacific long-term seismicity forecast. The solid black line – the strategy of random guess. The solid thick red diagonal line is a curve for the NW forecast. The blue line is the earthquake distribution from the GCMT catalog in 2004–2006 (forecast); the magenta line corresponds to the earthquake distribution from the GCMT catalog in 1977–2003; the cyan line is the earthquake distribution from the PDE catalog in 1968–2006. The thin green lines are ten simulations, displayed in Fig. 11.1, the first realization is shown by a thick green line. *Source:* Kagan (2009b), Fig. 10.

where ζ_i is a rate density for all the other point distributions. In fact, we displace all the curves horizontally by the abscissa difference between the forecast line and the unskilled forecast (the descending diagonal of the error diagram, cf. Fig. 11.4).

Figure 11.9 shows the difference between the forecast curve (red) and the earthquake curve (blue) better than Fig. 11.4. The plot also displays the curve for the 1977–2003 GCMT catalog. The numbers of events in the cell areas are shown in Table 11.1, column 8. Also shown is the curve for the PDE catalog for 1968–2006. Similarly to (11.7) we also calculate the empirical score I_1' for earthquakes which occurred in the training period. We obtain $I_1' = 3.5991$ bits/event for the 1977–2003 GCMT catalog and $I_1' = 2.9789$ bits for the PDE. These values are significantly larger than those forecast for 2004–2006. Therefore, our forecast predicts better locations of past earthquakes than those of future events. Why this paradox? In the forecast we use a broader smoothing kernel to capture the spread of seismicity with time (Section 11.2). Kagan and Jackson (1994, p. 13,696) performed two optimizations for the NW Pacific region: one by using the regular leave-one-out method (Silverman 1986), the other by subdividing the catalog into two parts. The standard density estimation method yielded the optimum value of the maximum distance of a smoothing kernel $R_{max} \simeq 180$ km as compared to the value $R_{max} \simeq 350$ km to predict the second part of the catalog based on the first part, that is, the leave-one-out procedure greatly under-smoothed the predicted future spatial distribution of earthquakes. A similar explanation is apparently valid for the PDE score value. Helmstetter *et al.* (2007, Table 1) obtained $G = 7.1$ (significantly higher than the G-values for predictive algorithms) when the same data were used to build the long-term seismicity model and to test it (see Subsection 11.3.1).

In Fig. 11.9 we also show several curves for the simulated earthquakes. These curves explain why the I_4-value (Eq. 11.23, see also row 5 in Table 11.3) is significantly larger than the other measures of the information score. The reason is twofold. First, the number of events in the three-year interval is relatively small and the curves often fluctuate around the expected value (the red curve). These fluctuations increase the sum value in (11.23). The curves are often below the red forecast line, which would usually cause the score value to increase. Second, the ED curve should be concave (Molchan 1997, 2003). I_4-values are calculated with the original curves shown in Figs. 11.4, 11.5 which have many convex sections. If we make a lower envelope of the curve points, this would decrease the I_4-value. However, our numerical experiments show that the decrease is not significant enough to bring the value sufficiently close to the I_0 score (see more explanations at and below Eq. 11.32).

The fluctuations of the synthetic curves also suggest that some strategies proposed to measure the performance of a prediction algorithm by considering the ED, like a sum of errors ($\nu + \tau$) or minimax errors (Molchan 1991; Molchan and Kagan 1992; Kossobokov 2006; Molchan and Keilis-Borok 2008) are biased for a small number of forecasted events. For western Pacific regions the number of predicted events (n_j) is relatively large; in many other applications of the ED (ibid.) this number is less than 10.

In Fig. 11.9 the forecast distribution curve is used as the template or the reference model (Zechar and Jordan 2008). Thus, we can measure the difference between this line and the other curves using many standard statistical techniques, like the Kolmogorov-Smirnov test or the Cramer-von Mises test (Stephens 1974), to infer whether these distributions are statistically different.

11.4 Tests and optimization for global high-resolution forecasts

Subsection 10.2.3 discussed the Fisher kernels that were applied for full Earth global earthquake forecast. Figure 11.10 demonstrates how the fixed Fisher kernel was optimized using earthquake history from 1969–2005 in the PDE catalog to forecast earthquake occurrence in 2006–2010. The red curve displays the theoretical gain of a forecast (specificity): the scores increase for the narrower kernels.

Scores I_1' calculated for earthquakes which occurred in the training period are significantly larger than those calculated for the forecast intervals. Therefore, as expected, our forecast "predicts" better the locations of past earthquakes than those of future (target) events (see Kagan 2009b, Fig. 10 and its explanations).

However, to select the most internally consistent kernel, we should compare these scores with those obtained for the test period (2006–2010). The latter scores can be calculated for the cell centers where the earthquakes occur, I_1, or for actual epicenter locations, I_2 (see Eqs. 11.7, 11.8). The intersection of I_0 and I_1 or I_2 scores suggests the best choice of kernel.

To test whether the initial estimate significantly influences the score values or not, as Silverman suggested (1986, his Ch. 5.3, see also Eq. 10.11), we produced two forecasts. In one the initial estimate for earthquake epicenter locations (10.8) was executed using $\omega = 10,000$ and in the second ω was taken as $100,000$. In both cases the actual forecast ω is selected as $100,000$. The resulting scores are

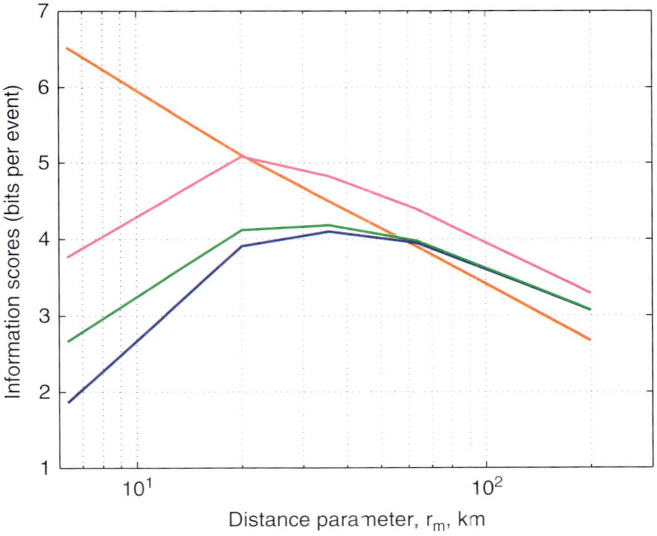

Fig. 11.10 Optimization of Fisher kernels: dependence of information scores on the smoothing scale parameter $r_m = R \times \rho_m$ in the Fisher distribution (10.14) for the 2006–2010 forecast based on the PDE catalog for 1969–2005. The abscissa r_m values correspond to $\omega =$ 1,000,000; 100,000; 10,000 $\times \sqrt{10}$; 10,000; 1000 in this order. The Red line is I_0 score, blue – I_1, green – I_2, and the magenta line is for I_1'. *Source:* Kagan and Jackson (2012), Fig. 3.

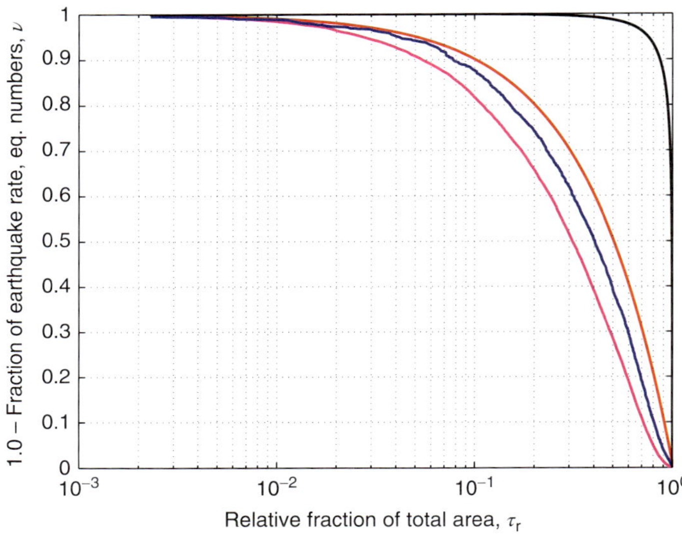

Fig. 11.11 Error diagram (τ, ν) for global long-term seismicity forecast relative to that of power-law forecast based on earthquakes in learning period, in the PDE catalog ($m \geq 5.0$) for 1969–2005. The solid black line, theoretical curve for spatially uniform rate density; the red line, theoretical results for reference power-law kernel (Eq. 10.1) with $r_s = 7.5$ km; the blue line, empirical results for test period 2006–2010; the magenta line, empirical results for retrospective application to training period, 1969–2005. *Source:* Kagan and Jackson (2012), Fig. 6.

$I_0 = 5.538$; 5.420, $I_1 = 3.865$; 3.859, $I_2 = 4.145$; 4.145, $I_1' = 4.940$; 4.972, respectively for both forecasts: differences connected with the change in the initial ω value are negligible.

Figure 11.11 shows two theoretical and two empirical error curves relevant to the fixed power-law forecast. To better show the curves on the graph, we use as a reference the forecast based on the power-law kernel (Eq. 10.1) with $r_s = 7.5$ km and the PDE earthquake catalog from 1969–2005. This display is equivalent to calculating the information scores by using forecast λ_i as a reference density shown in Eq. 11.28 (see Kagan 2009b, Fig. 10; Kagan and Jackson 2011, Fig. 7 and its explanations).

The information scores are still calculated with respect to the null hypothesis of uniform spatial rate density (black line). The red line describes the reference hypothesis. The black line, corresponding to the null hypothesis, lies above the reference curve, indicating that its performance is worse than that of the power law kernel. Also shown are the two empirical distributions, one based on the learning catalog (1969–2005) and the other on the test catalog (2006–2010).

Information scores for the curves shown in the plot are as follows: for the forecast (red curve) $I_0 = 3.13$; for earthquakes in the 2006–2010 test period (blue curve), $I_1 = 3.85$; for the backward forecast (magenta) $I_1' = 4.46$. The backward forecast does best as expected, because the test and learning data are the same (Kagan 2009b).

In principle, we could adjust our forecast so that $I_0 = I_1$, because for a correct forecast the empirical information score should converge to the theoretical one. In Fig. 11.11 the empirical score exceeds the theoretical score, indicating that the earthquakes are more concentrated than the forecast predicts. As shown in Fig. 10.4, this occurs when the kernel width is too large. Decreasing the value of r_s

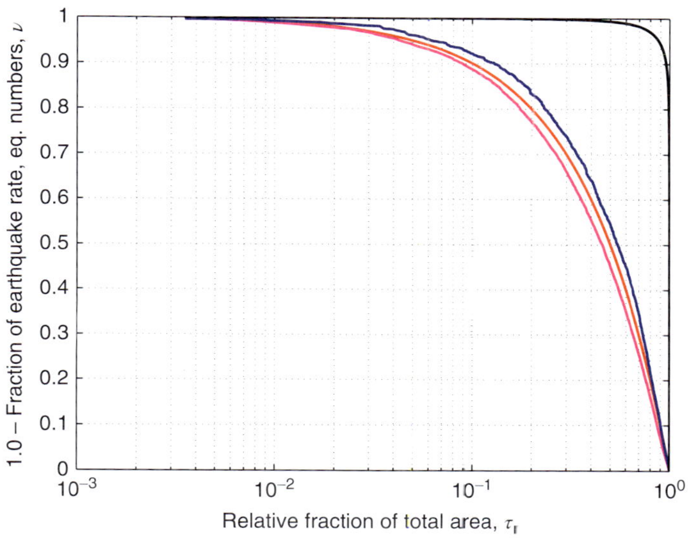

Fig. 11.12 Error diagram (τ, ν) for global long-term seismicity (PDE catalog, $m \geq 5.0$) forecast relative to that of the adaptive kernel forecast. Same format as in Fig. 11.11 is used. The solid black line, theoretical curve for spatially uniform rate density; the red line, theoretical results for reference forecast based on adaptive Fisher kernel (Eq. 10.6) with $\omega = 10,000$ and $\alpha = 0.5$; the blue line, empirical results for test period 2006–2010; the magenta line, empirical results for retrospective application to training period, 1969–2005. *Source:* Kagan and Jackson (2012), Fig. 7.

would bring closer agreement. However, the sample size for I_1 is fairly small, so the agreement between I_0 and I_1 is close enough.

The major part of the numerical difference between different forecasts in Fig. 11.11 generally corresponds to the curves behavior in the upper-left part of the plot; the higher I values are associated with curves closer to the vertical axis (Kagan 2009b). This is observed if only the upper-left corner of a diagram is plotted. In Fig. 11.11 the complete curves are shown, so this feature is not readily observable.

Figure 11.12 displays error diagrams for adaptive kernel smoothing. The diagram shows the adaptive Fisher kernels more effectively approximate earthquake productivity than the fixed power-law kernel does in Fig. 11.11: $I_0 = 4.26$, $I_1 = 4.04$, and $I_1' = 4.57$. In Figs. 11.13 and 11.14 we show the same error diagrams in a different, more regular format (cf. Fig. 11.5) to better display the approximation in low seismicity regions, that is, high values of τ. The adaptive kernel smoothing forecast is closer to the test period earthquake distribution in this part of the ED. Thus, Figs. 11.11–11.14 quantitatively confirm our observation in comparing the global fixed and adaptive kernel predictions (Figs. 10.4 and 10.5): the latter forecasts predict the earthquake rate better both in high and low seismicity areas.

Our forecasts do not predict well in regions where the previous earthquake rate was low. We have investigated whether our predictions can be improved by using the strain-based forecast (Bird *et al.* 2010). We normalized the seismicity and strain-based spatial densities, and tested a linear or geometric combination that preserves the normalization. Preliminary results show that a 50/50 percent linear combination works best.

The tests indicated above are "pseudo-prospective," in that except for I_1', the tests are performed on data not used in the forecast. However, they provide a good template for truly prospective tests.

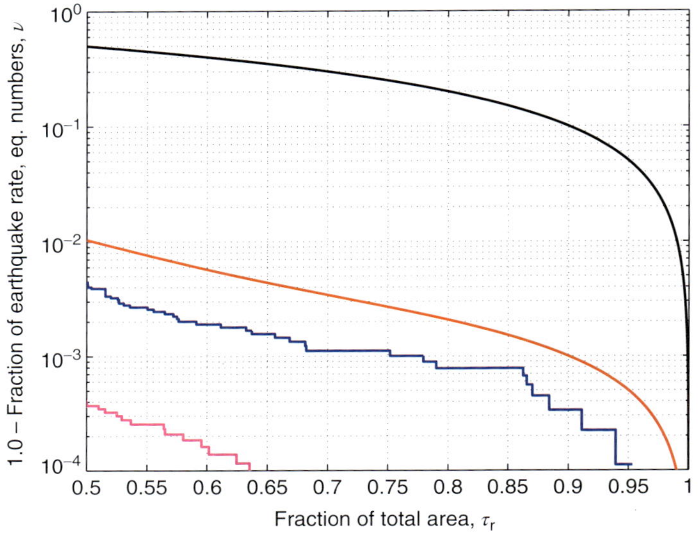

Fig. 11.13 Partial error diagram for fixed power-law kernel. Same as Fig. 11.11, except that reference forecast is the spatially uniform rate density model, the horizontal axis is plotted on the linear scale, and only the right half of the diagram is plotted. *Source:* Kagan and Jackson (2012), Fig. 8.

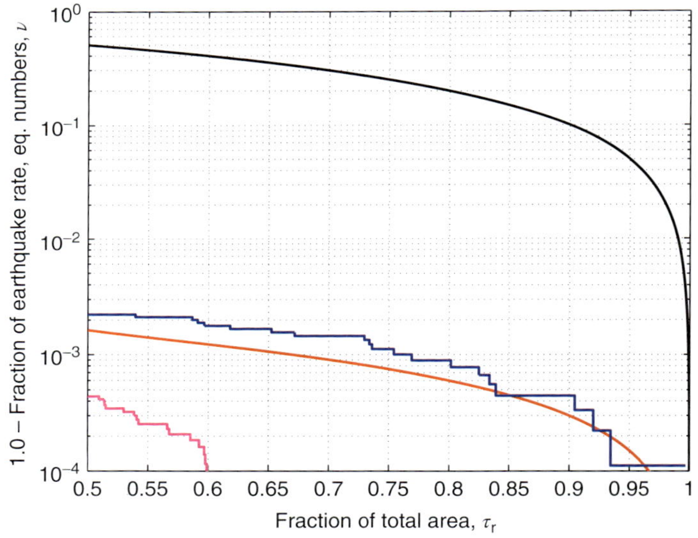

Fig. 11.14 Partial error diagram for adaptive Fisher kernel. Same as Fig. 11.12, except that similarly to Fig. 11.13 the reference forecast is the spatially uniform rate density model, the horizontal axis is plotted on linear scale, and only the right half of the diagram is plotted. *Source:* Kagan and Jackson (2012), Fig. 9.

11.5 Summary of testing results

Several information scores are displayed in Table 11.3. Although these scores appear different, the difference is caused either by the small event number or a small number of simulations. The following limits can easily be conjectured

$$I_0 = \lim_{\mathcal{N} \to \infty} < I_3 >, \tag{11.29}$$

(see Eq. 11.10). In Table 11.3 the difference between these two scores is small due to the large number ($\mathcal{N}$) of simulations. Similarly,

$$I = \lim_{|S_i| \to 0} I_0, \quad \text{or} \quad I = \lim_{N \to \infty} I_0, \tag{11.30}$$

where $|S_i|$ is the cell area, and N is the cell number (cf. Eq. 11.14). Also

$$I_2 = \lim_{|S_i| \to 0} I_1, \tag{11.31}$$

(see Eqs. 11.7, 11.8).

In addition, if the model of the long-term forecast is correct, then

$$I = \lim_{n \to \infty} I_1, \quad \text{and} \quad I = \lim_{n \to \infty} I_4, \tag{11.32}$$

where n is the event number (see Eqs. 11.7, 11.23 and Kagan 2007b).

As Table 11.3 demonstrates, the convergence of the I_4 score, shown in (11.32) is slow, even as tens or hundreds of events have been forecast in our examples. In many considerations and applications of error diagrams (see Section 11.1), behavior of the curves is considered only for a few observed events, thus random fluctuations of the observed distribution vs. the theoretical distribution should be much larger.

Our tests of forecast efficiency, described in Sections 10.2.4 and 11.4, are conducted mostly for $m5$ to $m6$ earthquakes. These earthquakes usually cause little damage. Preliminary tests for the forecast of higher magnitude earthquakes have shown that, although $m \geq 7.5$ are mostly concentrated in trenches, the forecast skill for these events is similar to that for medium-sized earthquakes.

Summarizing this chapter, we list the following major points:

- As with temporal stochastic processes (Kagan 2007b), we find forward and inverse relations between the information score and the error diagram curve for point spatial fields. The error diagram represents a more complete picture of the stochastic point process than does likelihood analysis.
- Since we are using a Poisson process to represent the long-term spatial point pattern, the resulting models are easier to visualize and calculate. However, the assumption of earthquake statistical independence and its influence on the information score value needs to be investigated.
- We extend our analysis for relatively small samples of events and show that for such samples we should modify some of the testing criteria proposed for error diagrams.
- The forecasting blueprint for estimating future earthquake point density differs from the standard methods of statistical density evaluation. Nevertheless, the connection between the likelihood score and error diagrams can be used in many density estimation problems.

- To test the long-term forecast, it is sufficient to process the forecast table to obtain the error diagram and most information scores. Thus, the simulation which requires significant computational resources, can be avoided in most cases.
- In this chapter we mostly considered relations between the theoretical EDs and likelihood scores (Eqs. 11.12 and 11.13) under the assumptions that the number of events is large and the theoretical model of the stochastic point process is true. Although some theoretical analysis has previously been published for the case when the above assumptions have been violated (see references in Section 11.1), and we discuss an extension of the analysis to more general situation (Figs. 11.7–11.9), a rigorous consideration of statistical behavior for both measures of prediction efficiency would require an extensive future effort.

12 Future prospects and problems

12.1 Community efforts for statistical seismicity analysis and earthquake forecast testing

12.1.1 Community Online Resource for Statistical Seismicity Analysis (CORSSA)

The topics discussed in this book have many points in common with the book(s) being prepared by the Community Online Resource for Statistical Seismicity Analysis (CORSSA) a web-based organization that provides software and tutorials on statistical seismology. CORSSA' goal is to promote excellence in statistical seismology by providing the knowledge and resources necessary to understand and implement best practices, in order that the users can apply these methods to their own research. For the CORSSA introductory material, see Michael and Wiemer (2010), Zechar *et al.* (2011), and http://www.corssa.org.

CORSSA is an educational resource and contains only methods that have already been published in established peer-reviewed journals. It does not contain new scientific results. The goal of each CORSSA article is to provide a tutorial based on the published literature. Each article covers a specific task or theme, including an explanation of why the topic is useful for research, a brief, referenced review of a theory, a list of methods and software relevant to the topic, a discussion of tradeoffs in analysis choices, pitfalls to be aware of, examples of results derived by applying the method to one of the CORSSA standard data sets, examples of applications in the scientific literature and pointers on further reading.

In contrast to CORSSA, which focuses on the development of statistical tools, this book emphasizes mathematical, statistical, and geometrical explanations of earthquake occurrence process. Many topics such as stable distributions or properties of focal mechanisms have not yet been discussed in the CORSSA documents. This book and the CORSSA documents will therefore complement one another.

Earthquakes: Models, Statistics, Testable Forecasts, First Edition. Yan Y. Kagan.

12.1.2 Collaboratory for the Study of Earthquake Predictability (CSEP): Global and regional forecast testing

Earthquake forecasts are important in estimating both hazard and risk and in making informed decisions in order to manage emergency response. How can we establish standards for reporting and testing earthquake forecasts? One significant effort to establish such standards began in California, where the Regional Earthquake Likelihood Models (RELM) project published a dozen models for earthquake rate density as well as a likelihood-based method for prospective testing (Field 2007; Schorlemmer and Gerstenberger 2007; Schorlemmer *et al.* 2007; Zechar *et al.* 2013). The Collaboratory for Study of Earthquake Predictability (CSEP), organized by the Southern California Earthquake Center (SCEC), is currently extending the tests to several natural laboratories around the Globe. A major focus of the CSEP activities (http://cseptesting.org/) is to develop international collaborations between regional testing centers and to accommodate a wide-ranging set of prediction experiments involving geographically distributed fault systems in different tectonic environments.

Many local forecasts are now produced by the researchers affiliated with CSEP infra-structure (Zechar *et al.* 2013). For these forecasts the maximum resolution is usually 0.1°; therefore, localized forecasts can be extracted from our forecasts and compared to those predictions. Furthermore, a combined forecast employing local and global data can be produced.

CSEP tests long- and short-term forecasts updated daily (Schorlemmer and Gerstenberger 2007; Nanjo *et al.* 2011; Marzocchi and Zechar 2011; Zechar *et al.* 2013). CSEP also applied our programs to replicate and test our forecasts for the western Pacific as well as the global forecast (see Chapter 10). These programs are available from CSEP. New testing methods are being developed by CSEP as well, for example, Clements *et al.* (2011); Rhoades *et al.* (2011).

The major goal for CSEP is developing testing routines and techniques. The test results should be both objective and reliably reproducible. Obviously, test routines ought to be specifically defined before the test has started in order to eliminate a possibility of a bias.

12.2 Results and challenges

As we show in this book, the earthquake process is controlled by scale-invariant, fractal distributions. Section 3.3 proposes that these scale-invariant distributions can be at least generally explained by simple assumptions regarding random stress behavior: (i) earthquake time behavior by the Brownian motion-like history of stress change (see also Section 6.4) and (ii) fault geometry by the Cauchy distribution of stress tensors due to randomly distributed defects in the rock medium (see Chapter 7 and Sections 8.1, 8.6.2). This Cauchy distribution induces the Cauchy 3-D rotation of focal mechanisms (8.54). Such a physical and mathematical explanation is a relatively rare case in the study of fractal distributions (Mandelbrot 1983).

However, it has not yet been explored whether simulated earthquakes (Section 3.3) are faithful representations of seismicity. There are several factors that complicate the comparison. The observational data, for example, are not sufficiently detailed, especially with regard to spatial and angular resolution (see Table 2.1). Furthermore, the mathematical and logical structure of the stochastic model needs to be explored in order to determine if it is consistent and can be extended to the continuum limit.

As we mentioned in several parts of this book, many of the mathematical techniques that are necessary to describe the earthquake geometry and its occurrence are still being developed: (i) the theory of stable distributions and their statistics, especially its multidimensional extension; (ii) statistics of 3-D rotations; and (iii) the random branching walk on non-commutative groups.

But developing a comprehensible theory of earthquake rupture may encounter serious mathematical difficulties. Earthquake faults, as shown for instance in Fig. 3.5, are stochastic fractal objects. The stress at the fractal boundary should not be a differentiable function. Thus, calculating the earthquake rupture criteria for points close to a "fault-tip" may be impossible to carry out effectively. For example, for the deterministic Mandelbrot set (Mandelbrot 1983, pp. 188–189) it has been shown (Blum *et al.* 1998, p. 55) that even if we use *real-number* arithmetic operations, no algorithm can decide in a finite number of steps whether the set contains an arbitrary point in a complex plane. The reason for the "undecidability" of the Mandelbrot set and many similar complex mathematical objects is that their boundary has a fractal Hausdorff dimension. This feature makes calculating the boundary of earthquake rupture faults effectively impossible (Kagan 2006).

In Chapter 1 we mentioned the similarity of two major problems in classical physics: the turbulence of fluids and fracture of solids (Kagan 1992a), and an apparent lack of interest by theoretical physicists in solving the former problem. This may have a simple explanation. Goldstein (1969, p. 23) remarks:

> *It was at a meeting of the British Association in London in 1932 that I remember that [Horace] Lamb remarked "I am an old man now, and when I die and go to Heaven there are two matters on which I hope for enlightenment. One is quantum electrodynamics, and the other is the turbulent motion of fluids. And about the former I am really rather optimistic."*

In other, undoubtedly apocryphal and more recent versions of the story Lamb is replaced by Einstein, or von Neumann, Heisenberg, Feynman, Fermi, or others (Yaglom 2001; Gleick 1987, pp. 121, 329; Mandelbrot 2012, p. 118). Do the multiple versions of this story genuinely reflect a general feeling among mathematical physicists that the turbulence problem may be unsolvable? The more complex problem of fracture in solids, including earthquake rupture process, may not be solvable either.

If we revisit our discussion in Chapter 1, a simple answer is that providing a theoretical explanation of earthquake occurrence is very difficult. It requires the application of mathematical methods that are outside of interest area of geophysicists and physicists. For example, many papers and a few monographs (e.g., Watson 1983, Fisher *et al.* 1987; Mardia and Jupp 2000) consider vector and axial statistics in 2-D and 3-D. There are, however, almost no publications, except for those few cited in Chapter 8, that deal with the statistics of 3-D rotations. Perhaps a recent development of statistical theory for topological manifolds (Small 1996; Kendall *et al.* 1999) could be adapted to describe the complex geometry of earthquake faulting, including 3-D rotations of focal mechanisms. Research techniques in crystallographic texture analysis (see Section 8.7) can also be applied to the problem of earthquake occurrence.

What else can be done? Clearly the level of mathematics employed in earthquake studies is inadequate. The mathematical tools in current use in seismological research can be traced back to the first half of nineteenth century and are therefore out of date. As we explained above, the level of mathematics needs to be raised by the order of a magnitude. Results in the forefront of modern mathematical research should be employed to describe earthquake occurrence and the geometry of earthquake faults in particular. Mathematical disciplines, such as modern differential geometry, matrix theory, group theory, topology, and the theory of stochastic processes must be involved in the solution. Otherwise significant progress is unlikely.

Although applied and pure mathematicians work in other geoscience disciplines for example atmospheric and plasma physics or geodynamo theory, until now, only statisticians have studied earthquake occurrence patterns. Even if we look at the development of earthquake science in the United States, we see that there are no professional statisticians involved in this research full-time. The results of the statistical research by D. Vere-Jones (N.Z.), G. Molchan (Russia) and Y. Ogata (Japan) have been extensively used in this book.

To summarize our discussion, we see that there are major, perhaps fundamental difficulties in creating a comprehensive physical/mathematical theory of brittle fracture and the earthquake rupture process. Yet developing the quantitative models of earthquake occurrence needed to evaluate probabilistic seismic maps is within our reach. It will require the combined effort of earth scientists, physicists, statisticians, as well as pure and applied mathematicians.

12.3 Future developments

Several concrete proposals to investigate stochastic and statistical properties of earthquake occurrence are scattered throughout the book. We summarize and list some of the major recommendations and challenges below.

1. In Section 3.3 we propose investigating analytically the continuum scale-invariant branching model of earthquake occurrence. The results described in this section are obtained through a multidimensional simulation in the size-time-space-focal mechanism manifold $M \times T \times R^3 \times SO(3)$. Such a model could simulate the source dynamic rupture with small-scale

complexity in fault geometry and time variations. Thus high-frequency synthetic seismograms can be generated and used in engineering applications and other studies. An analytical solution would be especially difficult because the focal mechanism orientation is defined by a non-commutative group $SO(3)$ that has a compact support. Thus, the self-similarity of the earthquake fault patterns is the complicated result of the branching random walk in the space-time-focal mechanism space that needs a concise mathematical treatment.

2. Chapter 4 analyzes earthquake number distributions. The discrete distributions we use to approximate the numbers were developed by statisticians in the mid-twentieth century. Many problems in estimating the parameters have still not been developed in a way that would be convenient for geophysicists to use. A new look by statisticians at these distributions and their applications in earthquake studies would be very useful.
3. In Section 5.3 we conjecture that brittle fracture and plastic deformation have the same exponent $\beta \equiv 1/2$ of their Pareto size distribution and the difference between them amounts only to the size of the maximum earthquake or dislocation avalanche each system allows. Thus, for example, a key dropped to the floor does not break, whereas a glass usually does. However, when a key is bent, it emits small cracking sounds. The reason a key does not break is that the maximum size of its cracks is much smaller than its thickness, allowing the object to deform plastically. It would be very helpful to study the maximum size distribution for dislocation avalanches in plastic objects of changing viscosity to determine whether or not our conjecture is correct. Similarly, a more profound investigation of earthquake size distribution, as well as a thorough study of the size distribution of tremors would certainly increase our understanding of the transition from brittle to plastic deformation in the Earth's interior.
4. Section 5.4 is dedicated to the study of approximating the sum of Pareto laws by stable distributions. It would be useful to obtain such approximations for the tapered Pareto (TGR) distribution and to develop a methodology for statistical modeling of the seismic moment release. The results reported in that section are obtained primarily through simulation. It may be possible to derive some analytical results useful to earthquake science and other scientific disciplines that study heavy-tailed distributions.
5. In Section 6.2.4 we investigated the relation between the moment release during an earthquake and the subsequent aftershock sequence. Such relations are typically studied in a retrospective analysis when the identification of a mainshock and its aftershocks is relatively easy. However, a major theoretical and practical issue is the real time prediction of the expected moment rate following any earthquake. For a retrospective consideration this rate is of the order $1/30$ of the maximum rate during a mainshock. We conjectured that in prospective studies the ratio would be of the order of 1.0

instead of 1/30. New investigations with good quality data are needed to resolve this question.

6. In Section 7.1 we consider multi-point spatial moments for earthquake hypocenters. Especially interesting is the 4-simplex or four-point distribution that shows that the hypocenter distribution is inversely proportional to the tetrahedron volume formed by these four points. This result negates the common plane model of the earthquake fault. We compared the observed distribution with a distribution for the Poisson set of points which was obtained by simulation. The simulation breaks down for small distances between the points, thus it would be valuable to obtain an analytical expression for the Poisson distribution.
7. Section 8.1 considers the distribution of stress in an elastic medium caused by random point-like defects, which can be regarded as micro-earthquakes. This distribution can be used to simulate further rupture of the infinitesimal defects from the assumed initial stage (Kagan 1990). Earthquakes, however, have a power-law size distribution (Chapter 5) and their spatial distribution is scale-invariant (Chapter 7). It would be therefore important to obtain the stress distribution function for more realistic conditions (see also Kagan, 1994b).
8. Subsection 8.2.2 describes the simulation of the CLVD source distribution. It is shown that the Γ-index distribution approaches the uniform law for the sum of rotated focal mechanisms. We propose to carry out a mathematical derivation of the results obtained by the simulations. We also show that the CLVD component of a complex earthquake source would be zero or close to zero for the suggested configurations of the geometrical barriers. It would be useful to show theoretically that the considered configurations are unique in this respect.
9. In Subsection 8.4.1 we propose to investigate the sign-change symmetry of a *DC* earthquake source. This symmetry may influence the properties of complex, extended earthquake sources.
10. In order to describe the focal mechanism orientation, in Subsection 8.7.3 we study the Rodrigues space statistics. This representation provides a new approach to focal mechanism investigation and, when fully developed, is likely to yield a more thorough understanding of this complex feature. More complete investigations of source orientations in the focal zone and the relation of focal mechanisms to the Earth's surface would be much more difficult to accomplish. They could, however, be carried out using research techniques developed in crystallographic texture analysis.
11. In Subsection 8.8.1 we considered a general representation of deformation sources in an infinite elastic space and a half-space. These solutions could prove useful when investigating complex earthquake sources and volcano deformation measurements. They can also be applied in other studies where detailed pictures of the deformation sources in elastic half-space are needed. Presently, such inversions are performed by approximating the deformation caused by simple sources and their combinations at the Earth's

surface. Such inversions are non-unique and may fail to represent important features of a source. On the other hand, using our representations make it possible to reconstruct a multipole image of an arbitrary complex deformation source. Such results would allow us to "see" the sources below the half-space boundary, which may constitute a major scientific achievement.

12. Subsection 8.8.2 considered a general representation of the earthquake focal zone by a higher-rank correlation tensor. As we suggested at the end of that subsection, this approach has significant potential advantages, though the mathematical challenges and difficulties with regard to observational investigations are considerable.
13. In Section 10.3 we point out a need for a more complete quantitative evaluation of the predictive skill of a short-term forecast. This would require an additional investigation of the local and global earthquake catalogs in order to estimate the predictive information rate for real-time forecasts. It would be also useful to include focal mechanism statistics in the short-term forecast as this may significantly increase its predictive score.
14. We suggest in Section 11.2 that further investigations should be carried out to better understand earthquake occurrence and the testing of earthquake forecasts. Relationships between proposed information scores should be explored further.
15. Similar to item 10, in Subsection 11.3.1 we indicate that due to the fractal nature of the earthquake spatial distribution, the definition of information gain for the long-term forecast requires a fundamental revision. The same point applies to the calculation of probability gain for short-term forecast (see above): because the aftershock rate is governed by a power-law (Omori's law) temporal distribution. Standard methods for the gain computation should be substantially modified to take into account the fractal properties of the time distribution.

Solving the above-mentioned problems will enhance our understanding of the stochastic nature of the earthquake process and will enable us to proceed from the point representation of earthquake occurrence to its continuum representation. A prediction of the continuum space process would represent a significant theoretical step forward and play an important practical role in the evaluation of the earthquake occurrence pattern.

References

Abercrombie, R. E. (1995). Earthquake source scaling relationships from −1 to 5 *mL* using seismograms recorded at 2.5-km depth, *J. Geophys. Res.*, 100, 24,015–24,036.

Abramowitz, M. and Stegun, I. A., eds. (1972). *Handbook of Mathematical Functions*, Dover, New York.

Advanced National Seismic System (ANSS) (2008). Catalog Search (2008) Northern California Earthquake Data Center (NCEDC), available at: http://www.ncedc.org/anss/catalog-search.html.

Agnew, D. C. (2005). Earthquakes: future shock in California, *Nature*, 435, 284–285.

Aki, K. and Richards, P. G. (2002). *Quantitative Seismology*, 2nd ed., University Science Books, Sausalito, CA.

Allen, R. M. (2007). Earthquake hazard mitigation: New directions and opportunities, in *Treatise on Geophysics, Earthquake Seismology*, 4(4.21), ed. H. Kanamori, pp. 607–648, Elsevier, Amsterdam.

Altmann, S. L. (1986). *Rotations, Quaternions and Double Groups*, Clarendon Press, Oxford.

Amelung, F. and King, G. (1997). Large-scale tectonic deformation inferred from small earthquakes, *Nature*, 386, 702–705.

Anderson, T. L. (2005). *Fracture Mechanics: Fundamentals and Applications*, 3rd ed., Taylor & Francis, Boca Raton, FL.

Anderson, T. W. (1958). *An Introduction to Multivariate Statistical Analysis*, John Wiley and Sons, New York.

Andrews, D. J. (1989). Mechanics of fault junctions, *J. Geophys. Res.*, 94, 9389–9397.

Angelier, J., Tarantola., A., Valette, B. and Manoussis, S. (1982). Inversion of field data in fault tectonics to obtain regional stress – I. Single phase fault populations: a new method of computing stress tensor, *Geophys. J. R. Astr. Soc.*, 69, 607–621.

Anraku, K. and Yanagimoto, T. (1990). Estimation for the negative binomial distribution based on the conditional likelihood, *Communications in Statistics: Simulation and Computation*, 19(3), 771–786.

Anscombe, F. J. (1950). Sampling Theory of the Negative Binomial and Logarithmic Series Distributions, *Biometrika*, 37(3/4), 358–382.

Antonioli, A., Cocco, M., Das, S. and Henry, C. (2002). Dynamic stress triggering during the great 25 March 1998 Antarctic Plate earthquake, *Bull. Seismol. Soc. Amer.*, 92(3), 896–903.

Apperson, K. D. (1991). Stress Fields of the Overriding Plate at Convergent Margins and Beneath Active Volcanic Arcs, *Science*, 254(5032), 670–678.

Athreya, K. B. and Ney, P. E. (1972). *Branching Processes*, Springer-Verlag, New York.

Backus, G. E. (1977a). Interpreting the seismic glut moments of total degree two or less, *Geophys. J. R. Astr. Soc.*, 51, 1–25.

Earthquakes: Models, Statistics, Testable Forecasts, First Edition. Yan Y. Kagan.

Backus, G. E. (1977b). Seismic sources with observable glut moments of spatial degree two, *Geophys. J. R. Astr. Soc.*, 51, 27–45.

Backus, G. and Mulcahy, M. (1976a). Moment tensor and other phenomenological descriptions of seismic sources – I. Continuous displacements. *Geophys. J. Roy. Astr. Soc.*, 46, 341–361.

Backus, G. and Mulcahy, M. (1976b). Moment tensor and other phenomenological descriptions of seismic sources – II. Discontinuous displacements. *Geophys. J. Roy. Astr. Soc.*, 47, 301–329.

Baddeley, A. (2007). Spatial point processes and their applications, *Lecture Notes in Mathematics*, 1892, pp. 1–75.

Baddeley, A., Turner, R., Moller, J. and Hazelton, M. (2005). Residual analysis for spatial point processes, *J. Royal Stat. Soc.*, B67(5), 617–651.

Baiesi, M. and Paczuski, M. (2005). Complex networks of earthquakes and aftershocks, *Nonlinear Processes Geophys.*, 12(1), 1–11.

Bailey, I. W., Becker, T. W. and Ben-Zion, Y. (2009). Patterns of co-seismic strain computed from southern California focal mechanisms, *Geophys. J. Int.*, 177(3), 1015–1036.

Bailey, I. W., Ben-Zion, Y., Becker, T. W. and Holschneider, M. (2010). Quantifying focal mechanism heterogeneity for fault zones in central and southern California, *Geophys. J. Int.*, 183(1), 433–450.

Bak, P., Christensen, K., Danon, L. and Scanlon, T. (2002). Unified scaling law for earthquakes, *Phys. Rev. Lett.*, 88, 178501, 1–4.

Bakun, W. H. and Lindh, A. G. (1985). The Parkfield, California, earthquake prediction experiment, *Science*, 229, 619–624.

Bakun, W. H. *et al.* (2005). Implications for prediction and hazard assessment from the 2004 Parkfield earthquake, *Nature*, 437, 969–974.

Bartlett, M. S. (1978). *An Introduction to Stochastic Processes with Special Reference to Methods and Applications*, 3rd ed., Cambridge University Press, Cambridge.

Ben-Menahem, A. (1995). A concise history of mainstream seismology – origins, legacy, and perspectives, *Bull. Seism. Soc. Am.*, 85, 1202–1225.

Ben-Zion, Y. (2008). Collective behavior of earthquakes and faults: continuum-discrete transitions, progressive evolutionary changes and different dynamic regimes, *Rev. Geophysics*, 46, RG4006, DOI: 10.1029/2008RG000260.

Ben-Zion, Y. and Sammis, C. G. (2003). Characterization of fault zones, *Pure Appl. Geophys.*, 160, 677–715.

Beran, M. J. (1968). *Statistical Continuum Theories*, Interscience Publishers, N.Y., pp. 424.

Beroza, G. C. and Ide, S. (2009). Deep tremors and slow quakes, *Science*, 324(5930), 1025–1026.

Bird, P. (2003). An updated digital model of plate boundaries, *Geochemistry, Geophysics, Geosystems*, 4(3), 1027.

Bird, P. and Kagan, Y. Y. (2004). Plate-tectonic analysis of shallow seismicity: apparent boundary width, beta, corner magnitude, coupled lithosphere thickness, and coupling in seven tectonic settings, *Bull. Seismol. Soc. Amer.*, 94(6), 2380–2399.

Bird, P. and Liu, Z. (2007). Seismic hazard inferred from tectonics: California, *Seism. Res Lett.*, 78(1), 37–48.

Bird, P., Kreemer, C. and Holt, W. E. (2010). A long-term forecast of shallow seismicity based on the Global Strain Rate Map, *Seismol. Res. Lett.*, 81(2), 184–194 (plus electronic supplement).

Blum, L., Cucker, F., Shub, M. and Smale, S. (1998). *Complexity and Real Computation*, Springer, New York.

Boettcher, M. and Jordan, T. H. (2004). Earthquake scaling relations for mid-ocean ridge transform faults, *J. Geophys. Res.*, 109(B12), Art. No. B12302.

Boettcher, M. S., McGarr, A. and Johnston, M. (2009). Extension of Gutenberg-Richter distribution to *Mw*-1.3, no lower limit in sight, *Geophys. Res. Lett.*, 36, L10307.

Bolt, B. A. (2003). *Earthquakes*, 5th ed., W. H. Freeman, New York.

Bouchbinder, E., Procaccia, I. and Sela, S. (2005). Disentangling scaling properties in anisotropic fracture, *Phys. Rev. Lett.*, 95(25), 255–303.

Bremaud, P. and Massoulie, L. (2001). Hawkes branching point processes without ancestors, *J. Applied Probability*, 38(1), 122–135.

Buehler, M. J. and Gao, H. J. (2006). Dynamical fracture instabilities due to local hyperelasticity at crack tips, *Nature*, 439(7074), 307–310.

Bugayevskiy, L. M. and Snyder, J. P. (1995). *Map Projections: A Reference Manual*, Taylor & Francis, London.

Bullen, K. E. (1979) *An Introduction to the Theory of Seismology*, 3rd ed., Cambridge University Press, Cambridge.

Burridge, R. and Knopoff, L. (1964). Body force equivalents for seismic dislocations, *Bull. Seismol. Soc. Amer.*, 54, 1875–1888.

Burridge, R. and Knopoff, L. (1967). Model and theoretical seismicity, *Bull. Seismol. Soc. Amer.*, 57, 341–371.

Cai, J. Q., Grafarend, E. W. and Schaffrin, B. (2005). Statistical inference of the eigenspace components of a two-dimensional, symmetric rank-two random tensor, *J. Geodesy*, 78(7–8), 425–436.

Calais, E., Freed, A. M., Van Arsdale, R. and Stein, S. (2010). Triggering of New Madrid seismicity by late-Pleistocene erosion, *Nature*, 466, 608–611.

Calais, E., Han, J. Y., DeMets, C., and Nocquet, J. M. (2006). Deformation of the North American plate interior from a decade of continuous GPS measurements, *J. Geophys. Res.*, 111(B6), B06402.

Castellaro, S., Mulargia, F. and Kagan, Y. Y. (2006). Regression problems for magnitudes, *Geophys. J. Int.*, 165(3), 913–930.

Chen, P., Jordan, T. H. and Zhao, L. (2005). Finite-moment tensor of the 3 September 2002 Yorba Linda earthquake, *Bull. Seismol. Soc. Amer.*, 95(3), 1170–1180.

Chen, P., Jordan, T. H. and Zhao, L. (2010). Resolving fault plane ambiguity for small earthquakes, *Geophys. J. Int.*, 181(1), 493–501.

Clauset, A., Shalizi, C. R. and Newman, M. E. J. (2009). Power-law distributions in empirical data, *SIAM Rev.*, 51, 661–703.

Clements, R. A., Schoenberg, F. P. and Schorlemmer, D. (2011). Residual analysis methods for spacetime point processes with applications to earthquake forecast models in California, *Annals Applied Statistics*, 5(4), 2549–2571.

Console, R., Lombardi, A. M., Murru, M. and Rhoades, D. (2003a). Båth's law and the self-similarity of earthquakes, *J. Geophys. Res.*, 108(B2), 2128.

Console, R. and Murru, M. (2001). A simple and testable model for earthquake clustering, *J. Geophys. Res.*, 106, 8699–8711.

Console, R., Murru, M. and Lombardi, A. M. (2003b). Refining earthquake clustering models, *J. Geophys. Res.*, 108(B10), Art. No. 2468.

Consul, P. C. (1989). *Generalized Poisson Distributions: Properties and Applications*, Dekker, New York.

Cornell, C. A. (1968). Engineering seismic risk analysis, *Bull. Seismol. Soc. Amer.*, 58, 1583–1606.

Corral, A. (2005). Renormalization-group transformations and correlations of seismicity, *Phys. Rev. Lett.*, 95(2), 028501.

Csikor, F. F., Motz, C., Weygand, D., Zaiser, M. and Zapperi, S. (2007). Dislocation avalanches, strain bursts, and the problem of plastic forming at the micrometer scale, *Science*, 318(5848), 251–254.

Dahmen, K. A., Ben-Zion, Y. and Uhl, J. T. (2009). Micromechanical model for deformation in solids with universal predictions for stress-strain curves and slip avalanches, *Phys. Rev. Lett.*, 102(17), Article Number: 175501.

Daley, D. J. and Vere-Jones, D. (2003). *An Introduction to the Theory of Point Processes*, 2nd ed., Vol. 1, Springer Verlag, New York.

Daley, D. J. and Vere-Jones, D. (2004). Scoring probability forecasts for point processes: the entropy score and information gain, *J. Applied Probability*, 41A, 297–312 (Special Issue).

Daley, D. J. and Vere-Jones, D. (2008). *An Introduction to the Theory of Point Processes*, 2nd ed., Vol. 2, Springer Verlag, New York.

Danilova, M. A., and Yunga. S. L. (1990). Statistical properties of matrices in problems of earthquake focal mechanism interpretation, *Izv. Acad. Sci. USSR, Phys. Solid Earth*, 26(2), 137–142 (English translation).

Davis, P. M., Jackson, D. D. and Kagan, Y. Y. (1989). The longer it has been since the last earthquake, the longer the expected time till the next?, *Bull. Seismol. Soc. Amer.*, 79(5), 1439–1456.

Dawson, T. E., McGill S. F. and Rockwell, T. K. (2003). Irregular recurrence of paleoearthquakes along the central Garlock fault near El Paso Peaks, California, *J. Geophys. Res.*, 108(B7), 2356.

De Luca, L., Lasocki, S., Luzio, D. and Vitale. M. (1999). Fractal dimension confidence interval estimation of epicentral distributions, *Annali di Geofisica*, 42(5), 911–925.

De Luca, L., Luzio. D. and Vitale. M. (2002). A ML estimator of the correlation dimension for left-hand truncated data samples, *Pure Appl. Geophys.*, 159, 2789–2803.

DelSole, T. and Tippett, M. K. (2007). Predictability: recent insights from information theory, *Rev. Geophys.*, 45, RG4002.

Dieterich, J. (1994). A constitutive law for rate of earthquake production and its application to earthquake clustering, *J. Geophys. Res.*, 99, 2601–2618.

Dimiduk, D. M., Woodward, C., LeSar, R., and Uchic, M. D. (2006). Scale-free intermittent flow in crystal plasticity, *Science*, 312, 1188–1190.

Dionysiou, D. D. and Papadopoulos, G. A. (1992). Poissonian and negative binomial modelling of earthquake time series in the Aegean area, *Phys. Earth Planet. Inter.*, 71(3–4), 154–165.

Dunn, F. and Parberry, I. (2011). *3D Math Primer for Graphics and Game Development*. A.K. Peters/CRC Press, Boca Raton, FL.

Dyson, F. (2004). A meeting with Enrico Fermi: How one intuitive physicist rescued a team from fruitless research, *Nature*, 427(6972), 297.

Dziewonski, A. M., Chou, T.-A. and Woodhouse, J. H. (1981). Determination of earthquake source parameters from waveform data for studies of global and regional seismicity, *J. Geophys. Res.*, 86, 2825–2852.

Dziewonski, A. M. and Woodhouse, J. H. (1983a). An experiment in systematic study of global seismicity: centroid-moment tensor solutions for 201 moderate and large earthquakes of 1981, *J. Geophys. Res.*, 88, 3247–3271.

Dziewonski, A. M. and Woodhouse, J. H. (1983b). Studies of the seismic source using normal mode theory, in *Earthquakes: Observation, Theory and Interpretation*, Proc. Int. School Phys. "Enrico Fermi," Course LXXXV, eds H. Kanamori and E. Boschi, North-Holland Publ., Amsterdam, pp. 45–137.

Ebel, J. E. (2009). Analysis of aftershock and foreshock activity in stable continental regions: implications for aftershock forecasting and the hazard of strong earthquakes, *Seismol. Res. Lett.*, 80(6), 1062–1068.

Ebel, J. E., Bonjer, K.-P. and Oncescu, M. C. (2000). Paleoseismicity: seismicity evidence for past large earthquakes, *Seismol. Res. Lett.*, 71(2), 283–294.

Ekström, G. (2007). Global seismicity: results from systematic waveform analyses, 1976–2005, in *Treatise on Geophysics, Earthquake Seismology*, 4(4.16), H. Kanamori, ed., Elsevier, Amsterdam, pp. 473–481.

Ekström, G. and Dziewonski, A. M. (1988). Evidence of bias in estimation of earthquake size, *Nature*, 332, 319–323.

Ekström, G., Nettles, M. and Dziewonski, A.M. (2012). The global CMT project 2004–2010: centroid moment tensors for 13,017 earthquakes, *Phys. Earth Planet. Inter.*, 200–201, 1–9.

Ellsworth, W. L. (1995). Characteristic earthquakes and long-term earthquake forecasts: implications of central California seismicity, in *Urban Disaster Mitigation: The Role of Science and Technology*, F. Y. Cheng and M. S. Sheu, eds., Elsevier Science Ltd., Oxford, pp. 1–14.

Ellsworth, W. L., Matthews, M. V., Nadeau, R. M., Nishenko, S. P., Reasenberg, P. A. and Simpson, R. W. (1999). A physically based earthquake recurrence model for estimation of long-term earthquake probabilities, *U.S. Geol. Surv. Open-File Rept.* 99–522.

Enescu, B., Mori, J., Miyazawa, M. and Kano, Y. (2009). Omori-Utsu law *c*-values associated with recent moderate earthquakes in Japan, *Bull. Seismol. Soc. Amer.*, 99(2A), 884–891.

England, P. and Jackson, J. (2011) Uncharted seismic risk, *Nature Geoscience*, 4, 348–349.

Evans, D. A. (1953). Experimental evidence concerning contagious distributions in ecology, *Biometrika*, 40(1–2), 186–211.

Evans, M., Hastings, N. and Peacock, B. (2000). *Statistical Distributions*, 3rd ed., John Wiley, New York.

Evans, R. (1997). Assessment of schemes for earthquake prediction: Editor's introduction, *Geophys. J. Int.*, 131, 413–420.

Fedotov, S. A. (1965). Regularities of the distribution of strong earthquakes in Kamchatka, the Kurile Islands and northeastern Japan, *Tr. Inst. Fiz. Zemli Akad. Nauk SSSR*, 36(203), 66–93 (in Russian).

Fedotov, S. A. (1968). On the seismic cycle, feasibility of quantitative seismic zoning and long-term seismic prediction, in *Seismic Zoning of the USSR*, pp. 121–150, Nauka, Moscow, (in Russian); English translation: Israel Program for Scientific Translations, Jerusalem, 1976.

Fedotov, S. A. (2005). Dolgosrochnyi Seismicheskii Prognoz dlya Kurilo-Kamchatskoi Dugi [Long-term Earthquake Prediction for the Kuril-Kamchatka Arc] (in Russian), Nauka / Moskva.

Feller, W. (1968). *An Introduction to Probability Theory and its Applications*, 1, 3rd ed., John Wiley, New York.

Feller, W. (1971). *An Introduction to Probability Theory and Its Applications*, 2, 2nd ed., John Wiley, New York.

Felzer, K., Becker, T. W., Abercrombie, R. E., Ekström, G. and Rice, J. R. (2002). Triggering of the 1999 *Mw* 7.1 Hector Mine earthquake by aftershocks of the 1992 *Mw* 7.3 Landers earthquake, *J. Geophys. Res.* 107(B9), 2190.

Field, E. H. (2007). Overview of the Working Group for the Development of Regional Earthquake Likelihood Models (RELM), *Seism. Res. Lett.*, 78(1), 7–16.

Fisher, N. I., Lewis, T. and Embleton, B. J. J. (1987). *Statistical Analysis of Spherical Data*, Cambridge University Press, Cambridge.

Fisher, R. A. (1928). General sampling distribution of the multiple correlation coefficient, *Proc. Roy. Soc. London*, ser. A, 121, 654–673.

Fisher, R. A. (1953). Dispersion on a sphere, *Proc. Roy. Soc. London*, ser. A, 271, 295–305.

Flinn, E. A., Engdahl, E. R. and Hill, A. R. (1974). Seismic and geographical regionalization, *Bull. Seismol. Soc. Amer.*, 64, 771–992.

Frank, F. C. (1988). Orientation mapping, *Metall. Trans. A*, 19, 403–408.

Frohlich, C. (1990). Note concerning non-double-couple source components from slip along surfaces of revolution, *J. Geophys. Res.*, 95(B5), 6861–6866.

Frohlich, C. (2001). Display and quantitative assessment of distributions of earthquake focal mechanisms, *Geophys. J. Int.*, 144, 300–308.

Frohlich, C. and Davis, S. D. (1999). How well constrained are well-constrained T, B, and P axes in moment tensor catalogs?, *J. Geophys. Res.*, 104, 4901–4910.

Frohlich, C., Riedesel, M. A. and Apperson, K. D. (1989). Note concerning possible mechanisms for non-double-couple earthquake sources, *Geophys. Res. Lett.*, 16(6), 523–526.

Frohlich, C. and Wetzel, L. R. (2007). Comparison of seismic moment release rates along different types of plate boundaries, *Geophys. J. Int.*, 171(2), 909–920.

Gabrielov, A. M. and Keilis-Borok, V. I. (1983). Patterns of stress corrosion: geometry of the principal stresses, *Pure Appl. Geophys.*, 121, 477–494.

Gabrielov, A., Keilis-Borok, V. and Jackson, D. D. (1996). Geometric incompatibility in a fault system, *P. Natl. Acad. Sci. USA*, 93, 3838–3842.

Garwood, F. (1947). The variance of the overlap of geometrical figures with reference to a bombing problem, *Biometrika*, 34, 1–17.

Gasperini, P. and Vannucci, G. (2003). FPSPACK: a package of FORTRAN subroutines to manage earthquake focal mechanism data, *Computers and Geosciences*, 29(7), 893–901.

Geist, E.L. and Parsons, T. (2005). Triggering of tsunamigenic aftershocks from large strike-slip earthquakes: analysis of the November 2000 New Ireland earthquake sequence, *Geochemistry Geophysics Geosystems*, 6, Article Number: Q10005.

Geller, R. J. (1997). Earthquake prediction: a critical review, *Geophys. J. Int.*, 131, 425–450.

Geller, R. J. (2011). Shake-up time for Japanese seismology, *Nature*, 472(7344), 407–409.

Geller, R. J., Jackson, D. D., Kagan, Y. Y. and Mulargia, F. (1997) Earthquakes cannot be predicted, *Science*, 275(5306), 1616–1617.

Gerstenberger, M. C., Wiemer, S., Jones, L. M. and Reasenberg, P. A. (2005). Real-time forecasts of tomorrow's earthquakes in California, *Nature*, 435, 328–331, DOI: 10.1038/nature03622.

Ghosh, P. (1951). Random distance within a rectangle and between two rectangles, *Bull. Calcutta Math. Soc.*, 43, 17–24.

Gilbert, F. and Dziewonski, A. M. (1975). An application of normal mode theory to the retrieval of structural parameters and source mechanisms from seismic spectra, *Phil. Trans. R. Soc. Lond. A*, 278, 187–269.

Gilbert, G. K. (1884). A theory of the earthquakes of the Great Basin, with a practical application, *Am. J. Sci.*, ser. 3, 27, no. 157, 49–53.

Girko, V. L. (1990). *Theory of Random Determinants*, Kluwer Academic Publishers, Boston.

Gleick, J. (1987). *Chaos: Making a New Science*, Viking, New York.

Goldstein, S. (1969). Fluid mechanics in the first half of this century, *Annual Rev. Fluid Mech.*, 1, 1–29.

Gomberg, J., Bodin, P. and Reasenberg, P. A. (2003). Observing earthquakes triggered in the near field by dynamic deformations, *Bull. Seismol. Soc. Amer.*, 93, 118–138.

Gradshteyn, I. S. and Ryzhik, I. M. (1980). *Table of Integrals, Series, and Products*, Academic Press, New York.

Grimmer, H. (1979). Distribution of disorientation angles if all relative orientations of neighboring grains are equally probable, *Scripta Metallurgica*, 13(2), 161–164.

Grunewald, E. D. and Stein, R. S. (2006). A new 1649–1884 catalog of destructive earthquakes near Tokyo and implications for the long-term seismic process, *J. Geophys. Res.*, 111, B12306.

Guo, Z. and Ogata, Y. (1997). Statistical relations between the parameters of aftershocks in time, space, and magnitude, *J. Geophys. Res.*, 102, 2857–2873.

Gutenberg, B. and Richter, C. F. (1944). Frequency of earthquakes in California, *Bull. Seism. Soc. Am.*, 34, 185–188.

Hammersley, J. M. (1950). The distribution of distance in a hypersphere. *Ann. Math. Stat.*, 21, 447–452.

Handscomb, D. C. (1958). On the random disorientation of two cubes, *Can. J. Math.*, 10, 85–88.

Hanks, T. C. (1992). Small earthquakes, tectonic forces, *Science*, 256, 1430–1432.

Hanks, T. C. and Kanamori, H. (1979). A moment magnitude scale, *J. Geophys. Res.*, 84, 2348–2350.

Hanson, A. J. (2006). *Visualizing Quaternions*, Elsevier, San Francisco.

Harris, R. A. (1998). Introduction to special section: stress triggers, stress shadows, and implications for seismic hazard, *J. Geophys. Res.*, 103, 24,347–24,358.

Harris, T. E. (1963). *The Theory of Branching Processes*, Springer, New York.

Harte, D. (1998). Dimension estimates of earthquake epicentres and hypocentres, *J. Nonlinear Science*, 8, 581–618.

Harte, D. (2001). *Multifractals: Theory and Applications*, Boca Raton, FL, Chapman and Hall.

Harte, D. (2013). Bias in fitting the ETAS model: a case study based on New Zealand seismicity, *Geophys. J. Int.*, 192, in press.

Haseltine, E. (2002). The 11 greatest unanswered questions of physics, *Discover*, 23(2).
Hauksson, E. and Shearer, P. (2005). Southern California hypocenter relocation with waveform cross-correlation, Part 1: Results using the double-difference method, *Bull. Seism. Soc. Am.*, 95(3), 896–903.
Hauksson, E., Yang, W. and Shearer, P. M. (2012). Waveform Relocated Earthquake Catalog for Southern California (1981 to June 2011), *Bull. Seismol. Soc. Amer.*, 102(5), 2239–2244.
Hawkes, A. G. (1971). Point spectra of some mutually exciting point processes, *J. Roy. Statist. Soc.*, B33, 438–443.
Hawkes, A. G. and Adamopoulos, L. (1973). Cluster models for earthquakes: regional comparisons, *Bull. Int. Statist. Inst.*, 45(3), 454–461.
Hawkes, A. G. and Oakes, D. (1974). A cluster process representation of a self-exciting process, *J. Appl. Prob.*, 11, 493–503.
Hayes, G. P., Earle, P. S., Benz, H. M., Wald, D. J., Briggs, R. W. and the USGS/NEIC Earthquake Response Team. (2011). 88 Hours: The U.S. Geological Survey National Earthquake Information Center Response to the 11 March 2011 Mw 9.0 Tohoku Earthquake, *Seismol. Res. Lett.*, 82(4), 481–493.
Headquarters for Earthquake Research Promotion (2005). National Seismic Hazard Maps for Japan available at http://www.jishin.go.jp/main/index-e.html.
Heaton, T. H. (1990). Evidence for and implications of self-healing pulses of slip in earthquake rupture, *Phys. Earth Planet. Inter.*, 64, 1–20.
Heinz, A. and Neumann, P. (1991). Representation of orientation and disorientation data for cubic, hexagonal, tetragonal and orthorhombic crystals, *Acta Cryst. A*, 47, 780–789.
Helmstetter, A., Kagan, Y. Y. and Jackson, D. D. (2005). Importance of small earthquakes for stress transfers and earthquake triggering, *J. Geophys. Res.*, 110(5), B05S08.
Helmstetter, A., Kagan, Y. Y. and Jackson, D. D. (2006). Comparison of short-term and time-independent earthquake forecast models for southern California, *Bull. Seismol. Soc. Amer.*, 96(1), 90–106.
Helmstetter, A., Kagan, Y. Y. and Jackson, D. D. (2007). High-resolution time-independent grid-based forecast for $m \geq 5$ earthquakes in California, *Seism. Res. Lett.*, 78(1), 78–86.
Helmstetter, A. and Sornette, D. (2003). Importance of direct and indirect triggered seismicity in the ETAS model of seismicity, *Geophys. Res. Lett.*, 30(11), 1576.
Helmstetter, A. and Sornette, D. (2004). Predictability in the epidemic-type aftershock sequence model of interacting triggered seismicity, *J. Geophys. Res.*, 108(B10), Art. No. 2482.
Hilbe, J. M. (2007). *Negative Binomial Regression*, Cambridge University Press, New York.
Hileman, J. A., Allen, C. R. and Nordquist, J. M. (1973). *Seismicity of the Southern California Region, 1 January 1932 to 31 December 1972*, Cal. Inst. Technology, Pasadena.
Hiramatsu, Y., Watanabe, T. and Obara, K. (2008). Deep low-frequency tremors as a proxy for slip monitoring at plate interface, *Geophys. Res. Lett.*, 35, L13304.
Hirose, T., Hiramatsu, Y. and Obara, K. (2010., Characteristics of short-term slow slip events estimated from deep low-frequency tremors in Shikoku, Japan, *J. Geophys. Res.*, 115, B10304.
Holt, W. E., Chamot-Rooke, N., Le Pichon, X., Haines, A. J., Shen-Tu, B. and Ren, J. (2000). Velocity field in Asia inferred from Quaternary fault slip rates and Global Positioning System observations, *J. Geophys. Res.*, 105, 19,185–19,209.
Horn, B. K. P. (1987). Closed-form solution of absolute orientation using unit quaternions, *J. Opt. Soc. Am. A*, 4(4), 629–642.
Hough, S. E. (2004). Scientific overview and historical context of the 1811–1812 New Madrid earthquake sequence, *Annals of Geophysics*, 47(23), 523–537.
Hough, S. E. (2009). *Predicting the Unpredictable: The Tumultuous Science of Earthquake Prediction*, Princeton University Press, Princeton, NJ.
Houston, H. (2001). Influence of depth, focal mechanism, and tectonic setting on the shape and duration of earthquake source time functions, *J. Geophys. Res.*, 106, 11137–11150.

Houston, H., Benz, H. M. and Vidale, J. E. (1998). Time functions of deep earthquakes from broadband and short-period stacks, *J. Geophys. Res.*, 103(B12), 29895–29913.

Huillet, T. and Raynaud, H. F. (2001). On rare and extreme events, *Chaos, Solutions and Fractals*, 12, 823–844.

Hutton, K., Hauksson, E., Clinton, J., Franck, J., Guarino, A., Scheckel, N., Given, D. and Young, A. (2006). Southern California Seismic Network update, *Seism. Res. Lett.*, 77(3), 389–395.

Hutton, K., Woessner, J. and Hauksson, E. (2010). Earthquake monitoring in southern California for seventy-seven years (1932–2008), *Bull. Seismol. Soc. Amer.*, 100, 423–446.

Hutton, L. K. and Jones, L. M. (1993). Local magnitudes and apparent variations in seismicity rates in Southern California, *Bull. Seismol. Soc. Am.*, 83, 313–329.

Ishimoto, M. and Iida, K. (1939). Observations sur les seismes enregistrés par le microsismographe construit dernièrement (1), *Bull. Earthquake Res. Inst. Tokyo Univ.*, 17, 443–478 (in Japanese).

Jackson, D. D. and Kagan, Y. Y. (1999). Testable earthquake forecasts for 1999, *Seism. Res. Lett.*, 70(4), 393–403.

Jackson, D. D. and Kagan, Y. Y. (2006). The 2004 Parkfield earthquake, the 1985 prediction, and characteristic earthquakes: lessons for the future, *Bull. Seismol. Soc. Amer.*, 96(4B), S397–S409.

Jackson, D. D. and Matsu'ura, M. (1985). A Bayesian approach to nonlinear inversion, *J. Geophys. Res.*, 90, 581–591.

Jaeger, J. C. and Cook, N. G. W. (1979). *Fundamentals of Rock Mechanics*, 3rd ed., Chapman and Hall, London.

Jirina, M. (1958). Stochastic branching processes with continuous state space, *Czech. Math. J.*, 8, 292–313.

Johnson, N. L., Kemp, A. W. and Kotz, S. (2005). *Univariate Discrete Distributions*, 3rd ed., Wiley, Hoboken, NJ.

Jolliffe, I. T. and Stephenson, D. B., eds. (2003). *Forecast Verification: A Practitioner's Guide in Atmospheric Science*, John Wiley, Chichester.

Jordan, T. H. (1997). Is the study of earthquakes a basic science?, *Seismol. Res. Lett.*, 68(2), 259–261.

Jordan, T. H., Chen, Y.-T., Gasparini, P., Madariaga, R., Main, I., Marzocchi, W., Papadopoulos, G., Sobolev, G., Yamaoka, K. and Zschau, J. (2011). Operational earthquake forecasting: state of knowledge and guidelines for utilization, *Annals Geophysics*, 54(4), 315–391.

Jordan, T. H. and Jones, L. M. (2010). Operational earthquake forecasting: some thoughts on why and how, *Seismol. Res. Lett.*, 81(4), 571–574.

Julian, B. R., Miller, A. D. and Foulger, G. R. (1998). Non-double-couple earthquakes, 1. Theory, *Rev. Geophys.*, 36(4), 525–549.

Kagan, Y. Y. (1973). Statistical methods in the study of the seismic process (with discussion: comments by M. S. Bartlett, A. G. Hawkes and J. W. Tukey), *Bull. Int. Statist. Inst.*, 45(3), 437–453.

Kagan, Y. Y. (1981a). 1 Spatial distribution of earthquakes: the three-point moment function, *Geophys. J. Roy. Astr. Soc.*, 67, 697–717.

Kagan, Y. Y. (1981b). Spatial distribution of earthquakes: the four-point moment function, *Geophys. J. Roy. Astr. Soc.*, 67, 719–733.

Kagan, Y. Y. (1982). Stochastic model of earthquake fault geometry, *Geophys. J. Roy. Astr. Soc.*, 71, 659–691.

Kagan, Y. Y. (1987). Point sources of elastic deformation: elementary sources, static displacements, *Geophys. J. Roy. Astr. Soc.*, 90, 1–34 (Errata, *Geophys. J. R. Astron. Soc.*, 93, 591, 1988).

Kagan, Y. Y. (1988). Static sources of elastic deformation in homogenous half-space, *J. Geophys. Res.*, 93(B9), 10,560–10,574.

Kagan, Y. Y. (1990). Random stress and earthquake statistics: spatial dependence, *Geophys. J. Int.*, 102, 573–583.

Kagan, Y. Y. (1991a). Fractal dimension of brittle fracture, *J. Nonlinear Sci.*, 1, 1–16.

Kagan, Y. Y. (1991b). Likelihood analysis of earthquake catalogues, *Geophys. J. Int.*, 106, 135–148.

Kagan, Y. Y. (1991c). 3-D rotation of double-couple earthquake sources, *Geophys. J. Int.*, 106, 709–716.

Kagan, Y. Y. (1992a). Seismicity: turbulence of solids, *Nonlinear Sci. Today*, 2, 1–13.

Kagan, Y. Y. (1992b). On the geometry of an earthquake fault system, *Phys. Earth Planet. Inter.*, 71, 15–35.

Kagan, Y. Y. (1992c). Correlations of earthquake focal mechanisms, *Geophys. J. Int.*, 110, 305–320.

Kagan, Y. Y. (1994a). Incremental stress and earthquakes, *Geophys. J. Int.*, 117, 345–364.

Kagan, Y. Y. (1994b). Distribution of incremental static stress caused by earthquakes, *Nonlinear Processes Geophys.*, 1(2/3), 172–181.

Kagan, Y. Y. (1996). Comment on "The Gutenberg-Richter or characteristic earthquake distribution, which is it?" by Steven G. Wesnousky, *Bull. Seismol. Soc. Amer.*, 86(1a), 274–285.

Kagan, Y. Y. (1997a). Seismic moment-frequency relation for shallow earthquakes: regional comparison, *J. Geophys. Res.*, 102(B2), 2835–2852.

Kagan, Y. Y. (1997b). Are earthquakes predictable?, *Geophys. J. Int.*, 131(3), 505–525.

Kagan, Y. Y. (1999). Universality of the seismic moment-frequency relation, *Pure Appl. Geoph.*, 155(24), 537–573.

Kagan, Y. Y. (2000). Temporal correlations of earthquake focal mechanisms, *Geophys. J. Int.*, 143, 881–897.

Kagan, Y. Y. (2002a). Seismic moment distribution revisited: I. Statistical results, *Geophys. J. Int.*, 148, 520–541.

Kagan, Y. Y. (2002b). Aftershock zone scaling, *Bull. Seismol. Soc. Amer.*, 92(2), 641–655.

Kagan, Y. Y. (2002c). Seismic moment distribution revisited: II. Moment conservation principle, *Geophys. J. Int.*, 149, 731–754.

Kagan, Y. Y. (2003). Accuracy of modern global earthquake catalogs, *Phys. Earth Planet. Inter.*, 135(23), 173–209.

Kagan, Y. Y. (2004). Short-term properties of earthquake catalogs and models of earthquake source, *Bull. Seismol. Soc. Amer.*, 94(4), 1207–1228.

Kagan, Y. Y. (2005a). Earthquake slip distribution: a statistical model, *J. Geophys. Res.*, 110(5), B05S11.

Kagan, Y. Y. (2005b). Double-couple earthquake focal mechanism: random rotation and display, *Geophys. J. Int.*, 163(3), 1065–1072.

Kagan, Y. Y. (2006). Why does theoretical physics fail to explain and predict earthquake occurrence?, in *Modelling Critical and Catastrophic Phenomena in Geoscience: A Statistical Physics Approach, Lecture Notes in Physics*, 705, P. Bhattacharyya and B. K. Chakrabarti, eds., Springer Verlag, Berlin, pp. 303–359.

Kagan, Y. Y. (2007a). Earthquake spatial distribution: the correlation dimension, *Geophys. J. Int.*, 168(3), 1175–1194.

Kagan, Y. Y. (2007b). On earthquake predictability measurement: information score and error diagram, *Pure Appl. Geoph.*, 164(10), 1947–1962.

Kagan, Y. Y. (2007c). Simplified algorithms for calculating double-couple rotation, *Geophys. J. Int.*, 171(1), 411–418.

Kagan, Y. Y. (2009a). On the geometric complexity of earthquake focal zone and fault systems: a statistical study, *Phys. Earth Planet. Inter.*, 173(3–4), 254–268.

Kagan, Y. Y. (2009b). Testing long-term earthquake forecasts: likelihood methods and error diagrams, *Geophys. J. Int.*, 177(2), 532–542.

Kagan, Y. Y. (2010a). Statistical distributions of earthquake numbers: consequence of branching process, *Geophys. J. Int.*, 180(3), 1313–1328.

Kagan, Y. Y. (2010b). Earthquake size distribution: power-law with exponent $\beta \equiv 1/2$?, *Tectonophysics*, 490(1–2), 103–114.

Kagan, Y. Y. (2011). Random stress and Omori's law, *Geophys. J. Int.*, 186(3), 1347–1364.

Kagan, Y. Y. (2013). Double-couple earthquake source: symmetry and rotation, *Geophys. J. Int.*, 194(2), 1167–1179.

Kagan, Y. Y., Bird, P. and Jackson, D. D. (2010). Earthquake patterns in diverse tectonic zones of the Globe, *Pure Appl. Geoph. (The Frank Evison Volume)*, 167(6/7), 721–741.

Kagan, Y. Y. and Houston, H. (2005). Relation between mainshock rupture process and Omori's law for aftershock moment release rate, *Geophys. J. Int.*, 163(3), 1039–1048.

Kagan, Y. Y. and Jackson, D. D. (1991a). Long-term earthquake clustering, *Geophys. J. Int.*, 104, 117–133.

Kagan, Y. Y. and Jackson, D. D. (1991b). Seismic gap hypothesis: ten years after, *J. Geophys. Res.*, 96(B13), 21,419–21,431.

Kagan, Y. Y. and Jackson, D. D. (1994). Long-term probabilistic forecasting of earthquakes, *J. Geophys. Res.*, 99(B7), 13,685–13,700.

Kagan, Y. Y. and Jackson, D. D. (1995). New seismic gap hypothesis: five years after, *J. Geophys. Res.*, 100(B3), 3943–3959.

Kagan, Y. Y. and Jackson, D. D. (1998). Spatial aftershock distribution: effect of normal stress, *J. Geophys. Res.*, 103(B10), 24,453–24,467.

Kagan, Y. Y. and Jackson, D. D. (1999). Worldwide doublets of large shallow earthquakes, *Bull. Seismol. Soc. Amer.*, 89(5), 1147–1155.

Kagan, Y. Y. and Jackson, D. D. (2000). Probabilistic forecasting of earthquakes, *Geophys. J. Int.*, 143, 438–453.

Kagan, Y. Y. and Jackson, D. D. (2006). Comment on 'Testing earthquake prediction methods: "The West Pacific short-term forecast of earthquakes with magnitude MwHRV $\geq$ 5.8"' by V. G. Kossobokov, *Tectonophysics*, 413(1–2), 33–38.

Kagan, Y. Y. and Jackson, D. D. (2011). Global earthquake forecasts, *Geophys. J. Int.*, 184(2), 759–776.

Kagan, Y. Y. and Jackson, D. D. (2012). Whole Earth high-resolution earthquake forecasts, *Geophys. J. Int.*, 190(1), 677–686.

Kagan, Y. Y. and Jackson, D. D. (2013). Tohoku earthquake: a surprise? *Bull. Seismol. Soc. Amer.*, 103(2B), 1181–1194, DOI: 10.1785/0120120110.

Kagan, Y. Y., Jackson, D. D. and Geller, R. J. (2012). Characteristic earthquake model, 1884–2011, R.I.P., *Seismol. Res. Lett.*, 83(6), 951–953.

Kagan, Y. Y., Jackson, D. D. and Rong, Y. F. (2006). A new catalog of southern California earthquakes, 1800–2005, *Seism. Res. Lett.*, 77(1), 30–38.

Kagan, Y. Y., Jackson, D. D. and Rong, Y. F. (2007). A testable five-year forecast of moderate and large earthquakes in southern California based on smoothed seismicity, *Seism. Res. Lett.*, 78(1), 94–98.

Kagan, Y. Y. and Knopoff, L. (1976). Statistical search for non-random features of the seismicity of strong earthquakes, *Phys. Earth Planet. Inter.*, 12(4), 291–318.

Kagan, Y. Y. and Knopoff, L. (1977). Earthquake risk prediction as a stochastic process, *Phys. Earth Planet. Inter.*, 14(2), 97–108.

Kagan, Y. Y. and Knopoff, L. (1980). Spatial distribution of earthquakes: the two-point correlation function, *Geophys. J. Roy. Astr. Soc.*, 62, 303–320.

Kagan, Y. Y. and Knopoff, L. (1981). Stochastic synthesis of earthquake catalogs, *J. Geophys. Res.*, 86, 2853–2862.

Kagan, Y. Y. and Knopoff, L. (1984). A stochastic model of earthquake occurrence, *Proc. 8th Int. Conf. Earthq. Eng.*, San Francisco, 1, 295–302.

Kagan, Y. Y. and Knopoff, L. (1985a). The first-order statistical moment of the seismic moment tensor, *Geophys. J. Roy. Astr. Soc.*, 81, 429–444.

Kagan, Y. Y. and Knopoff, L. (1985b). The two-point correlation function of the seismic moment tensor, *Geophys. J. Roy. Astr. Soc.*, 83, 637–656.

Kagan, Y. Y. and Knopoff, L. (1987a). Random stress and earthquake statistics: time dependence, *Geophys. J. Roy. Astr. Soc.*, 88, 723–731.

Kagan, Y. Y. and Knopoff, L. (1987b). Statistical short-term earthquake prediction, *Science*, 236, 1563–1567.

Kagan, Y. Y. and Vere-Jones, D. (1996). Problems in the modelling and statistical analysis of earthquakes in: *Lecture Notes in Statistics (Athens Conference on Applied Probability and Time Series)* 114, C. C. Heyde, Yu. V. Prohorov, R. Pyke and S. T. Rachev, eds., New York, Springer, pp. 398–425, DOI: 10.1007/978-1-4612-0749-8 29.

Kanamori, H. (1977). The energy release in great earthquakes, *J. Geophys. Res.*, 82, 2981–2987.

Kanamori, H. and Brodsky, E. E. (2004). The physics of earthquakes, *Rep. Prog. Phys.*, 67, 1429–1496.

Kendall, D. G., Barden, D., Carn, T. K. and Le, H. (1999). *Shape and Shape Theory*, Wiley, New York.

Kendall, M. G. and Moran, P. A. P. (1963). *Geometrical Probabilities*, Hafner, NY.

Ken-Tor, R., Agnon, A., Enzel, Y., Stein, M., Marco, S. and Negendank, J. F.W. (2001). High-resolution geological record of historic earthquakes in the Dead Sea basin, *J. Geophys. Res.*, 106(B2), 2221–2234.

Kilb, D., Gomberg, J. and Bodin, P. (2002) Aftershock triggering by complete Coulomb stress changes, *J. Geophys. Res.*, 107(B4), 2060.

King, G. (1983). The accommodation of large strains in the upper lithosphere of the Earth and other solids by self-similar fault systems: the geometrical origin of *b*-value, *Pure Appl. Geophys.*, 121, 761–815.

King, G. C. P. (1986). Speculations on the geometry of the initiation and termination processes of earthquake rupture and its relation to morphology and geological structure, *Pure Appl. Geophys.*, 124(3), 567–585.

Klein, F. (1932). *Elementary Mathematics from an Advanced Standpoint*. Vol. I. *Arithmetic, Algebra, Analysis*, trans. E. R. Hedrick and C. A. Noble, Macmillan, London.

Knopoff, L. and Randall, M. J. (1970). The compensated linear vector dipole: a possible mechanism for deep earthquakes, *J. Geophys. Res.*, 75, 4957–4963.

Kossobokov, V. G. (2006). Testing earthquake prediction methods: "The West Pacific short-term forecast of earthquakes with magnitude MwHRV ≥ 5.8", *Tectonophysics*, 413(1–2), 25–31.

Kostrov, B. V. (1974). Seismic moment and energy of earthquakes, and seismic flow of rock, *Izv. Acad. Sci. USSR, Phys. Solid Earth*, January, 13–21.

Kostrov, B. V. and Das, S. (1988). *Principles of Earthquake Source Mechanics*, Cambridge University Press, New York.

Kotz, S., Balakrishnan, N, Read, C. and Vidakovic, B. (2006). *Encyclopedia of Statistical Sciences*, 2nd ed., Wiley-Interscience, Hoboken, NJ, 16 vols.

Kreemer, C., Holt, W. E., and Haines, A. J. (2003). An integrated global model of present-day plate motions and plate boundary deformation, *Geophys. J. Int.*, 154, 8–34.

Krieger, L. and Heimann, S. (2012). MoPaD – Moment tensor plotting and decomposition: a tool for graphical and numerical analysis of seismic moment tensors, *Seismol. Res. Lett.*, 83(3), 589–595.

Kuhn, T. S. (1965). Logic of discovery or psychology of research?, in *Criticism and the Growth of Knowledge*, I. Lakatos and A. Musgrave eds., Cambridge University Press, Cambridge. pp. 1–23.

Kuipers, J. B. (1999). *Quaternions and Rotation Sequences: A Primer with Applications to Orbits, Aerospace and Virtual Reality*, Princeton University Press, Princeton, NJ.

Kwiatek, G., Plenkers, K. Dresen, G. and JAGUARS Research Group. (2011) Source parameters of picoseismicity recorded at Mponeng Deep Gold Mine, South Africa: implications for scaling relations, *Bull. Seismol. Soc. Amer.*, 101(6), 2592–2608.

Langer, J. S., Carlson, J. M., Myers, C. R. and Shaw, B. E. (1996). Slip complexity in dynamic models of earthquake faults, *Proc. Nat. Acad. Sci. USA*, 93, 3825–3829.

Lavallée, D. (2008). On the random nature of earthquake sources and ground motions: a unified theory, *Advances in Geophysics*, 50, 427–461.

Lay, T. and Wallace, T. C. (1995). *Modern Global Seismology*, San Diego, Academic Press.

Lee, W. H. K., Kanamori, H., Jennings, P. C. and Kisslinger, C. eds. (2002) *IASPEI Handbook of Earthquake and Engineering Seismology, Part A*, Academic Press, Boston.

Libicki, E. and Ben-Zion, Y. (2005). Stochastic branching models of fault surfaces and estimated fractal dimension, *Pure Appl. Geophys.*, 162(6–7), 1077–1111.

Lockwood, E. H. and Macmillan, R. H. (1978). *Geometric Symmetry*, Cambridge, Cambridge Univ. Press, 228 pp.

Lombardi, A. M. and Marzocchi, W. (2007). Evidence of clustering and nonstationarity in the time distribution of large worldwide earthquakes, *J. Geophys. Res.*, 112, B02303.

Lomnitz, C. (1994). *Fundamentals of Earthquake Prediction*, Wiley, New York.

Lumley, J. L. (ed.) (1990). *Whither Turbulence?: Turbulence at the Crossroads*, Berlin, Springer-Verlag.

Lyakhovsky, V., Ben-Zion, Y. and Agnon, A. (2005). A viscoelastic damage rheology and rate- and state dependent friction, *Geophys. J. Int.*, 161(1), 179–190; Correction, *Geophys. J. Int.*, 161(2), 420, (2005).

Mackenzie, J. K. (1958). Second paper on statistics associated with the random disorientation of cubes, *Biometrika*, 45, 229–240.

Mackenzie, J. K. (1964). Distribution of rotation axes in random aggregate of cubic crystals, *Acta Metallurgica*, 12(2), 223–225.

Mandelbrot, B. B. (1983). *The Fractal Geometry of Nature*, 2nd ed., W. H. Freeman, San Francisco.

Mandelbrot, B. B. (2012). *The Fractalist: Memoir of a Scientific Maverick*, Pantheon Books, New York.

Marco, S., Stein, M., Agnon, A. and Ron, H. (1996). Long-term earthquake clustering: a 50,000-year paleoseismic record in the Dead Sea Graben, *J. Geophys. Res.*, 101, 6179–6191.

Marder, M. (1998). Computational science: unlocking dislocation secrets, *Nature*, 391, 637–638.

Mardia, K. V. and Jupp, P. E. (2000). *Directional Statistics*, Wiley, Chichester.

Marsaglia, G. (1972). Choosing a point from the surface of a sphere, *Ann. Math. Stat.*, 43, 645–646.

Marsan, D. (2005). The role of small earthquakes in redistributing crustal elastic stress, *Geophys. J. Int.*, 163(1), 141–151.

Marsan, D. and Lengliné, O. (2008). Extending earthquakes' reach through cascading, *Science*, 319, 1076–1079.

Marzocchi, W. and Lombardi, A. M. (2008). A double branching model for earthquake occurrence, *J. Geophys. Res.*, 113, B08317.

Marzocchi W. and Zechar, J. D. (2011). Earthquake forecasting and earthquake prediction: different approaches for obtaining the best model, *Seismol. Res. Lett.*, 82(3), 442–448.

Mason, I. B. (2003). Binary events, in *Forecast Verification: A Practitioner's Guide in Atmospheric Science*, I. T. Jolliffe and D. B. Stephenson, eds, John Wiley, Chichester, pp. 37–76.

Mason, J. K. and Schuh, C. A. (2009). The generalized Mackenzie distribution: Disorientation angle distributions for arbitrary textures, *Acta Materialia*, 57, 4186–4197.

Matheron, G. (1971). *The Theory of Regionalized Variables and its Applications*, Cahiers du Centre de Morphologie Mathematique de Fontainebleau, No. 5.

Matthews, M. V., Ellsworth, W. L. and Reasenberg, P. A. (2002). A Brownian model for recurrent earthquakes, *Bull. Seismol. Soc. Amer.*, 92, 2233–2250.

McCaffrey, R. (2007). The next great earthquake, *Science*, 315, 1675.

McCaffrey, R. (2008). Global frequency of magnitude 9 earthquakes, *Geology*, 36(3), 263–266, (GSA Data Repository item 2008063, Table DR1).

McCann, W. R., Nishenko, S. P., Sykes, L. R. and Krause, J. (1979). Seismic gaps and plate tectonics: seismic potential for major boundaries, *Pure Appl. Geophys.*, 117, 1082–1147.

McCulloch, J. H. and Panton, D. B. (1997). Precise tabulation of the maximally-skewed stable distributions and densities, *Comput. Statist. Data Anal.*, 23, 307–320; Erratum, 26, 101.

McGuire, J. J. and Beroza, G. C. (2012). A rogue earthquake off Sumatra, *Science*, 336(6085), 1118–1119.

McGuire, J. J., Boettcher, M. S. and Jordan, T. H. (2005). Foreshock sequences and short-term earthquake predictability on East Pacific Rise transform faults, *Nature*, 434(7032), 457–461; Correction – *Nature*, 435(7041), 528.

McGuire, J. J., Li Zhao and Jordan, T. H. (2001). Teleseismic inversion for the second-degree moments of earthquake space-time distributions, *Geophys. J. Int.*, 145, 661–678.

McKenzie, D. P. and Morgan, W. J. (1969). Evolution of triple junctions, *Nature*, 224(5215), 125–133.

Meade, B. J. and Hager, B. H. (2005). Spatial localization of moment deficits in southern California, *J. Geophys. Res.*, 110(4), B04402.

Mehta, M. L., (1991). *Random Matrices*, 2nd ed., Boston, Academic Press.

Meister, L. and Schaeben, H. (2005). A concise quaternion geometry of rotations, *Math. Meth. Appl. Sci.*, 28, 101–126.

Michael, A. J. (2011). Random variability explains apparent global clustering of large earthquakes, *Geophys. Res. Lett.*, 38, L21301.

Michael A.J. and Wiemer, S. (2010). CORSSA: The Community Online Resource for Statistical Seismicity Analysis, DOI: 10.5078/corssa-39071657. Available at: http://www.corssa.org.

Miguel, M.-C., Vespignani, A., Zapperi, S., Weiss, J. and Grasso, J.-R. (2001). Intermittent dislocation flow in viscoplastic deformation, *Nature*, 410, 667–671.

Mindlin, R. D. (1936). Force at a point in the interior of a semi-infinite solid, *Physics*, 7, 195–202.

Mogi, K. (1995). Earthquake prediction research in Japan, *J. Phys. Earth*, 43, 533–561.

Molchan, G. M. (1990). Strategies in strong earthquake prediction, *Phys. Earth Planet. Inter.*, 61(1–2), 84–98.

Molchan, G. M. (1991). Structure of optimal strategies in earthquake prediction, *Tectonophysics*, 193(4), 267–276.

Molchan, G. M. (1997). Earthquake prediction as a decision-making problem, *Pure Appl. Geophys.*, 149(1), 233–247.

Molchan, G. M. (2003). Earthquake prediction strategies: a theoretical analysis, in V. I. Keilis-Borok and A. A. Soloviev, eds., *Nonlinear Dynamics of the Lithosphere and Earthquake Prediction*, Springer, Heidelberg, pp. 208–237.

Molchan, G. M. (2010). Space-time earthquake prediction: the error diagrams, *Pure Appl. Geoph. (The Frank Evison Volume)*, 167(8/9), 907–917.

Molchan, G. M. and Kagan, Y. Y. (1992). Earthquake prediction and its optimization, *J. Geophys. Res.*, 97(B4), 4823–4838.

Molchan, G. and Keilis-Borok. V. (2008). Earthquake prediction: probabilistic aspect, *Geophys. J. Int.*, 173(3), 1012–1017.

Molchan, G. and Kronrod, T. (2005). On the spatial scaling of seismicity rate, *Geophys. J. Int.*, 162(3), 899–909.

Molchan, G. and Kronrod, T. (2009). The fractal description of seismicity, *Geophys. J. Int.*, 179(3), 1787–1799.

Molchan, G. M. and Podgaetskaya, V. M. (1973). Parameters of global seismicity, I, in V. I. Keilis-Borok, ed., *Computational Seismology*, 6, Nauka, Moscow, 44–66 (in Russian).

Monin, A. S. and Yaglom, A. M. (1971) and (1975). *Statistical Fluid Mechanics*, Vols. 1 and 2, Cambridge, Mass.: MIT Press.

Moran, P. A. P. (1975). Quaternions, Haar measure and estimation of paleomagnetic rotation, in *Perspectives in Probability and Statistics*, J. Gani, ed., Academic Press, San Diego, pp. 295–301.

Morawiec, A. (1995). Misorientation-angle distribution of randomly oriented symmetrical objects, *J. Applied Crystallography*, 28(3), 289–293.

Morawiec, A. (1996). Distributions of rotation axes for randomly oriented symmetric objects, *J. Applied Crystallography*, 29(2), 164–169.

Morawiec, A. (2004). *Orientations and Rotations: Computations in Crystallographic Textures*, Springer, Berlin.

Morawiec, A. and Field, D. P. (1996). Rodrigues parameterization for orientation and misorientation distributions, *Philos. Mag. A*, 73(4), 1113–1130.

Nadeau, R. M., Foxall, W. and McEvilly, T. V. (1995). Clustering and periodic recurrence of microearthquakes on the San Andreas fault at Parkfield, California, *Science*, 267, 503–507.

Nadeau, R. M. and Johnson, L. R. (1998). Seismological studies at Parkfield VI: moment release rates and estimates of source parameters for small repeating earthquakes, *Bull. Seismol. Soc. Amer.*, 88, 790–814.

Nanjo, K. Z., Tsuruoka, H., Hirata, N. and Jordan, T. H. (2011). Overview of the first earthquake forecast testing experiment in Japan, *Earth, Planets, Space*, 63(3), 159–169.

Nature (1999). Debate on earthquake prediction, February–April. Available at: http://www.nature.com/nature/-debates/earthquake/equake frameset.html.

Nechad, H., Helmstetter, A., El Guerjouma, R. and Sornette, D. (2005). Andrade and critical time to-failure laws in fiber-matrix composites: experiments and model, *J. Mech. Phys. Solids*, 53, 1099–1127.

Nerenberg, M. A. H. and Essex, C. (1990). Correlation dimension and systematic geometric effects, *Phys. Rev.*, A42, 7065–7074.

Neumann, P. (1992). The role of geodesic and stereographic projections for the visualization of directions, rotations, and textures, *Phys. Stat. Sol.* (a), 131, 555–567.

Nishenko, S. P. (1991). Circum-Pacific seismic potential – 1989–1999, *Pure Appl. Geophys.*, 135, 169–259.

O'Brien, J. F. and Hodgins, J. K. (1999). Graphical modeling and animation of brittle fracture, *Proceedings of Assoc. Computing Machinery (ACM) SIGGRAPH*, 99, 137–146.

Ogata, Y. (1983). Estimation of the parameters in the modified Omori formula for aftershock frequencies by the maximum likelihood procedure, *Journal of Physics of the Earth*, 31, 115–124.

Ogata, Y. (1988). Statistical models for earthquake occurrence and residual analysis for point processes, *J. Amer. Statist. Assoc.*, 83, 9–27.

Ogata, Y. (1998). Space-time point-process models for earthquake occurrences, *Ann. Inst. Statist. Math.*, 50(2), 379–402.

Ogata, Y. (2004). Space-time model for regional seismicity and detection of crustal stress changes, *J. Geophys. Res.*, 109(B3), Art. No. B03308. Correction, J. Geophys. Res., 109(B6), Art. No. B06308 (2004).

Ogata, Y., Jones, L. M. and Toda, S. (2003). When and where the aftershock activity was depressed: contrasting decay patterns of the proximate large earthquakes in southern California, *J. Geophys. Res.*, 108(B6), 2318.

Ogata, Y. and Katsura, K. (1991). Maximum likelihood estimates of the fractal dimension for random point patterns, *Biometrika*, 78, 463–474.

Ogata, Y. and Zhuang. J. C. (2006). Space-time ETAS models and an improved extension, *Tectonophysics*, 413(1–2), 13–23.

Ojala, I. O., Main, I. G. and Ngwenya, B. T. (2004). Strain rate and temperature dependence of Omori law scaling constants of AE data: implications for earthquake foreshock-aftershock sequences, *Geophys. Res. Lett.*, 31, L24617.

Okal, E. A. (2013). Earthquake: focal mechanism, *Encyclopedia of Earth Sciences Series 2011*, pp. 194–199, DOI: 10.1007/978-90-481-8702-7 158.

Okal, E. A. and Romanowicz, B. A. (1994) On the variation of *b*-values with earthquake size, *Phys. Earth Planet. Inter.*, 87, 55–76.

Okal, E. A., Borrero, J. C. and Chagui-Goff, C. (2011). Tsunamigenic predecessors to the 2009 Samoa earthquake, *Earth-Sci. Rev.*, 107, 128–140.

Omori, F. (1894). On the after-shocks of earthquakes, *J. College Sci., Imp. Univ. Tokyo*, 7, 111–200 (with Plates IV–XIX).

Otter, R. (1949). The multiplicative process, *Annals Math. Statistics*. 20(2), 206–224.

Palmer, T. N. and Hagedorn, R. (2006). *Predictability of Weather and Climate*, Cambridge University Press, New York.

Pareto, V. (1897). *Œuvres Complètes*, Publ. by de Giovanni Busino, Droz, Geneva, vol. II (1964).

Park, S.C. and Mori, J. (2007). Triggering of earthquakes during the 2000 Papua New Guinea earthquake sequence, *J. Geophys. Res.*, 112(B3), Article Number: B03302.

Parsons, T. (2002). Global Omori law decay of triggered earthquakes: large aftershocks outside the classical aftershock zone, *J. Geophys. Res.*, 107, 2199.

Parsons, T. (2009). Lasting earthquake legacy, *Nature* 462(7269), 42–43.

Patil, G. P. (1962). Some methods of estimation for logarithmic series distribution, *Biometrics*, 18(1), 68–75.

Patil, G. P. and Wani, J. K. (1965). Maximum likelihood estimation for the complete and truncated logarithmic series distributions, *Sankhya*, 27A(2/4), 281–292.

Penrose, R. (2005). *The Road to Reality: A Complete Guide to the Laws of the Universe*, New York, Knopf.

Perez-Campos, X., Singh, S. K. and Beroza, G. C. (2003). Reconciling teleseismic and regional estimates of seismic energy, *Bull. Seismol. Soc. Amer.*, 93(5), 2123–2130.

Pisarenko, D. V. and Pisarenko, V. F. (1995). Statistical estimation of the correlation dimension, *Physics Letters A*, 197, 31–39.

Pisarenko, V. and Rodkin, M. (2010). *Heavy-Tailed Distributions in Disaster Analysis*, Springer, New York.

Pisarenko, V. F. (1998). Non-linear growth of cumulative flood losses with time, *Hydrological Processes*, 12, 461–470.

Pondrelli, S., Salimbeni, S., Morelli, A., Ekström, G., Boschi, E. (2007). European-Mediterranean regional centroid moment tensor catalog: Solutions for years 2003 and 2004, *Phys. Earth Planet. Inter.*, 164(1–2), 90–112.

Popper, K. R. (1980). *Logic of Scientific Discovery*, 2nd ed., Hutchinson, London.

Powers, P. M. and Jordan, T. H. (2010). Distribution of seismicity across strike-slip faults in California, *J. Geophys. Res.*, 115(B05), Article Number B05305.

Press, W. H., Teukolsky, S. A., Vetterling, W. T. and Flannery, B. P. (1992). *Numerical Recipes in FORTRAN*, 2nd ed., Cambridge University Press, Cambridge.

Reasenberg, P. A. (1999). Foreshock occurrence rates before large earthquakes worldwide, *Pure Appl. Geophys.*, 155(2–4), 355–379.

Reasenberg, P. A. and Jones, L. M. (1989). Earthquake hazard after a mainshock in California, *Science*, 243, 1173–1176.

Reasenberg, P. A. and Jones, L. M. (1994). Earthquake aftershocks: update, *Science*, 265, 1251–1252.

Rhoades, D. A. (1996). Estimation of the Gutenberg-Richter relation allowing for individual earthquake magnitude uncertainties, *Tectonophysics*, 258, 71–83.

Rhoades, D. A. (1997). Estimation of attenuation relations for strong-motion data allowing for individual earthquake magnitude uncertainties, *Bull. Seismol. Soc. Amer.*, 87(6), 1674–1678.

Rhoades, D. A. (2007). Application of the EEPAS model to forecasting earthquakes of moderate magnitude in southern California, *Seism. Res. Lett.*, 78(1), 110–115.

Rhoades, D. A. and Dowrick, D. J. (2000). Effects of magnitude uncertainties on seismic hazard estimates, in Proceedings of the 12th World Conference on Earthquake Engineering, Auckland, New Zealand, 30th January 4th February 2000, Paper No. 1179. New Zealand Society for Earthquake Engineering, Upper Hutt, New Zealand, *Bull. N.Z. Soc. Earthqu. Eng.*, 33(3).

Rhoades, D. A. and Evison, F. F. (2005). Test of the EEPAS forecasting model on the Japan earthquake catalogue, *Pure Appl. Geophys.*, 162(67), 1271–1290.

Rhoades, D. A. and Evison, F. F. (2006). The EEPAS forecasting model and the probability of moderate-to-large earthquakes in central Japan, *Tectonophysics*, 417(1–2), 119–130.

Rhoades, D. A., Schorlemmer, D., Gerstenberger, M. C., Christophersen, A., Zechar, J. D. and Imoto, M. (2011). Efficient testing of earthquake forecasting models, *Acta Geophysica*, 59(4), 728–747.

Rice, J. R. and Ben-Zion, Y. (1996). Slip complexity in earthquake fault models, *Proc. Nat. Acad. Sci. USA*, 93, 3811–3818.

Richards-Dinger, K. B. and Shearer, P. M. (2000). Earthquake locations in southern California obtained using source-specific station terms, *J. Geophys. Res.*, 105(B5), 10939–10960.

Richardson, E. and Jordan, T. H. (2002). Low-frequency properties of intermediate-focus earthquakes, *Bull. Seismol. Soc. Amer.*, 92(6), 2434–2448.

Richeton, T., Weiss, J., and Louchet F. (2005). Breakdown of avalanche critical behaviour in polycrystalline plasticity, *Nature Materials*, 4(6), 465–469.

Robertson, M. C., Sammis, C. G., Sahimi, M. and Martin, A. J. (1995). Fractal analysis of three-dimensional spatial distributions of earthquakes with a percolation interpretation, *J. Geophys. Res.*, 100, 609–620.

Rockwell, T. K., Lindvall, S., Herzberg, M., Murbach, D. Dawson, T. and Berger, G. (2000). Paleoseismology of the Johnson Valley, Kickapoo, and Homestead Valley faults: clustering of earthquakes in the eastern California shear zone, *Bull. Seismol. Soc. Amer.*, 90, 1200–1236.

Roeloffs, E. and Langbein, J. (1994). The earthquake prediction experiment at Parkfield, California, *Rev. Geophys.*, 32, 315–336.

Romashkova, L. L. (2009). Global-scale analysis of seismic activity prior to 2004 Sumatra-Andaman mega-earthquake, *Tectonophysics*, 470, 329–344.

Rong, Y.-F. and Jackson, D. D. (2002). Earthquake potential in and around China: estimated from past earthquakes, *Geophys. Res. Lett.*, 29(16): art. no. 1780.

Rong, Y.-F., Jackson, D. D. and Kagan, Y. Y. (2003). Seismic gaps and earthquakes, *J. Geophys. Res.*, 108(B10), 2471, ESE-6.

Rundle, J. B. (1989). Derivation of the complete Gutenberg-Richter magnitude-frequency relation using the principle of scale invariance, *J. Geophys. Res.*, 94, 12,337–12,342.

Rundle, J. B., Turcotte, D. L., Shcherbakov, R., Klein, W. and Sammis, C. (2003). Statistical physics approach to understanding the multiscale dynamics of earthquake fault systems, *Rev. Geophys.*, 41(4), Article Number 1019.

Samorodnitsky, G. and Taqqu, M. S. (1994). *Stable Non-Gaussian Random Processes: Stochastic Models with Infinite Variance*, Chapman and Hall, New York.

Schaeben, H. (1996). Texture approximation or texture modelling with components represented by the von Mises-Fisher matrix distribution on *SO(3)* and the Bingham distribution on S^4_+, *Appl. Cryst.*, 29, 516–525.

Schaeben, H. (2010). Special issue on spherical mathematics and statistics, *Math Geosci.*, 42, 727–730.

Schoenberg, F. P., Kumar, S., Zaliapin, I., and Kagan, Y. (2006). Statistical modeling of seismic moment release, *Eos Trans. AGU*, 87(52), Fall Meet. Suppl., Abstract S13A-0212.

Scholz, C. (1997). Whatever happened to earthquake prediction?, *Geotimes*, 42(3), 16–19.

Scholz, C. H. (1996). Faults without friction?, *Nature*, 381, 556–557.

Scholz, C. H. (2002) *The Mechanics of Earthquakes and Faulting*, 2nd ed., Cambridge University Press, Cambridge.

Schorlemmer, D. and Gerstenberger, M. C. (2007). RELM testing Center, *Seism. Res. Lett.*, 78(1), 30–35.

Schorlemmer, D., Gerstenberger, M. C. Wiemer, S., Jackson, D. D. and Rhoades, D. A. (2007). Earthquake likelihood model testing, *Seism. Res. Lett.*, 78(1), 17–29.

Schorlemmer D., Wiemer, S. and Wyss, M. (2005). Variations in earthquake-size distribution across different stress regimes, *Nature*, 437, 539–542.

Schwartz, D. P. and Coppersmith, K. J. (1984). Fault behavior and characteristic earthquakes: examples from Wasatch and San Andreas fault zones, *J. Geophys. Res.*, 89, 5681–5698.

Schwartz, S. Y. and Rokosky, J. M. (2007). Slow slip events and seismic tremor at circum-Pacific subduction zones, *Reviews Geophysics*, 45(3), Art. No. RG3004.

Sharon, E. and Fineberg, J. (1996). Microbranching instability and the dynamic fracture of brittle materials, *Physical Review B*, 54, 7128–7139.

Sharon, E. and Fineberg, J. (1999). Confirming the continuum theory of dynamic brittle fracture for fast cracks, *Nature*, 397, 333–335.

Shearer, P., Hauksson E., and Lin, G. Q. (2005). Southern California hypocenter relocation with waveform cross-correlation, Part 2: Results using source-specific station terms and cluster analysis, *Bull. Seismol. Soc. Am.*, 95(3), 904–915.

Shearer, P. and Stark, P. B. (2012). Global risk of big earthquakes has not recently increased, *Proceedings of PNAS*, 109(3), 717–721.

Shelly, D. R., Beroza, G. C. and Ide, S. (2007). Non-volcanic tremor and low frequency earthquake swarms, *Nature*, 446, 305–307.

Shenton, L. R. and Myers, R. (1963). Comments on estimation for the negative binomial distribution, in *Classical and Contagious Discrete Distributions*, G. P. Patil, Ed., *Stat. Publ. Soc.*, Calcutta, pp. 241–262.

Shepperd, S. W. (1978). Quaternion from rotation matrix, *J. Guidance and Control*, 1, 223–224.

Shimazaki, K. and Nakata, T. (1980). Time-predictable recurrence model for large earthquakes, *Geophys. Res. Lett.*, 7, 279–282.

Shlien, S. and Toksöz, M. N. (1970). A clustering model for earthquake occurrences. *Bull. Seismol. Soc. Amer.*, 60(6), 1765–1788.

Silver, P. G. and Jordan, T. H. (1983). Total-moment spectra of fourteen large earthquakes, *J. Geophys. Res.*, 88, 3273–3293.

Silver, P. and Masuda, T. (1985). Source extent analysis of the Imperial Valley earthquake of October 15, 1979, and the Victoria earthquake of June 9, 1980, *J. Geophys. Res.*, 90, 7639–7651.

Silverman, B. W. (1986). *Density Estimation for Statistics and Data Analysis*, Chapman and Hall, London.

Simons, M., Minson, S. E., Sladen, A., Ortega, F., Jiang, J. L., Owen, S. E., Meng, L. S., Ampuero, J. P., Wei, S. J., Chu, R. S., Helmberger, D. V., Kanamori, H., Hetland, E., Moore, A.W., and Webb, F. H. (2011) The 2011 magnitude 9.0 Tohoku-Oki earthquake: mosaicking the megathrust from seconds to centuries, *Science*, 332(6036), 1421–1425.

Small, C. G. (1996). *The Statistical Theory of Shape*, Springer, New York.

Smith, L. A. (1988). Intrinsic limits on dimension calculations, *Physics Letters A*, 133, 283–288.

Snieder, R., and van Eck, T. (1997). Earthquake prediction: a political problem?, *Geologische Rundschau*, 86, 446–463.

Snoke, J. A. (2003). FOCMEC: FOcal MEChanism determinations, in *International Handbook of Earthquake and Engineering Seismology* W. H. K. Lee, H. Kanamori, P. C. Jennings and C. Kisslinger, eds., Academic Press, San Diego, pp. 1629–1630.

Soler, T. and van Gelder, B. H. W. (1991). On covariances of eigenvalues and eigenvectors of second-rank symmetric tensors, *Geophys. J. Int.*, 105, 537–546.

Sornette, D. (2003). *Critical Phenomena in Natural Sciences (Chaos, Fractals, Self-organization, and Disorder: Concepts and Tools)*, 2nd edn, Springer, New York.

Sornette, D. and Werner, M. J. (2005). Apparent clustering and apparent background earthquakes biased by undetected seismicity, *J. Geophys. Res.*, 110(9), B09303.

Sornette, D. and Werner, M. J. (2008). Statistical physics approaches to seismicity, in *Encyclopedia of Complexity and System Science*, Vol. 9, R. Meyer, ed., Springer, New York, pp. 7872–7891.

Stark, P. B. and Freedman, D. A. (2003). What is the chance of an earthquake?, in *Earth-Quake Science and Seismic Risk Reduction*, F. Mulargia and R. J. Geller, eds., Kluwer, Dordrecht pp. 201–213.

Steacy, S., Gomberg, J. and Cocco, M. (2005). Introduction to special section: stress transfer, earthquake triggering, and time-dependent seismic hazard, *J. Geophys. Res.*, 110(B5), B05S01.

Stein, R. S. (1999). The role of stress transfer in earthquake occurrence, *Nature*, 402, 605–609.

Stein, S., Geller, R., and Liu, M. (2011). Bad assumptions or bad luck: why earthquake hazard maps need objective testing, *Seismol. Res. Lett.*, 82(5), 623–626.

Stein, S. and Liu, M. (2009). Long aftershock sequences within continents and implications for earthquake hazard assessment, *Nature* 462(7269), 87–89.

Stein, R. S., Toda, S., Parsons, T. and Grunewald, D. E. (2006). A new probabilistic seismic hazard assessment for greater Tokyo, *Phil. Trans. R. Soc. A*, 364, 1965–1988.

Stephens, M. A. (1974). EDF statistics for goodness of fit and some comparisons, *J. Amer. Statist. Assoc. (JASA)*, 69, 730–737.

Swets, J. A. (1973). The relative operating characteristic in psychology, *Science*, 182(4116), 990–1000.

Tanioka, Y. and Ruff, L. (1997). Source time functions, *Seismol. Res. Lett.*, 68(3), 386–397.

Tape, W. and Tape, C. (2012). Angle between principal axis triples, *Geophys. J. Int.* 191(2), 813–831.

Thio, H. K. and Kanamori, H. (1996). Source complexity of the 1994 Northridge earthquake and its relation to aftershock mechanisms, *Bull. Seismol. Soc. Amer.*, 86, S84–S92.

Thomson Reuters Scientific/ISI (2012). Available at http://www.isinet.com/; ISI Web of Science, 2006. The Thomson Corporation, Thomson Reuters' Web of Science available at: http://portal.isiknowledge.com/portal.cgi (last accessed September 2013).

Thorne, K. S. (1980). Multipole expansions of gravitational radiation, *Rev. Mod. Phys.*, 52, 299–339.

Tinti, S. and Mulargia, F. (1985). Effects of magnitude uncertainties on estimating the parameters in the Gutenberg-Richter frequency-magnitude law, *Bull. Seismol. Soc. Amer.*, 75, 1681–1697.

Touati, S., Naylor, M. and Main, A. M. (2009). Origin and nonuniversality of the earthquake interevent time distribution, *Phys. Rev. Lett.*, 102, 168501.

Tripathi, R. C. (2006). Negative binomial distribution, in *Encyclopedia of Statistical Sciences*, Kotz, S., N. Balakrishnan, C. Read and B. Vidakovic, eds., 2nd ed., Wiley-Interscience, Hoboken, NJ, vol. 8, pp. 5413–5420.

Tsai, V.C., Nettles, M., Ekström, G., and Dziewonski, A.M. (2005). Multiple CMT source analysis of the 2004 Sumatra earthquake, *Geophys. Res. Lett.*, 32, L17304, DOI: 10.1029/2005GL023813.

Turcotte, D. L. (1986). A fractal model for crustal deformation, *Tectonophysics*, 132, 261–269.

Turcotte, D. L. (1997). *Fractals and Chaos in Geology and Geophysics*, 2nd ed., Cambridge University Press, Cambridge.

Uchaikin, V. V. and Zolotarev, V, M. (1999). *Chance and Stability: Stable Distributions and Their Applications*, VSP International Science Publishers, Utrecht.

U.S. Geological Survey, Preliminary Determinations of Epicenters (PDE) (2012). U.S. Dep. of Inter., Natl. Earthquake Inf. Cent. available at: http://neic.usgs.gov/neis/epic/epic.html.

Utsu, T. (1961). A statistical study on the occurrence of aftershocks, *Geoph. Magazine*, 30, 521–605.

Utsu, T. (1999). Representation and analysis of the earthquake size distribution: a historical review and some new approaches, *Pure Appl. Geophys.*, 155, 509–535.

Utsu, T., Ogata, Y. and Matsu'ura, R. S. (1995). The centenary of the Omori formula for a decay law of aftershock activity, *J. Phys. Earth*, 43, 1–33.

van Stiphout, T., Wiemer, S. and Marzocchi, W. (2010). Are short-term evacuations warranted? Case of the 2009 L'Aquila earthquake, *Geophys. Res. Lett.*, 37, L06306.

Vere-Jones, D. (1970). Stochastic models for earthquake occurrence (with discussion), *J. Roy. Stat. Soc.*, B32, 1–62.

Vere-Jones, D. (1976). A branching model for crack propagation, *Pure Appl. Geophys.*, 114, 711–725.

Vere-Jones, D. (1988). Statistical aspects of the analysis of historical earthquake catalogues, in *Historical Seismicity of Central-Eastern Mediterranean Region*, C. Margottini, ed., pp. 271–295.

Vere-Jones, D. (1998). Probabilities and information gain for earthquake forecasting, *Computational Seismology*, 30, *Geos, Moscow*, 248–263.

Vere-Jones, D. (1999). On the fractal dimensions of point patterns, *Adv. Appl. Probab.*, 31, 643–663.

Vere-Jones, D. (2009). Stochastic models for earthquake occurrence and mechanisms, *Encyclopedia of Complexity and Systems Science*, Springer, New York, pp. 2555–2581.

Vere-Jones, D. (2010). Foundations of statistical seismology, *Pure Appl. Geophys.*, 167(6/7), 645–653.

Vere-Jones, D., Davies, R. B., Harte, D., Mikosch, T. and Wang, Q. (1997). Problems and examples in the estimation of fractal dimension from meteorological and earthquake data, in *Proc. Int. Conf. on Application of Time Series in Physics, Astronomy and Meteorology*, T. Subba, M. Rao, B. Priestley and O. Lessi, eds., Chapman & Hall, London, pp. 359–375.

Vere-Jones, D., Robinson, R. and Yang, W. Z. (2001). Remarks on the accelerated moment release model: problems of model formulation, simulation and estimation, *Geophys. J. Int.*, 144, 517–531.

Wang, Q., Jackson, D. D. and Kagan, Y. Y. (2009). California earthquakes, 1800–2007: a unified catalog with moment magnitudes, uncertainties, and focal mechanisms, *Seism. Res. Lett.*, 80(3), 446–457.

Wang, Q., Schoenberg, F. P. and Jackson, D. D. (2010). Standard errors of parameter estimates in the ETAS model, *Bull. Seismol. Soc. Amer.*, 100(5A), 1989–2001.

Ward, J. P. (1997). *Quaternions and Cayley Numbers: Algebra and Applications*, Kluwer Academic Publishers, London.

Watson, G. S. (1983). *Statistics on Spheres*, John Wiley, New York.

Weeks, J. R. (2002). *The Shape of Space*, 2nd ed., CRC Press, Boca Raton, FL.

Weiss, J. and Marsan, D. (2003). Three-dimensional mapping of dislocation avalanches: clustering and space/time coupling, *Science*, 299, 89–92.

Wells, D. L. and Coppersmith, K. J. (1994). New empirical relationships among magnitude, rupture length, rupture width, rupture area, and surface displacement, *Bull. Seismol. Soc. Amer.*, 84, 974–1002.

Werner, M. J., Helmstetter, A., Jackson, D. D. and Kagan, Y. Y. (2011). High resolution long- and short-term earthquake forecasts for California, *Bull. Seismol. Soc. Amer.*, 101(4), 1630–1648.

Werner, M. J. and Sornette, D. (2008). Magnitude uncertainties impact seismic rate estimates, forecasts and predictability experiments, *J. Geophys. Res.*, 113(B8), Article Number: B08302.

Wilks, S. S. (1962). *Mathematical Statistics*, John Wiley, New York.

Wolfram, S. (1999). *The Mathematica Book*, 4th ed., Wolfram Media, Champaign, IL, Cambridge University Press, New York.

Wood, H. O. and Gutenberg, B. (1935). Earthquake prediction, *Science*, 82, 219–220.

Working Group on California Earthquake Probabilities (WG02) (2003). *Earthquakes Probabilities in the San Francisco Bay Region: 2002 to 2031*, USGS, Open-file Rept. 03 214; available at: http://pubs.usgs.gov/of/2003/of03-214.

Xu, P. L. (1999). Spectral theory of constrained second-rank symmetric random tensors, *Geophys. J. Int.*, 138, 1–24.

Xu, P. L. (2002). Isotropic probabilistic models for directions, planes, and referential systems, *Proc. R. Soc. London*, 458A(2024), 2017–2038.

Yaglom, A. (2001). The century of turbulence theory: The main achievements and unsolved problems, in *New Trends in Turbulence*, M. Lesieur, A. Yaglom and F. David, eds., NATO ASI, Session LXXIV, Springer, Berlin., pp. 1–52.

Young, J. B., Presgrave, B. W., Aichele, H., Wiens, D. A. and Flinn, E. A. (1996). The Flinn-Engdahl regionalisation scheme: the 1995 revision, *Phys. Earth Planet. Inter.*, 96, 223–297.

Zaiser, M. (2006). Scale invariance in plastic flow of crystalline solids, *Advances in Physics*, 55(1–2), 185–245.

Zaiser, M. and Moretti, P. (2005). Fluctuation phenomena in crystal plasticity: a continuum model, *J. Statistical Mechanics*. P08004, 1–19.

Zaiser, M. and Nikitas, N. (2007). Slip avalanches in crystal plasticity: scaling of the avalanche cut-off, *J. Stat. Mech.*, P04013.

Zaiser, M., Schwerdtfeger, J., Schneider, A. S., Frick, C. P., Clark, B. G., Gruber, P. A. and Arzt, E. (2008). Strain bursts in plastically deforming molybdenum micro- and nanopillars, *Philosophical Magazine*, 88(30), 3861–3874.

Zaliapin, I. V., Kagan Y. Y. and Schoenberg, F. P. (2005a). Approximating the distribution of Pareto sums, *Pure Appl. Geoph.*, 162(6–7), 1187–1228.

Zaliapin, I., Kagan Y. Y. and Schoenberg, F. P. (2005b). Estimation of seismic moment release with implications for regional hazard assessment, *Eos Trans. AGU*, 86(52), Fall Meet. Suppl., Abstract S53B-1096.

Zechar, J. D., Hardebeck, J. L., Michael, A. J., Naylor, M., Steacy, S., Wiemer, S., Zhuang, J. and the CORSSA Working Group (2011). Community Online Resource for Statistical Seismicity Analysis, *Seismol. Res. Lett.*, 82(5), 686–690.

Zechar, J. D. and Jordan, T. H. (2008). Testing alarm-based earthquake predictions, *Geophys. J. Int.*, 172(2), 715–724.

Zechar, J. D., Schorlemmer, D., Werner, M. J., Gerstenberger, M.,C., Rhoades, D. A. and Jordan, T. H. (2013). Regional earthquake likelihood models I: first-order results, *Bull. Seismol. Soc. Amer.*, 103(2a), 787–798.

Zhuang, J. (2011). Next-day earthquake forecasts for the Japan region generated by the ETAS model, *Earth Planets Space*, 63, 207–216.

Zhuang, J.C., Chang, C.-P., Ogata, Y. and Chen, Y.-I. (2005). A study on the background and clustering seismicity in the Taiwan region by using point process models, *J. Geophys. Res.*, 110, B05S18.

Zhuang, J. C., Ogata, Y. and Vere-Jones, D. (2004). Analyzing earthquake clustering features by using stochastic reconstruction, *J. Geophys. Res.*, 109(B5), Art. No. B05301.

Zolotarev, V. M. (1986). *One-Dimensional Stable Distributions*, Amer. Math. Soc., Providence, RI.

Zolotarev, V. M. and Strunin, B. M. (1971). Internal-stress distribution for a random distribution of point defects, *Soviet Phys. Solid State*, 13, 481–482 (English translation).

Index

References to figures are given in *italic* type. Reference to tables are given in **bold** type.